Smart Innovation, Systems and Technologies

Volume 72

About this Series

The Smart Innovation, Systems and Technologies book series encompasses the topics of knowledge, intelligence, innovation and sustainability. The aim of the series is to make available a platform for the publication of books on all aspects of single and multi-disciplinary research on these themes in order to make the latest results available in a readily-accessible form. Volumes on interdisciplinary research combining two or more of these areas is particularly sought.

The series covers systems and paradigms that employ knowledge and intelligence in a broad sense. Its scope is systems having embedded knowledge and intelligence, which may be applied to the solution of world problems in industry, the environment and the community. It also focusses on the knowledge-transfer methodologies and innovation strategies employed to make this happen effectively. The combination of intelligent systems tools and a broad range of applications introduces a need for a synergy of disciplines from science, technology, business and the humanities. The series will include conference proceedings, edited collections, monographs, handbooks, reference books, and other relevant types of book in areas of science and technology where smart systems and technologies can offer innovative solutions.

High quality content is an essential feature for all book proposals accepted for the series. It is expected that editors of all accepted volumes will ensure that contributions are subjected to an appropriate level of reviewing process and adhere to KES quality principles.

More information about this series at http://www.springer.com/series/8767

Ireneusz Czarnowski · Robert J. Howlett
Lakhmi C. Jain
Editors

Intelligent Decision Technologies 2017

Proceedings of the 9th KES International Conference on Intelligent Decision Technologies (KES-IDT 2017) – Part I

Editors
Ireneusz Czarnowski
Maritime University
Gdynia
Poland

Robert J. Howlett
Bournemouth University
Poole
UK

and

KES International
Shoreham-by-Sea
UK

Lakhmi C. Jain
University of Canberra
Canberra, ACT
Australia

and

Bournemouth University
Poole
UK

and

KES International
Shoreham-by-Sea
UK

ISSN 2190-3018 ISSN 2190-3026 (electronic)
Smart Innovation, Systems and Technologies
ISBN 978-3-319-86621-5 ISBN 978-3-319-59421-7 (eBook)
DOI 10.1007/978-3-319-59421-7

Printed on acid-free paper

This Springer imprint is published by Springer Nature
The registered company is Springer International Publishing AG
The registered company address is: Gewerbestrasse 11, 6330 Cham, Switzerland

9th International KES Conference On Intelligent Decision Technologies (KES-IDT 2017), Proceedings, Part I

Preface

This volume contains the proceedings (Part II) of the 9th International KES Conference on Intelligent Decision Technologies (KES-IDT 2017), which will be held in Algarve, Portugal, on June 21–23, 2017.

The KES-IDT is a well-established international annual conference organized by KES International. The KES-IDT conference is a sub-series of the KES Conference series.

The KES-IDT is an interdisciplinary conference and provides excellent opportunities for the presentation of interesting new research results and discussion about them, leading to knowledge transfer and generation of new ideas.

This edition, KES-IDT 2017, attracted a number of researchers and practitioners from all over the world. The KES-IDT 2017 Program Committee received papers for the main track and 11 special sessions. Each paper has been reviewed by 2–3 members of the International Program Committee and International Reviewer Board. Following a review process, only the highest quality submissions were accepted for inclusion in the conference. The 63 best papers have been selected for oral presentation and publication in the two volumes of the KES-IDT 2017 proceedings.

We are very satisfied with the quality of the program and would like to thank the authors for choosing KES-IDT as the forum for presentation of their work. Also, we gratefully acknowledge the hard work of the KES-IDT international Program Committee members and of the additional reviewers for taking the time to review the submitted papers and selecting the best among them for presentation at the conference and inclusion in its proceedings.

We hope and intend that KES-IDT 2017 significantly contributes to the fulfillment of the academic excellence and leads to even greater successes of KES-IDT events in the future.

June 2017

Ireneusz Czarnowski
Robert J. Howlett
Lakhmi C. Jain

KES-IDT 2017 Conference Organization

Honorary Chairs

Lakhmi C. Jain	University of Canberra, Australia and Bournemouth University, UK
Gloria Wren-Phillips	Loyola University, USA
Junzo Watada	Waseda University, Japan

General Chair

Ireneusz Czarnowski	Gdynia Maritime University, Poland

Executive Chair

Robert J. Howlett	KES International and Bournemouth University, UK

Program Chair

Alfonso Mateos Caballero	Universidad Politécnica de Madrid, Spain

Publicity Chair

Izabela Wierzbowska	Gdynia Maritime University, Poland

Special Sessions

Decision Making Theory for Economics

Eizo Kinoshita	Meijo University, Japan
Takao Ohya	Kokushikan University, Japan

Advances of Soft Computing in Industrial and Management Engineering: New Trends and Applications

Shing Chiang Tan	Multimedia University, Malaysia
Junzo Watada	Universiti Teknologi Petronas, Malaysia
Chee Peng Lim	Deakin University, Australia

Digital Architecture and Decision Management

Alfred Zimmermann	Reutlingen University, Germany
Rainer Schmidt	Munich University, Germany

Specialized Decision Techniques for Data Mining, Transportation and Project Management

Piotr Jędrzejowicz	Gdynia Maritime University, Poland
Ireneusz Czarnowski	Gdynia Maritime University, Poland

Interdisciplinary Approaches in Business Intelligence Research and Practice

Ivan Luković	University of Novi Sad, Serbia
Ralf-Christian Härting	Hochschule Aalen, Germany

Eye Movement Data Processing and Analysis

Katarzyna Harezlak	Silesian University of Technology, Poland
Paweł Kasprowski	Silesian University of Technology, Poland

Decision Support Systems

Wojciech Froelich	University of Silesia, Poland

Pattern Recognition for Decision Making Systems

Paolo Crippa	Università Politecnica delle Marche, Italy
Claudio Turchetti	Università Politecnica delle Marche, Italy

Reasoning-Based Intelligent Systems

Kazumi Nakamatsu	University of Hyogo, Japan
Jair M. Abe	Paulista University, Brazil

Intelligent Data Analysis and Applications

Urszula Stańczyk	Silesian University of Technology, Gliwice, Poland
Beata Zielosko	University of Silesia, Katowice, Poland

Social Media Analysis and Mining

Clara Pizzuti	National Research Council of Italy (CNR), Italy

International Program Committee

Jair Minoro Abe	Paulista University and University of São Paulo, Brazil
Witold Abramowicz	Poznan University Economics, Poland
Stefan Aier	University of St Gallen, Germany
Rainer Alt	Universitat Leipzig, Germany
Colin Atkinson	University of Mannheim, Germany
Ahmad Taher Azar	Benha University, Egypt
Dariusz Barbucha	Gdynia Maritime University, Poland
Alina Barbulescu	Higher Colleges of Technology, Sharjah, UAE
Monica Bianchini	Dipartimento di Ingegneria dell'Informazione e Scienze Matematiche, Italy
Andreas Behrend	University of Bonn, Germany
Mokhtar Beldjehem	University of Ottawa, Canada
Gloria Bordogna	CNR IDPA, Italy
Oliver Bossert	McKinsey & Co. Inc., Germany

János Botzheim	Tokyo Metropolitan University, Japan
Lars Brehm	University of Munich, Germany
Wei Cao	School of Economics, HeFei University of Technology, China
Frantisek Capkovic	Slovak Academy of Sciences, Slovakia
Mario Giovanni C.A. Cimino	University of Pisa, Italy
Marco Cococcioni	University of Pisa, Italy
Angela Consoli	Defence Science and Technology Group, Australia
Paulo Cortez	University of Minho, Portugal
Paolo Crippa	Università Politecnica delle Marche, Ancona, Italy
Alfredo Cuzzocrea	University of Trieste, Italy
Ireneusz Czarnowski	Gdynia Maritime University, Poland
Eman El-Sheikh	University of West Florida, USA
Margarita N. Favorskaya	Siberian State Aerospace University, Russia
Michael Fellmann	University of Rostock, Germany
Raquel Florez-Lopez	University Pablo Olavide of Seville, Spain
Bogdan Franczyk	Universitat Leipzig, Germany
Ulrik Franke	KTH Royal Institute of Technology, Sweden
Wojciech Froelich	University of Silesia, Poland
Mauro Gaggero	National Research Council of Italy, Italy
Mauro Gaspari	Universita' di Bologna, Italy
Marina Gavrilova	University of Calgary, Canada
Michael Gebhart	Iteratec GmbH, Germany
Raffaele Gravina	University of Calabria, Italy
Christos Grecos	Central Washington University, USA
Foteini Grivokostopoulou	University of Patras, Greece
Jerzy W. Grzymala-Busse	University of Kansas, USA
Ralf-Christian Härting	Hochschule Aalen, Germany
Katarzyna Harezlak	Silesian University of Technology, Poland
Ioannis Hatzilygeroudis	University of Patras, Greece
Robert Hirschfeld	University of Potsdam, Germany
Dawn E.Holmes	University of California, USA
Katsuhiro Honda	Osaka Prefecture University, Japan
Daocheng Hong	Fudan University and Victoria University, China
Tzung-Pei Hong	National University of Kaohsiung, Taiwan
Yuh-Jong Hu	National Chengchi University, Taipei, Taiwan
Yuji Iwahori	Chubu University, Japan
Joanna Jedrzejowicz	University of Gdansk, Poland
Piotr Jędrzejowicz	Gdynia Maritime University, Poland
Björn Johansson	Lund University, Sweden
Nikos Karacapilidis	University of Patras, Greece
Dimitris Karagiannis	Universitat Wien, Germany
Pawel Kasprowski	Silesian University of Technology, Poland

Radosław Katarzyniak	Wrocław University of Technology and Science, Poland
Eizo Kinoshita	Meijo University, Japan
Frank Klawonn	Ostfalia University, Germany
Petia Koprinkova-Hristova	Bulgarian Academy of Sciences
Marek Kretowski	Bialystok University of Technology, Poland
Vladimir Kurbalija	University of Novi Sad, Serbia
Kazuhiro Kuwabara	Ritsumeikan University, Japan
Birger Lantow	University of Rostock, Germany
Frank Leymann	Universitat Stuttgart, Germany
Chee Peng Lim	Deakin University, Australia
Ivan Luković	University of Novi Sad, Serbia
Neel Mani	Dublin City University, Dublin, Ireland
Alfonso Mateos-Caballero	Universidad Politécnica de Madrid, Spain
Shimpei Matsumoto	Hiroshima Institute of Technology, Japan
Raimundas Matulevicius	University of Tartu, Estonia
Mohamed Arezki MELLAL	M'Hamed Bougara University, Algeria
Lyudmila Mihaylova	University of Sheffield, UK
Yasser Mohammad	ASSIUT University and EJUST, Egypt
Masoud Mohammadian	University of Canberra, Australia
Michael Möhring	University of Munich, Germany
Stefania Montani	University of Piemonte Orientale, Italy
Daniel Moldt	University of Hamburg, Germany
Mikhail Moshkov	KAUST, Saudi Arabia
Kazumi Nakamatsu	University of Hyogo, Japan
Selmin Nurcan	University of Paris, France
Marek Ogiela	AGH University of Science and Technology, Krakow, Poland
Takao Ohya	Kokushikan University, Japan
Isidoros Perikos	University of Patras, Greece
Petra Perner	Institute of Computer Vision and applied Computer Sciences IBaI, Germany
James F. Peters	University of Manitoba, Winnipeg, Canada
Gunther Piller	University of Mainz, Germany
Camelia-M. Pintea	UT Cluj-Napoca, Romania
Clara Pizzuti	National Research Council of Italy (CNR), Italy
Bhanu Prasad	Florida A&M University, USA
Jim Prentzas	Democritus University of Thrace, Greece
Radu-Emil Precup	Politehnica University of Timisoara, Romania
Georg Peters	Munich University of Applied Sciences
Marcos Quiles	Federal University of São Paulo - UNIFESP
Miloš Radovanović	University of Novi Sad, Serbia
Sheela Ramanna	The University of Winnipeg, Canada
Ewa Ratajczak-Ropel	Gdynia Maritime University, Poland
Manfred Reichert	Universitat Uulm, Germany

John Ronczka	SCOTTYNCC Independent research Scientist, Australia
Kurt Sandkuhl	University of Rostock, Germany
Mika Sato-Ilic	University of Tsukuba, Japan
Miloš Savić	University of Novi Sad, Serbia
Rafał Scherer	Częstochowa University of Technology, Poland
Rainer Schmidt	Munich University, Germany
Christian Schweda	Technical University Munich, Germany
Hirosato Seki	Osaka Institute of Technology, Japan
Bharat Singh	Big Data Labs, Hamburg, Germany
Urszula Stańczyk	Silesian University of Technology, Gliwice, Poland
Ulrike Steffens	Hamburg University of Applied Sciences, Germany
Janis Stirna	Stockholm University, Sweden
Catalin Stoean	University of Craiova, Romania
Ruxandra Stoean	University of Craiova, Romania
Mika Sulkava	Natural Resources Institute Finland
Keith Swenson	Fujitsu America Inc., USA
Shing Chiang Tan	Multimedia University
Mieko Tanaka-Yamawaki	Tottori University, Japan
Dilhan J. Thilakarathne	VU University Amsterdam, Netherlands
Claudio Turchetti	Università Politecnica delle Marche, Italy
Eiji Uchino	Yamaguchi University, Japan
Pandian Vasant	Universiti Teknologi PETRONAS, Tronoh, Malaysia
Ljubo Vlacic	Griffith University, Gold Coast Australia
Zeev Volkovich	Ort Braude College, Karmiel, Israel
Gottfried Vossen	WWU Munster, Germany
Junzo Watada	Waseda University, Japan
Fen Wang	Central Washington University, USA
Matthias Wissotzki	University of Rostock, Germany
Dmitry Zaitsev	International Humanitarian University, Odessa, Ukraine
Gian Piero Zarri	University Paris sorbonne, France
Beata Zielosko	University of Silesia, Katowice, Poland
Alfred Zimmermann	Reutlingen University, Germany

International Referee Board

Contents

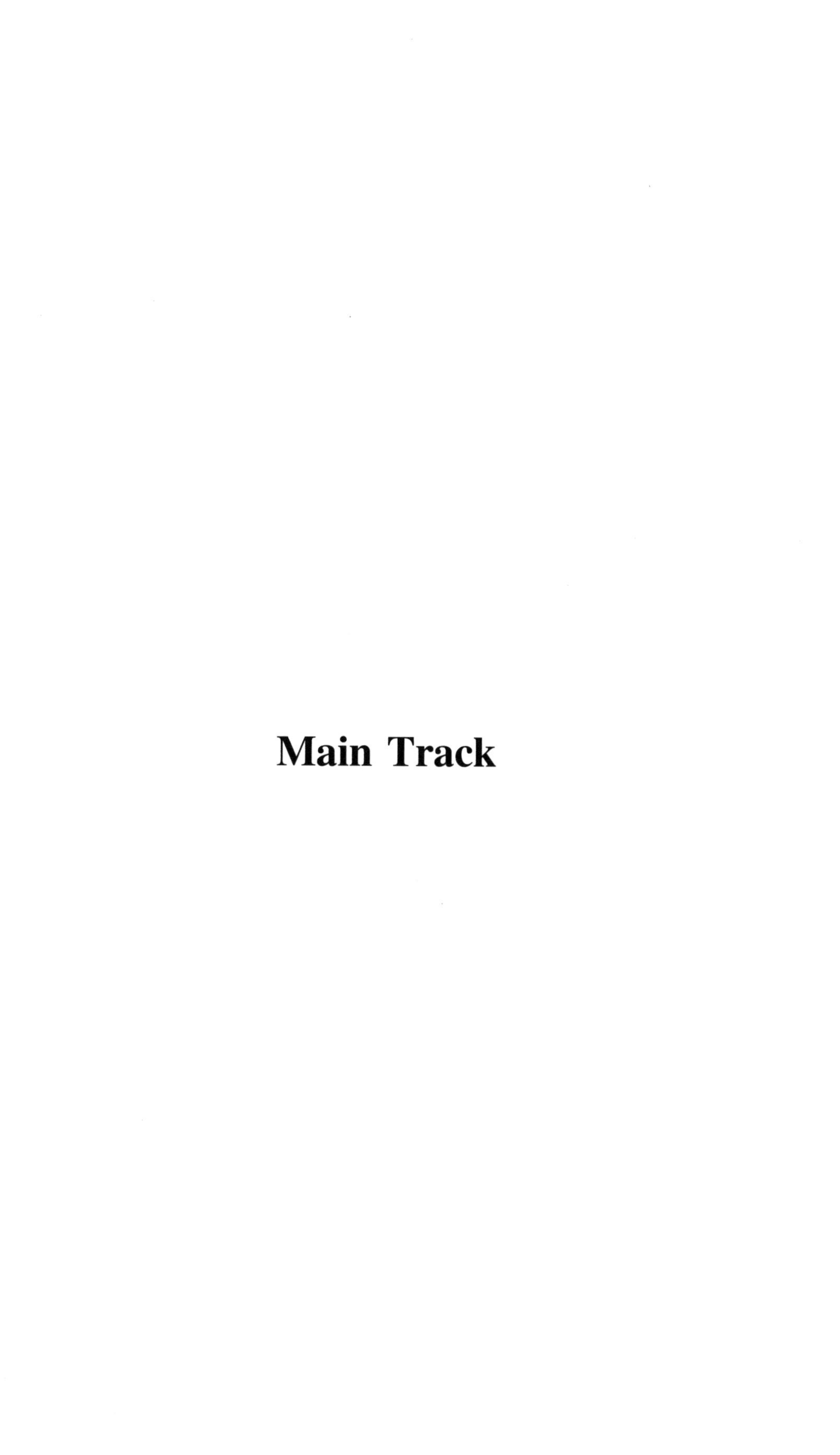

Main Track

The Shapley Value and Consistency Axioms of Cooperative Games Under Incomplete Information

Satoshi Masuya(✉)

Faculty of Business Administration, Daito Bunka University,
1-9-1, Takashimadaira Itabashi-ku, Tokyo 175-8571, Japan
masuya@ic.daito.ac.jp

Abstract. In this paper, we study cooperative TU games in which the worths of some coalitions are not known. We investigate superadditive games and the Shapley values for general partially defined cooperative games. We show that the set of the superadditive complete games and the set of the Shapley values which can be obtained from a given incomplete game. Furthermore, we propose selection methods of the one-point solution from the set of the Shapley values and axiomatize one of the proposed solutions.

Keywords: Cooperative game · Partially defined cooperative game · Superadditive game · Shapley value

1 Introduction

The cooperative game theory provides useful tools to analyze various cost and/or surplus allocation problems, the distribution of voting power in a parliaments or a country, and so on. The problems to be analyzed by the cooperative game theory include n entities called players and are usually expressed by characteristic functions which map each subset of players to a real number. The solutions to the problems are given by a set of n-dimensional real numbers or value functions which assign a real number to each player. Such a real number can show the cost borne by the player, power of influence, an allocation of the shared profits, and so on. Several solution concepts for cooperative games have been proposed. As representative examples of solution concepts, the core, the Shapley value [5] and the nucleolus [4] are well-known. The core can be represented by a set of solutions while the Shapley value and the nucleolus are one-point solutions.

A classical approach of von Neumann and Morgenstern [6] to cooperative games assumes that the worths of all coalitions are given. However, in the real world problems, there may exist situations in which the worths of some coalitions are unknown. Such cooperative games under incomplete information have not yet investigated considerably.

Cooperative games under incomplete information are first considered by Willson [7] which are called partially defined cooperative games. In Willson [7] he

I. Czarnowski et al. (eds.), *Intelligent Decision Technologies 2017*,
Smart Innovation, Systems and Technologies 72, DOI 10.1007/978-3-319-59421-7_1

proposed the generalized Shapley value which is obtained by using only known coalitional worths of a game and he axiomatized the proposed Shapley value. After that, Housman [1] continued the study of Willson [7]. Housman [1] characterized the generalized Shapley value by Willson [7]. However, the simplest interpretation of the generalized Shapley value is that it coincides with the ordinary Shapley value of a game whose coalitional worths take zero if they are unknown and given values otherwise. Such a game usually dissatisfies natural properties such as superadditivity.

In our previous paper [2,3], we propose an approach to partially defined cooperative games from a different angle from Willson [7] and Housman [1]. We have assumed that some coalitional worths are known but the others are unknown. For the sake of simplicity, in those papers, a partially defined cooperative game, i.e., a cooperative game under incomplete information is called "an incomplete game" while a cooperative game with complete information is called "a complete game". In Masuya and Inuiguchi [2], although we have defined the lower and upper games associated with the given general incomplete game which is assumed the superadditivity, we have considered the simplest case when only worths of singleton coalitions and the grand coalition are known for the investigation of the solution concepts to incomplete games. Then, in Masuya [3], we investigated the Shapley values for more general incomplete games such that the set of coalitions whose cardinalities are one to k is known.

Then, in this paper, we investigate superadditive games and the Shapley values for completely general incomplete games. In this case, we show that the set of the superadditive complete games which can be obtained from a given incomplete game is a polytope. Further, we show the set of the Shapley values which can be obtained from a given incomplete game assumed the superadditivity. Furthermore, we propose selection methods of the one-point solution from the set of the Shapley values and axiomatize one of the proposed solutions. These results of incomplete games are generalized results of those obtained in Masuya [3]. In addition to these results, we investigate properties which we call consistency axioms.

This paper is organized as follows. In Sect. 2, we introduce the classical cooperative game and well-known solution concepts. Further, we present partially defined cooperative games, the lower and upper games and their properties by [2]. In Sect. 3, we investigate the set of superadditive games and the set of the Shapley values which can be obtained from a given incomplete game. In Sect. 4, we propose two selection methods of the one-point solution from the set of the Shapley values. Further, we investigate properties which we call consistency axioms. In Sect. 5, concluding remarks are given.

2 Classical Cooperative Games and Partially Defined Cooperative Games

Let $N = \{1, 2, \dots, n\}$ be the set of players and $v\colon 2^N \to \mathbb{R}$ such that $v(\emptyset) = 0$. A classical cooperative game, i.e., a coalitional game with transferable utility (a TU game) is characterized by a pair (N, v). A set $S \subseteq N$ is regarded as a coalition of players and the number $v(S)$ represents a collective payoff that players in S can gain by forming coalition S. For arbitrary coalition S, the number $v(S)$ is called the worth of coalition S.

A game (N, v) is said to be superadditive if and only if

$$v(S \cup T) \geq v(S) + v(T), \ \forall S, \ T \subseteq N \text{ such that } S \cap T = \emptyset. \tag{1}$$

Superadditivity is a natural property that gives each player an incentive to form a larger coalition.

Now let us introduce basic solution concepts in cooperative games. In cooperative games, it is assumed that the grand coalition N forms. The problem is how to allocate the collective payoff $v(N)$ to all players. A solution is a vector $x = (x_1, x_2, \ldots, x_n) \in \mathbb{R}^n$ where each component $x_i \in \mathbb{R}$ represents the payoff to player i. Many solution concepts have been proposed. We describe the Shapley value.

A solution x is efficient in a game (N, v) iff $\sum_{i \in N} x_i = v(N)$. A set of requirements $x_i \geq v(\{i\}), \forall i \in N$ is called the individual rationality. Let $I(N, v)$ denote the set of payoff vectors which satisfy efficiency and individual rationality in (N, v), or "imputations".

The Shapley value is a well-known one-point solution concept. It selects an imputation when the game is superadditive.

Let $G(N)$ be the set of all cooperative games with the player set N. For convenience, because the set of players is fixed as N, cooperative game (N, v) is denoted simply by v. Let π be a vector function from $G(N)$ to $\mathbb{R}^n$ specifying a payoff vector to a cooperative game. The i-th component of π is denoted by π_i.

The Shapley value is characterized by four axioms; axioms of null player, symmetry, efficiency and additivity. Player i is said to be a *null player* if and only if $v(S) - v(S \setminus i) = 0$ for all $S \subseteq N$ such that $i \in S$. Then the axiom of null player means $\pi_i(v) = 0$ if i is a null player. The axiom of symmetry means if players i and j are equivalent in the sense that $v(S \setminus i) = v(S \setminus j)$ for all $S \subseteq N$ such that $\{i, j\} \subseteq S$ then $\pi_i(v) = \pi_j(v)$. The axiom of efficiency means the satisfaction of the efficiency, i.e., $\sum_{i \in N} \pi_i(v) = v(N)$. The axiom of additivity means $\pi(v + w) = \pi(v) + \pi(w)$ for all $v, w \in G(N)$, where the sum of games $v + w \in G(N)$ is defined by $(v + w)(S) = v(S) + w(S), \ \forall S \subseteq N$.

The Shapley value is known as the unique function $\phi : G(N) \to \mathbb{R}^n$ satisfying these four axioms (see Shapley [5]) and its explicit form is

$$\phi_i(v) = \sum_{\substack{S \subseteq N \\ S \ni i}} \frac{(|S| - 1)!(n - |S|)!}{n!} (v(S) - v(S \setminus i)), \ \forall i \in N. \tag{2}$$

In the rest of this section, we present the results on partially defined cooperative games by Masuya and Inuiguchi [2]. In classical cooperative games, we assume that worths of all coalitions are known. However, in the real world problems, there may exist situations in which worths of some coalitions are unknown.

To avoid the confusion, we call such games "incomplete games" and the conventional cooperative games "complete games".

The incomplete games can be characterized by a set of players $N = \{1, 2, \dots, n\}$, a set of coalitions whose worths are known, say $\mathcal{K} \subseteq 2^N$, and a function $\nu : \mathcal{K} \to \mathbb{R}$, with $\emptyset \in \mathcal{K}$ and $\nu(\emptyset) = 0$. We assume that worths of singleton coalitions and the grand coalition are known and the worths of singleton coalitions are nonnegative, i.e., $\{i\} \in \mathcal{K}$ and $\nu(\{i\}) \geq 0$, $i \in N$ and $N \in \mathcal{K}$. Moreover, we assume that ν is superadditive in the following sense,

$$\nu(S) \geq \sum_{i=1}^{s} \nu(T_i),\ \forall S, T_i \in \mathcal{K},\ i = 1, 2, \dots, s \text{ such that } \bigcup_{i=1,2,\dots,s} T_i = S$$
$$\text{and } T_i,\ i = 1, 2, \dots, s \text{ are disjoint.} \tag{3}$$

A triple $(N, \mathcal{K}, \nu)$ identifies an incomplete game. When we consider only games under fixed N and $\mathcal{K}$, incomplete game $(N, \mathcal{K}, \nu)$ is simply written as ν.

Given an incomplete game $(N, \mathcal{K}, \nu)$, we define two associated complete games $(N, \underline{\nu})$ and $(N, \overline{\nu})$:

$$\underline{\nu}(S) = \max_{\substack{T_i \in \mathcal{K},\ i=1,2,\dots,s \\ \cup_i T_i = S,\ T_i \text{ are disjoint}}} \sum_{i=1}^{s} \nu(T_i), \tag{4}$$

$$\overline{\nu}(S) = \min_{\hat{S} \in \mathcal{K},\ \hat{S} \supseteq S} \left(\nu(\hat{S}) - \underline{\nu}(\hat{S} \setminus S) \right) \tag{5}$$

From the superadditivity of ν, we have $\underline{\nu}(S) = \nu(S)$ and $\overline{\nu}(S) = \nu(S)\ \forall S \in \mathcal{K}$.

A complete game (N, v) such that $v(T) = \nu(T)$, $\forall T \in \mathcal{K}$ is called a complete extension of $(N, \mathcal{K}, \nu)$, or simply a complete extension of ν. As shown in the following theorem, $\underline{\nu}(S)$ is the minimal payoff of coalition S among superadditive complete extensions of $(N, \mathcal{K}, \nu)$. On the other hand, $\overline{\nu}(S)$ is the maximal payoff of coalition S among superadditive complete extensions of $(N, \mathcal{K}, \nu)$.

Theorem 1 ([2]). *Let $(N, \mathcal{K}, \nu)$ be an incomplete game, and $(N, \underline{\nu})$ and $(N, \overline{\nu})$ the complete extensions defined by (4) and (5). For an arbitrary superadditive complete extension (N, v) of $(N, \mathcal{K}, \nu)$, we obtain*

$$\underline{\nu}(S) \leq v(S),\ \forall S \subseteq N, \tag{6}$$
$$\overline{\nu}(S) \geq v(S),\ \forall S \subseteq N. \tag{7}$$

Therefore, we call complete games $(N, \underline{\nu})$ and $(N, \overline{\nu})$ "lower game" and "upper game" associated with $(N, \mathcal{K}, \nu)$, respectively. When there is no confusion of the underlying incomplete game, those games are simply called the lower game and the upper game.

Using the incomplete information expressed by ν, we may consider the set $V(\nu)$ of possible complete games. More explicitly let the set of superadditive completions of ν be

$$V(\nu) = \{v : 2^N \to \mathbb{R} \mid v \text{ is superadditive},\ v(S) = \nu(S),\ \forall S \in \mathcal{K}\}. \tag{8}$$

3 The Set of Superadditive Complete Extensions of Partially Defined Cooperative Games

In this section, we investigate the set of superadditive complete games $V(\nu)$ which can be obtained from a given general incomplete game $(N, \mathcal{K}, \nu)$.

Consider a finite set of coalitions $\mathcal{T} = \{T_1, \ldots, T_m\}$ whose coalitional worths are not known and each coalition in the set is not included each other. That is, we consider $\mathcal{T}$, for $T_p \in \mathcal{T}$, $T_p \notin \mathcal{K} \setminus \{N\}$ holds and for arbitrary $T_p, T_s \in \mathcal{T}$ $(p \neq s)$, $T_p \not\subseteq T_s$ and $T_s \not\subseteq T_p$ holds. Note that the cardinality of $\mathcal{T}$ which is denoted by m is not constant but varies corresponding to $\mathcal{T}$. Although a number of the set $\mathcal{T}$ exists, the number of $\mathcal{T}$ is finite. The set of all $\mathcal{T}$ is denoted by $\Gamma'(N, \mathcal{K})$. We define the complete game dependent on $\mathcal{T} \in \Gamma'(N, \mathcal{K})$ as follows:

$$v^{\mathcal{T}}(S) = \begin{cases} \overline{\nu}(S), & \text{if } \exists T \in \mathcal{T},\ S \supseteq T, \text{ where } S \notin \mathcal{K}, \\ \underline{\nu}(S), & \text{if } \forall T \in \mathcal{T}, S \not\supseteq T, \text{ where } S \notin \mathcal{K}, \\ \nu(S), & \text{otherwise.} \end{cases} \tag{9}$$

The complete game $v^{\mathcal{T}}$ which is given by (9) takes the given value if its coalitional worths are known, otherwise, takes the value of the upper game when there are some coalitions in $\mathcal{T}$ which are included in the coalition S, and that of the lower game when there are no coalitions in $\mathcal{T}$ which are included in the coalition S. In other words, the complete game $v^{\mathcal{T}}$ is the game which takes the highest coalitional worths when the all members of some coalition in $\mathcal{T}$ are joined in the coalition S. That is, $\mathcal{T}$ is the list of coalitions which make a significant contribution.

Then, the next lemma is obtained.

Lemma 1. *Let $\mathcal{T} \in \Gamma'(N, \mathcal{K})$ satisfying $|T_k| \geq \lceil \frac{n}{2} \rceil$ $\forall k$. Then, $v^{\mathcal{T}}$ is superadditive.*

Let Γ^1 be the set of $\mathcal{T} \in \Gamma'(N, \mathcal{K})$ satisfying $|T_k| \geq \lceil \frac{n}{2} \rceil$ $\forall k$ and let Γ^2 be the set of $\mathcal{T} \in \Gamma'(N, \mathcal{K})$ such that $|T_k| < \lceil \frac{n}{2} \rceil$ holds at least one T_k and $v^{\mathcal{T}}$ is superadditive. Moreover, let $\Gamma(N, \mathcal{K}) = \Gamma^1 \cup \Gamma^2$. Clearly, $\Gamma(N, \mathcal{K}) \subseteq \Gamma'(N, \mathcal{K})$ holds. $\Gamma(N, \mathcal{K})$ is the set of $\mathcal{T}$ where $v^{\mathcal{T}}$ is superadditive.

From the above lemma, the next theorem is obtained.

Theorem 2. *Let ν be a general incomplete game. Then, the set of superadditive complete games $V(\nu)$ is the polytope whose extreme points are $v^{\mathcal{T}}$, $\forall \mathcal{T} \in \Gamma(N, \mathcal{K})$. That is, the next expression holds.*

$$V(\nu) = \left\{ v : 2^N \to \mathbb{R} \;\middle|\; v = \sum_{\mathcal{T} \in \Gamma(N,\mathcal{K})} c_{\mathcal{T}} v^{\mathcal{T}}, \sum_{\mathcal{T} \in \Gamma(N,\mathcal{K})} c_{\mathcal{T}} = 1, c_{\mathcal{T}} \geq 0,\ \forall \mathcal{T} \in \Gamma(N, \mathcal{K}) \right\}. \tag{10}$$

Theorem 2 means that $v^{\mathcal{T}}, \forall \mathcal{T} \in \Gamma(N, \mathcal{K})$ are the extreme points of $V(\nu)$ and all superadditive complete games can be obtained by the convex combinations of those extreme points.

Finally, we investigate the set of all the Shapley values which can be obtained from a general incomplete game ν. This set is denoted by $\Phi(\nu)$.

From the linearity of the Shapley value and Theorem 2, the set of all the Shapley values which can be obtained from an incomplete game ν is the polytope whose extreme points are $\phi(v^{\mathcal{T}}), \forall \mathcal{T} \in \Gamma(N, \mathcal{K})$.

Example 1. Let $N = \{1, 2, 3, 4\}$, and let $\mathcal{K} = \{\{1\}, \{2\}, \{3\}, \{4\}, \{1, 2\}, \{1, 3\}, \{2, 3\}, \{1, 2, 3\}, \{1, 2, 3, 4\}\}$. We consider an incomplete game $\nu : \mathcal{K} \to \mathbb{R}$ with

$$
\begin{aligned}
&\nu(\{1\}) = 8,\ \nu(\{2\}) = 7,\ \nu(\{3\}) = 3,\\
&\nu(\{4\}) = 1,\\
&\nu(\{1, 2\}) = 20,\ \nu(\{1, 3\}) = 16, \nu(\{2, 3\}) = 11,\\
&\nu(\{1, 2, 3\}) = 29, \nu(\{1, 2, 3, 4\}) = 40.
\end{aligned}
$$

Then, using (4) and (5), the lower game $\underline{\nu}$ and the upper game $\overline{\nu}$ are obtained as follows:

$$
\begin{aligned}
&\underline{\nu}(\{1\}) = 8,\ \underline{\nu}(\{2\}) = 7,\ \underline{\nu}(\{3\}) = 3,\ \underline{\nu}(\{4\}) = 1,\\
&\underline{\nu}(\{1, 2, 3, 4\}) = 40,\\
&\underline{\nu}(\{1, 2\}) = 20,\ \underline{\nu}(\{1, 3\}) = 16,\\
&\underline{\nu}(\{2, 3\}) = 11,\ \underline{\nu}(\{1, 2, 3\}) = 29,\\
&\underline{\nu}(\{1, 4\}) = 9,\ \underline{\nu}(\{2, 4\}) = 8,\ \underline{\nu}(\{3, 4\}) = 4,\\
&\underline{\nu}(\{1, 2, 4\}) = 21,\ \underline{\nu}(\{1, 3, 4\}) = 17,\ \underline{\nu}(\{2, 3, 4\}) = 12.
\end{aligned}
$$

$$
\begin{aligned}
&\overline{\nu}(\{1\}) = 8,\ \overline{\nu}(\{2\}) = 7,\ \overline{\nu}(\{3\}) = 3,\ \overline{\nu}(\{4\}) = 1,\\
&\overline{\nu}(\{1, 2, 3, 4\}) = 40,\\
&\overline{\nu}(\{1, 2\}) = 20,\ \overline{\nu}(\{1, 3\}) = 16,\\
&\overline{\nu}(\{2, 3\}) = 11,\ \overline{\nu}(\{1, 2, 3\}) = 29,\\
&\overline{\nu}(\{1, 4\}) = 29,\ \overline{\nu}(\{2, 4\}) = 24,\ \overline{\nu}(\{3, 4\}) = 20,\\
&\overline{\nu}(\{1, 2, 4\}) = 37,\ \overline{\nu}(\{1, 3, 4\}) = 33,\ \overline{\nu}(\{2, 3, 4\}) = 32.
\end{aligned}
$$

4 Selection Methods of the One-Point Solution from the Set of the Shapley Values and Consistency Axioms

Considering the applications of incomplete games, it is necessary to select the rational one-point solution in some sense from the set of all the Shapley values.

In this section, we propose two selection methods of the one-point solution from the set of all the Shapley values $\Phi(\nu)$. These selection methods are generalized methods of those we proposed in [3]. After that, we axiomatize one of the two proposed solutions and investigate some consistency axioms which can be developed from the axiom system of the Shapley value which is described in Sect. 2.

First selection method is to select the center of gravity. Since the set of the Shapley values which can be obtained from ν is a polytope, we propose the center of gravity of $\Phi(\nu)$ as the rational Shapley value $\tilde{\phi}(\nu)$. That is, $\tilde{\phi}(\nu)$ is defined as follows.

$$\tilde{\phi}(\nu) = \frac{\sum_{\mathcal{T}\in\Gamma(N,\mathcal{K})} \phi(v^{\mathcal{T}})}{|\Gamma(N,\mathcal{K})|} \tag{11}$$

Second selection method is the way of the minimization of the maximal excess. This is the application of the idea from the definition of the nucleolus. Particularly, we define the excess by the difference between an interior point of the maximal and the minimal Shapley values and the payoff of each player. Further, we propose the payoff vector of the minimization of the maximal excess as the rational Shapley value.

Given an incomplete game ν, let $x = (x_1, \ldots, x_n)$ be an n-dimensional payoff vector where $x \in \Phi(\nu)$. Let $\overline{\phi}_i(\nu)$ be the maximal Shapley value of player i, and $\underline{\phi}_i(\nu)$ be the minimal Shapley value of player i. Let α be a real number which satisfies $0 \leq \alpha \leq 1$. Then, the excess of player i to payoff vector x is defined by

$$e_i^{\alpha}(x) = \alpha\overline{\phi}_i(\nu) + (1-\alpha)\underline{\phi}_i(\nu) - x_i. \tag{12}$$

Further, we arrange $e_i^{\alpha}(x)$ for all players $i \in N$ in nonincreasing order. Then the arranged vector is denoted by $\theta(x)$.

Moreover, when two payoff vectors x, y satisfies the next condition, we call x is lexicographically smaller than y.

$$\begin{array}{c} \theta_1(x) < \theta_1(y), \text{ or} \\ \exists h \in \{1, 2, \ldots, n\} \text{ such that} \\ \theta_i(x) = \theta_i(y), \forall i < h \text{ and } \theta_h(x) < \theta_h(y) \end{array} \tag{13}$$

Finally, we call the Shapley value which is lexicographically minimized the minimized maximal excess Shapley value $\hat{\phi}^{\alpha}(\nu)$ or the min-max excess Shapley value in short.

We obtain the following theorem.

Theorem 3. *Let ν be an incomplete game. Let $\overline{\phi}_i(\nu)$ be the maximal Shapley value which player i can obtain and $\underline{\phi}_i(\nu)$ be the minimal Shapley value. Then, $\hat{\phi}^{\alpha}(\nu)$ is given by*

$$\hat{\phi}_i^{\alpha}(\nu) = \alpha\overline{\phi}_i(\nu) + (1-\alpha)\underline{\phi}_i(\nu) \ \forall i \in N,$$

$$\textit{where } \alpha = \frac{\nu(N) - \sum_{i\in N} \underline{\phi}_i(\nu)}{\sum_{i\in N} \left(\overline{\phi}_i(\nu) - \underline{\phi}_i(\nu)\right)}. \tag{14}$$

Note that the number $\alpha (0 \leq \alpha \leq 1)$ is given uniquely for arbitrary ν. Therefore, from Theorem 3, it can be said that the min-max excess Shapley value can be obtained easily.

Next, we axiomatize the gravity center of the Shapley values $\tilde{\phi}(\nu)$.

Definition 1. *Let ν be an incomplete game. Then we consider two players $i, j \in N$ and an arbitrary $S \subseteq N \setminus \{i, j\}$ satisfying $S \cup \{i\} \in \mathcal{K}$ and $S \cup \{j\} \in \mathcal{K}$. Then, if $\nu(S \cup i) = \nu(S \cup j)$ $\forall S \cup \{i\} \in \mathcal{K}$ and $\forall S \cup \{j\} \in \mathcal{K}$ holds, we call players i, j are symmetric in ν.*

Axiom 1 (ν-consistent symmetry). *Let ν be an incomplete game. We assume that two players $i, j \in N$ are symmetric in ν. Then, we call that the Shapley value $\pi(\nu) \in \Phi(\nu)$ satisfies ν-consistent symmetry if the following holds.*

$$\pi_i(\nu) = \pi_j(\nu) \tag{15}$$

Axiom 1 means that if two players i and j are symmetric in an incomplete game, the two players are symmetric in the superadditive complete extension of the game and the Shapley value of player i coincides with that of player j.

Axiom 2 (the principle of insufficient reason). *Let ν be an incomplete game and the Shapley value $\pi \in \Phi(\nu)$ be $\pi = \sum_{\mathcal{T} \in \Gamma(N, \mathcal{K})} \phi(c_{\mathcal{T}} v^{\mathcal{T}})$ where $\sum_{\mathcal{T} \in \Gamma(N, \mathcal{K})} c_{\mathcal{T}} = 1$ and $c_{\mathcal{T}} \geq 0$ $\forall \mathcal{T} \in \Gamma(N, \mathcal{K})$. Then, with respect to two families of coalitions $\mathcal{T}_1, \mathcal{T}_2 \in \Gamma(N, \mathcal{K})$, if $\mathcal{T}_2$ is not an exchange of symmetric players belonged to $\mathcal{T}_1$, the following holds:*

$$c_{\mathcal{T}_1} = c_{\mathcal{T}_2}.$$

$c_{\mathcal{T}}$ can be interpreted as the reward by forming the family of coalitions $\mathcal{T}$. Therefore, if there is no additional information on the relation of two families of coalitions such that one family can be obtained by an exchange of symmetric players belonged to the other family, there is no sufficient reason why one family can have more reward than the other family.

Then the following theorem is obtained.

Theorem 4. *$\tilde{\phi}(\nu)$ is the unique Shapley value in $\Phi(\nu)$ satisfying Axioms 1 and 2.*

From Theorem 4, the gravity center of the Shapley values $\tilde{\phi}(\nu)$ is axiomatized.

Here, we consider the meaning of ν-consistent symmetry axiom as a consistency axiom which works as a bridge from incomplete games to complete games. We developed ν-consistent symmetry axiom inspired from symmetry axiom which is belonged to the axiom system of the Shapley value described in Sect. 2. As well as developing the ν-consistent symmetry axiom, we can develop consistency axioms of other axioms belonged to the axiom system of the Shapley value, that is, the null player axiom and the additivity axiom. In the rest of this section, we develop ν-consistent null player axiom and ν-consistent additivity axiom and investigate whether the gravity center of the Shapley values satisfies these axioms or not.

Definition 2. *Let ν be an incomplete game and $i \in N$. Then if $\nu(S \cup i) - \nu(S) = 0$ holds for an arbitrary $S \subseteq N \setminus \{i\}$ satisfying $S \cup \{i\} \in \mathcal{K}$ and $S \in \mathcal{K}$, we call player i is a null player in ν.*

Axiom 3 (ν-consistent null player). *Let ν be an incomplete game and $i \in N$. We assume that player i is a null player in ν. Then, we call that the Shapley value $\pi(\nu)$ satisfies ν-consistent null player if the following holds.*

$$\pi_i(\nu) = 0. \tag{16}$$

Axiom 4 (ν-consistent additivity). *Let ν_1 and ν_2 be two incomplete games. The sum of two incomplete games $\nu_1 + \nu_2$ is defined by $(\nu_1 + \nu_2)(S) = \nu_1(S) + \nu_2(S)$, $\forall S \in \mathcal{K}$. Then, we call that the Shapley value π satisfies ν-consistent additivity if the following holds.*

$$\pi(\nu_1 + \nu_2) = \pi(\nu_1) + \pi(\nu_2). \tag{17}$$

Theorem 5. *The gravity center of the Shapley values $\tilde{\phi}(\nu)$ does not satisfy Axiom 3 or 4.*

Table 1. The set of coalitions $\mathcal{T}$, the Shapley values of each player $\phi(v^{\mathcal{T}})$ and $\tilde{\phi}(\nu)$

$\mathcal{T}$	$\phi_1(v^{\mathcal{T}})$	$\phi_2(v^{\mathcal{T}})$	$\phi_3(v^{\mathcal{T}})$	$\phi_4(v^{\mathcal{T}})$
{1,2,3,4}	15.5	12.5	8.5	3.5
{1,2,4}	16.83	13.83	4.5	4.83
{1,3,4}	16.83	8.5	9.83	4.83
{2,3,4}	10.5	14.16	10.16	5.16
{1,2,4}, {1,3,4}	18.16	9.83	5.83	6.16
{1,2,4}, {2,3,4}	11.83	15.5	6.16	6.5
{1,3,4}, {2,3,4}	11.83	10.16	11.5	6.5
{1,2,4}, {1,3,4}, {2,3,4}	13.16	11.5	7.5	7.83
{1,4}	19.83	8.16	4.16	7.83
{2,4}	10.5	16.83	4.83	7.83
{3,4}	10.5	8.83	12.83	7.83
{1,4}, {2,4}	13.5	11.16	4.5	10.83
{1,4}, {3,4}	13.5	8.5	7.16	10.83
{2,4}, {3,4}	10.5	11.5	7.5	10.5
{1,4}, {2,4},{3,4}	12.16	9.83	5.83	12.16
{1,4}, {2,3,4}	14.83	9.83	5.83	9.5
{2,4}, {1,3,4}	11.83	12.83	6.16	9.16
{3,4}, {1,2,4}	11.83	10.16	8.83	9.16
$\tilde{\phi}(\nu)$	13.53	11.31	7.31	7.83

From this theorem, we can conclude that the gravity center of the Shapley values satisfies ν-consistent symmetry axiom only among the three consistency axioms.

Example 2. Using the result of Example 1, we obtain the Shapley values of each v^T and the gravity center $\tilde{\phi}(\nu)$ as Table 1.

Moreover, using Theorem 3, the min-max excess Shapley value $\hat{\phi}^\alpha(\nu)$ can be obtained as follows:

$$\hat{\phi}^\alpha(\nu) = (14.11, 11.51, 7.51, 6.85), \text{ where } \alpha = 0.38.$$

5 Concluding Remarks

In this paper, we investigated the superadditive complete extensions and the Shapley values of general partially defined cooperative games. We have shown that the set of superadditive complete extensions and the set of the Shapley values of a general incomplete game are polytopes respectively. Further, we proposed two selection methods of the one-point solution from the set of the Shapley values and axiomatized one of them, which can be interpreted as generalized methods of those proposed in Masuya [3]. Finally, we considered consistency axioms which work as bridges from incomplete games to complete games. That is, we developed ν-consistent symmetry axiom, ν-consistent null player axiom and ν-consistent additivity axiom. Finally, we have shown that the gravity center of the Shapley values satisfies ν-consistent symmetry axiom only among these three axioms.

References

1. Housman, D.: Linear and symmetric allocation methods for partially defined cooperative games. Int. J. Game Theor. **30**, 377–404 (2001)
2. Masuya, S., Inuiguchi, M.: A fundamental study for partially defined cooperative games. Fuzzy Optim. Decis. Making **15**, 281–306 (2016)
3. Masuya, S.: The Shapley value on a class of cooperative games under incomplete information. In: Czarnowski, I. et al. (eds.) SIST, vol. 56, pp. 129–139. Springer (2016)
4. Schmeidler, D.: The nucleolus of a characteristic function game. SIAM J. Appl. Math. **17**, 1163–1170 (1969)
5. Shapley, L.S.: A value for n-person games. In: Kuhn, H., Tucker, A. (eds.) Contributions to the Theory of Games II, pp. 307–317. Princeton (1953)
6. Von Neumann, J., Morgenstern, O.: Theory of Games and Economic Behavior. Princeton University Press, Princeton (1944)
7. Willson, S.J.: A value for partially defined cooperative games. Int. J. Game Theor. **21**, 371–384 (1993)

Patients' EEG Data Analysis via Spectrogram Image with a Convolution Neural Network

Longhao Yuan and Jianting Cao(✉)

Graduate School of Engineering, Saitama Institute of Technology,
Fusaiji 1690, Fukaya-shi, Saitama 3690293, Japan
e60051mu@sit.ac.jp

Abstract. Electroencephalogram (EEG) recording is relatively safe for the patients who are in deep coma or quasi brain death, so it is often used to verify the diagnosis of brain death in clinical practice. The objective of this paper is to apply deep learning method to EEG signal analysis in order to confirm clinical brain death diagnosis. A novel approach using spectrogram images produced from EEG signals as the input dataset of Convolution Neural Network (CNN) is proposed in this paper. A deep CNN was trained to obtain the similarity degree of the patients' EEG signals with the clinical diagnosed symptoms. This method can evaluate the condition of the brain damage patients and can be a reliable reference of quasi brain death diagnosis.

Keywords: Deep learning · CNN · EEG · Spectrogram image · Brain death diagnosis

1 Introduction

Deep learning is a new method of training multi-layer neural network. Despite of the insufficiencies of shallow learning that are optimization difficulty and short in feature expression ability, deep learning has the unique hierarchical structure and the capability of extracting high-level features from low-level features which can solve these problems of shallow learning. [1] Back propagation (BP) is a typical algorithm of traditional shallow learning, it appears bad performance when the number of hidden layers increases [2]. In 2006, Hinton [3] and Bengio Y [4] proposed the unsupervised greedy layer-wise training algorithm based on Deep Belief Networks (DBN) which brings hope to solving multi-layer neural network optimization problems. From that time, deep learning has become a new area of machine learning and applied in the fields of speech recognition, computer vision, nature language processing and information retrieval. Lecun et al. [5, 6] proposed Convolutional Neural Networks (CNN) algorithm and applied it in MNIST handwritten digits recognition. Several methods were applied to the CNN to reduce the number of weights so the deep hierarchical structure can be trained in an acceptable time.

Electroencephalogram (EEG) is a recording of voltage fluctuations produced by ionic current flows in the neurons of brain and refers to the recording of the brain's spontaneous electrical activity over a period of time. EEG signal is applied to many fields such

I. Czarnowski et al. (eds.), *Intelligent Decision Technologies 2017*,
Smart Innovation, Systems and Technologies 72, DOI 10.1007/978-3-319-59421-7_2

as brain-computer interface (BCI) [7], diagnosis of brain-related diseases like Epilepsy [8] and Alzheimer [9]. In our previous work, many algorithms were proposed to analyze EEG signal and evaluate the state of patients' brain activity [10–12]. EEG signal is acquired from human scalp by measuring electrical activities at different electrode positions. Raw EEG signals are the record of amplified voltage varies with time. Like other electric signals, EEG signals can be characterized by amplitude and frequency. Human brain EEG signals are classified according to four different frequency bands, Delta (0.5 to 4 Hz), Theta (4 to 8 Hz), Alpha (8 to 13 Hz) and Beta (13 to 30 Hz) [13]. The signal features of each band can reflect human's physical conditions. Raw EEG signal is time-domain signal format and needs to be processed in order to obtain useful features. It is usually analyzed by three methods, time-domain analysis, frequency-domain analysis and time-frequency analysis. For frequency-based analyze, Fourier Transform (FT) is often used to transform raw EEG signal into frequency-domain signal.

Spectrogram image is a visual representation of signals. The frequency spectrum of spectrogram image varies with time and different colors on the image represent different energy values. Spectrogram image is another form of raw EEG signal's feature representation. Comparing to using some feature extraction methods, spectrogram image contains more unknown features of EEG signals and may have a better performance in a classification network. In order to produce spectrogram images from EEG signals to represent the features of EEG, Short Time Fourier Transform (STFT) technique was applied as a time-frequency analysis method.

With the aim to using EEG to help clinical brain death diagnosis, this paper applied a novel method of using EEG signals to train a CNN to accomplish EEG signal classification work. Firstly the study of deep learning, EEG and spectrogram image was briefly introduced. Then several works of brain death diagnosis, signal classification and time-frequency analysis which are related to this paper's study was illustrated. Next, the basic principle of CNN was explained. Finally, the experiment method and result was elaborated, and the conclusion and future work was proposed.

2 Related Work

Chen Z, Cao J, Cao Y, et al. [12] did a series of work related to brain death diagnosis. They firstly applied independent component analysis (ICA) and Fourier and time-frequency analysis to EEG signals processing. Then they used several methods to do statistical complexity measures in order to evaluate the difference between coma EEG and brain death EEG. Significant differences of the two kinds of samples were obtained from the preliminary experimental results.

There are several studies related to EEG-based signal classification. Koelstra et al. [14] applied a support vector machine (SVM) classifier to realize emotion classification by single trail EEG signals. Power spectrum density (PSD) of EEG signals was used as input features of the classifier and two levels of valence states and two levels of arousal states were classified. Junhua Li and Andrzej Cichocki [15] used a multi-fractal attributes extraction method to extract useful features from EEG signals. Then the extracted features were put into a deep network initialized by a block of denoising auto encoder (DAE) to recognize the subjects' motor imagery.

Although time-frequency analysis is applied to EEG signal processing [16] and other areas [17], very few are in image processing area. Mustafa [18] trained an Artificial Neuron Network (ANN) for brainwave balancing application. Spectrogram images were produced by spectrum data of EEG signals and Gray Level Co-occurrence Matrix (GLCM) features of the images was extracted as the input of the ANN.

3 Convolution Neural Network

Convolution Neural Network (CNN) is a deep learning algorithm which achieves high performance especially in image classification area. CNN uses relative space position relations to reduce the number of training parameters by a large margin in order to increase training speed and training performance. As shown in Fig. 1, a normal CNN consists of image input layer, convolution layer, sub-sample layer, and classification layer. Different CNNs vary in different algorithms of convolution layer and sub-sample layer and different structures of the network.

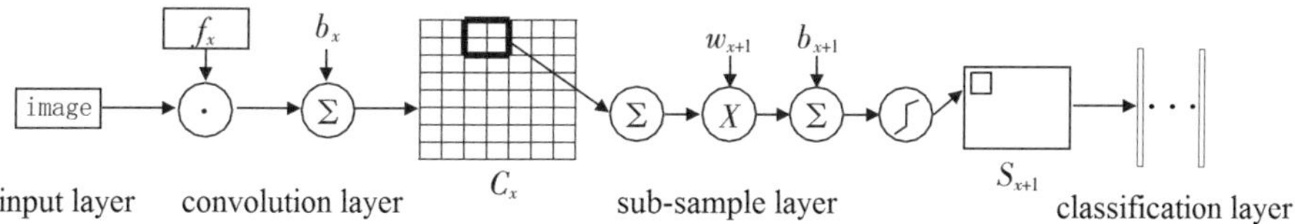

Fig. 1. Structure diagram of a typical CNN

3.1 Image Input Layer

Image input layer receives raw images from training samples and transforms the data into a unified form in order to deliver the data into next layer correctly. This layer also defines the initial parameters such as the scale of the local receptive fields and different filters.

3.2 Convolution Layer

Convolution layer processes the input data by convolution algorithm and produces several layers called feature map which consist of the convolution calculation results from the previous layers. The output equation of the convolution calculation is as follows:

$$x_j^l = f\left(\sum\nolimits_{i \in M_j} x_i^{l-1} * k_{ij}^l + b_j^l\right) \quad (1)$$

$$f(x) = \frac{1}{1 + e^{-x}} \quad (2)$$

Where x is the output value of the convolution layer, k is the kernel (or called the filter), l is the number of output layers which is decided by the number of kernels, i is the stride that the kernel moves in every step of calculation, M_j is the jth feature map produced by different kernels, b is the bias and f is an activation function usually defined as a sigmoid function showed in Eq. (2). Though every output neuron has different receptive fields, every neuron of the same feature map shares the same weights and bias. In this way, training parameters are greatly decreased.

3.3 Sub-sample Layer

This layer sub-samples every feature map from the previous convolution layer. The sub-sample method is weighted summation calculation or taking the maximum value in an $n \times n$ area of every feature map. The output of sub-sample layer is as follows:

$$x_j^l = f\left(\beta_j^l downsample\left(x_j^{l-1}\right) + b_j^l\right) \tag{3}$$

Where x^l is the output value of the l th sub-sample layer, *downsample* is the sub-sample function, β is the bias of the sub-sample function, f and b are the activation function and the bias respectively. Sub-sample layer reduces the number of training parameters, filters noises and avoids over-fitting of the network.

3.4 Classification Layer

After the data goes through several convolution layers and sub-sample layers, the size of output feature maps continuously decreases. For the classification layer, every feature map consists of only one neuron and becomes a 1D feature vetor. The vector is fully connected with a classifier. Usually the classifier is a traditional fully connected neural network.

4 Experiment Method and Results

The EEG signals used in this paper were obtained from the brain damage patients of a hospital in Shanghai. Each of the patients was diagnosed by clinical doctor. As for the experiment, firstly Short Time Fourier Transform (STFT) technique was applied as a time-frequency analysis method and spectrogram images were produced to represent the features of EEG signals. Then the images were labeled by patients' symptoms and used as the training samples in a deep convolutional neural network proposed in paper [19]. Finally the trained network was used to evaluate the similarity degree of the other patients' EEG signal with the two symptoms.

The experiment procedure is shown in Fig. 2 Firstly raw EEG signals of the patients were extracted from the data file by EEGLab Toolbox of Matlab software. The EEG data were pre-processed to prepare for producing spectrogram images. Then spectrogram images were labeled by the clinical diagnosed symptoms and the labeled images were used as the dataset to train a deep CNN. Finally, a set of raw EEG data from several other patients was put into the trained network to evaluate the state of the patients' conditions.

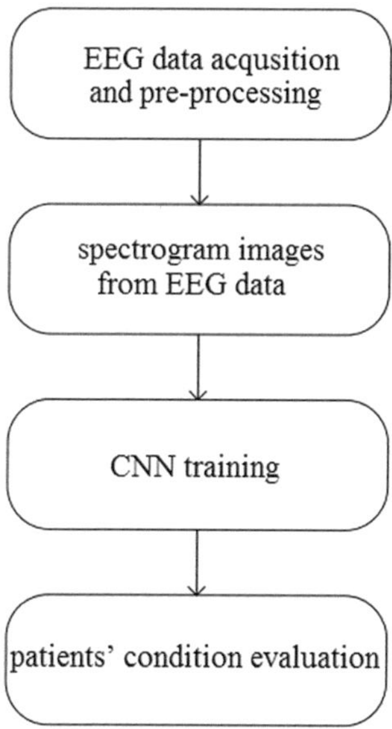

Fig. 2. Experiment method

4.1 EEG Data Acquisition and Pre-processing

The EEG data used in this paper was obtained from the brain damage patients in the intensive care unit (ICU) of a hospital in Shanghai. And the symptoms of the patients had been clinical diagnosed by the doctors before the record of the data. A total of 36 patients including 19 coma diagnosed patients and 17 brain death diagnosed patients with their age ranging from 18 to 85 years old were examined.

A portable EEG device named NEUROSCAN ESI was applied to measure EEG signals of the patients. Nine electrodes including six exploring electrodes (Fp1, Fp2, F3, F4, F7 and F8), one ground electrode (GND) and two reference electrodes (A1, A2) were used to record the EEG signals. The sampling rate was set as 1000 Hz and the resistance of the electrodes was set as less than 1000 kΩ. The recoding time of every patient ranged from 314 s to 1576 s.

The raw EEG signals used in this paper were obtained from the above method. Pre-processing including filtering specified voltage range (–150 μV to 150 μV) and frequency range (0.5 Hz to 30 Hz was done to filter the noise.

4.2 Spectrogram Images from EEG Data

The spectrogram images were produced from pre-processed EEG signals using STFT method at Matlab software platform. According to time sequence, every 20 s of the EEG signals was sampled and produced one spectrogram image. In order to increase the number of produced images, six channels of the EEG signals were used to produce spectrogram images respectively. The individual differences of the patients were ignored so the images produced from different patients were taken as one dataset. And to make the most use of the data, every window of STFT overlapped 20% with the adjacent windows. Finally every produced spectrogram image was resized to 256 × 256 pixels for the purpose of matching the input format of the deep CNN.

Figure 3 (a) and (b) are the spectrogram images of coma patient and brain death patient respectively. However, the EEG signals used to produce spectrogram images are raw signals so they may contain some noises. Obviously some images like Fig. 4 (a) and (b) were produced by noise signals so they were removed manually. Totally, the EEG data of 15 coma patients and 15 brain death patients was used and 6000 spectrogram images including 2400 images produced from coma patients and 3600 images produced from brain death patients were produced, labeled and disorganized to be used as the dataset of the deep CNN.

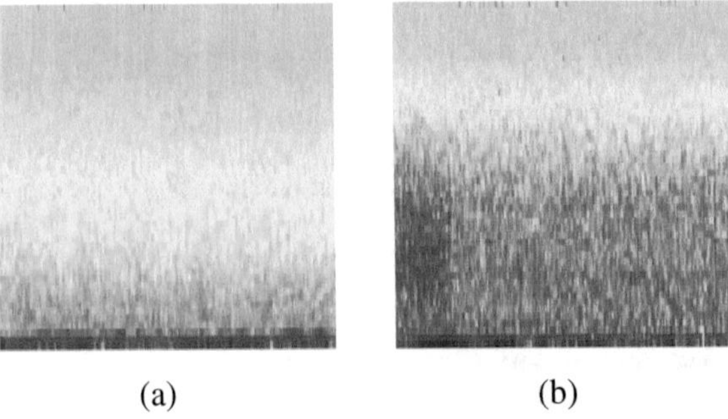

(a) (b)

Fig. 3. EEG spectrogram images for coma patient (a) and brain death patient (b)

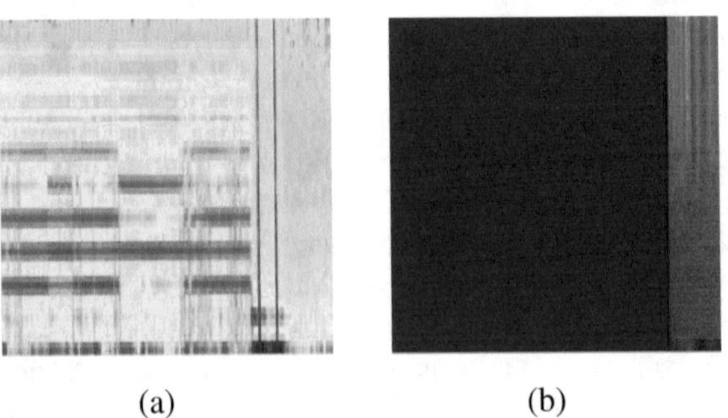

(a) (b)

Fig. 4. Spectrogram images (a) and (b) produced from noise signals

4.3 CNN Training

In the training step, a computer with a Core (TM) i7-2600 K (3.40 GHz), a 16.0 GB DDR3 and a GTX 1060 Graphics Processing Unit (GPU) was used. The operating system is Linux-Ubuntu 14.04.

A deep learning framework named Caffe which is developed by Yangqing Jia [20] was used to build the deep convolution neural network. The deep CNN from paper [12] in ImageNet classification and we used this network directly to train our spectrogram image dataset. This deep CNN has many convolution layers and has a large-scale input which can receive data from a high pixel image. By using this network, every single image can contain the data of a long period of time of EEG signals, and this is good for feature expression of time sequence data. The number of output unit of the classifier should be equal to the category of the class label so only the classification layer of the network was modified.

As for the dataset, 80% was used as training samples and 20% was used as testing samples. The max iteration was set as 12000 and GPU was used for training the network. The training time was 26 min and the test accuracy of this network is 99.8%.

4.4 Patients' Condition Evaluation

The EEG data from the other 6 brain damage patients which was not used in training the deep CNN model was used to evaluate the trained network. For every patient, 100 spectrogram images were extracted from the EEG data. Then according to the clinical diagnosis results, the images were put into the trained network to validate the recognition accuracy.

Table 1 shows the accuracy of every patient's validation result. Different from the test accuracy obtained from the training process, this accuracy is the recognition accuracy of samples from every single patient. Hence this can be regarded as the similarity degree of a patient's EEG signal with coma or brain death symptom.

Table 1. Experiment results

Diagnosed results	Samples	Accuracy (%)
Coma	100	95
Coma	100	89
Coma	100	96
Coma	100	88
Brain death	100	94
Brain death	100	92

5 Conclusion and Future Work

According to the experiment results, based on the trained deep CNN, the highest accuracy of coma and brain death diagnosed patients' samples are 96% and 94% respectively.

It proves the method this paper proposed is feasible and the method of converting a time-domain signal into several discrete spectrogram images is a good way of applying EEG signal to deep learning classification.

Using spectrogram images as the feature expression of EEG is a novel approach of EEG signal processing. Compared to the former brain death diagnosis studies which used many signal processing and feature extraction methods, we used raw signals to express more EEG features. Instead of working out some quantitative analysis results from processed EEG signals, we obtained the similarity between the symptoms and the patients' samples, which is more reliable in symptom evaluation. The method proposed in this paper can help evaluate the state of quasi brain death patient and be a reliable reference of the clinical diagnosis of brain-related illness.

In future work, instead of training the existing network which is initially used in other fields, we will build and test a specialized deep neural network of brain death diagnosis in order to raise calculation speed and classification accuracy. And an automatical denoising method should be proposed instead of picking out noising spectrogram images manually in case of processing mass data. In addition, more patient samples should be tested by the trained network in order to raise the reliability of the method.

References

1. Bengio, Y., Delalleau, O.: On the expressive power of deep architectures. In: International Conference on Algorithmic Learning Theory, pp. 18–36. Springer, Heidelberg (2011)
2. Bengio, Y.: Learning deep architectures for AI. Found. Trends® Mach. Learn. **2**(1), 1–127 (2009)
3. Hinton, G.E., Osindero, S., Teh, Y.W.: A fast learning algorithm for deep belief nets. Neural Comput. **18**(7), 1527–1554 (2006)
4. Bengio, Y., Lamblin, P., Popovici, D., et al.: Greedy layer-wise training of deep networks. Adv. Neural. Inf. Process. Syst. **19**, 153 (2007)
5. LeCun, Y., Bottou, L., Bengio, Y., et al.: Gradient-based learning applied to document recognition. Proc. IEEE **86**(11), 2278–2324 (1998)
6. Sermanet, P., Chintala, S., LeCun, Y.: Convolutional neural networks applied to house numbers digit classification. In: 21st International Conference on Pattern Recognition (ICPR), pp. 3288–3291. IEEE (2012)
7. Wolpaw, J.R., et al.: Brain–computer interfaces for communication and control. Clin. Neurophysiol. **113**(6), 767–791 (2002)
8. Ocak, H.: Automatic detection of epileptic seizures in EEG using discrete wavelet transform and approximate entropy. Expert Syst. Appl. **36**(2), 2027–2036 (2009)
9. Morabito, F.C., Labate, D., La Foresta, F., et al.: Multivariate multi-scale permutation entropy for complexity analysis of Alzheimer's disease EEG. Entropy **14**(7), 1186–1202 (2012)
10. Cao, J., Chen, Z.: Advanced EEG signal processing in brain death diagnosis. In: Signal Processing Techniques for Knowledge Extraction and Information Fusion, pp. 275–298. Springer, US (2008)
11. Cao, J.: Analysis of the quasi-brain-death EEG data based on a robust ICA approach. In: International Conference on Knowledge-Based and Intelligent Information and Engineering Systems. Springer (2006)
12. Chen, Z., Cao, J., Cao, Y., et al.: An empirical EEG analysis in brain death diagnosis for adults. Cogn. Neurodyn. **2**(3), 257–271 (2008)

13. Teplan, M.: Fundamentals of EEG measurement. Measur. Sci. Rev. **2**(2), 1–11 (2002)
14. Koelstra, S., Yazdani, A., Soleymani, M., et al.: Single trial classification of EEG and peripheral physiological signals for recognition of emotions induced by music videos. In: Brain Informatics, pp. 89–100. Springer, Heidelberg (2010)
15. Li, J., Cichocki, A.: Deep learning of multifractal attributes from motor imagery induced EEG. In: International Conference on Neural Information Processing, pp. 503–510. Springer (2014)
16. Roach, B.J., Mathalon, D.H.: Event-related EEG time-frequency analysis: an overview of measures and an analysis of early gamma band phase locking in schizophrenia. Schizophr. Bull. **34**(5), 907–926 (2008)
17. Goren, Y., Davrath, L.R., Pinhas, I., et al.: Individual time-dependent spectral boundaries for improved accuracy in time-frequency analysis of heart rate variability. IEEE Trans. Biomed. Eng. **53**(1), 35–42 (2006)
18. Mustafa, M., Taib, M.N., Murat, Z.H., et al.: The analysis of EEG spectrogram image for brainwave balancing application using ANN. In: UkSim 13th International Conference on Computer Modelling and Simulation (UKSim), pp. 64–68. IEEE (2011)
19. Krizhevsky, A., Sutskever, I., Hinton, G.E.: Imagenet classification with deep convolutional neural networks. In: Advances in Neural Information Processing Systems, pp. 1097–1105 (2012)
20. Jia, Y., Shelhamer, E., Donahue, J., et al.: Caffe: convolutional architecture for fast feature embedding. In: Proceedings of the 22nd ACM International Conference on Multimedia, pp. 675–678. ACM (2014)

Decremental Subset Construction

Gianfranco Lamperti[1(✉)] and Xiangfu Zhao[2]

[1] Department of Information Engineering, University of Brescia, Brescia, Italy
gianfranco.lamperti@unibs.it

[2] College of Maths, Physics and Information Engineering, Zhejiang Normal University, Jinhua, China

Abstract. Finite automata are exploited in several domains, including language processing, artificial intelligence, automatic control, and software engineering. Let $\mathcal{N}$ be a nondeterministic finite automaton (NFA) and $\mathcal{D}$ the deterministic finite automaton (DFA) equivalent to $\mathcal{N}$. Assume to decrement $\mathcal{N}$ by $\Delta\mathcal{N}$, thus obtaining the decremented NFA $\mathcal{N}' = \mathcal{N} \setminus \Delta\mathcal{N}$, where some states and/or transitions are missing. Consider determinizing $\mathcal{N}'$. Instead of determinizing $\mathcal{N}'$ from scratch using the classical *Subset Construction* algorithm, we propose *Decremental Subset Construction*, an algorithm which generates the DFA $\mathcal{D}'$ equivalent to $\mathcal{N}'$ by updating $\mathcal{D}$ based on $\mathcal{N}$ and $\Delta\mathcal{N}$. This way, only the actions necessary to transform $\mathcal{D}$ into $\mathcal{D}'$ are applied. Although evidence from worst-case complexity analysis indicates that *Decremental Subset Construction* is not better than *Subset Construction*, in practice, when $\mathcal{N}$ is large and $\Delta\mathcal{N}$ relatively small, *Decremental Subset Construction* may outperform *Subset Construction* significantly.

1 Introduction

A finite automaton (FA) [12] can be either deterministic (DFA) or nondeterministic (NFA). For practical reasons, it is often convenient to transform an NFA into an equivalent DFA. This determinization process is typically performed by *Subset Construction* [22]. On the other hand, there are some application domains, such as diagnosis of active systems in artificial intelligence [16–18,20,21], or model-based testing in software engineering [1,2], where an NFA expands over discrete time in a sequence $\mathcal{A} = [\mathcal{N}_0, \mathcal{N}_1, \ldots, \mathcal{N}_q]$, where each NFA $\mathcal{N}_i$, $i \in [1 \mathinner{..} q]$, is an augmentation (a set of additional states and transitions) of the previous NFA $\mathcal{N}_{i-1}$, and determinization is required for each NFA in $\mathcal{A}$ in the given order. Basically, given an NFA $\mathcal{N}$, the DFA $\mathcal{D}$ equivalent to $\mathcal{N}$, and an augmentation $\Delta\mathcal{N}$ of $\mathcal{N}$, the problem consists in generating the DFA $\mathcal{D}'$ equivalent to $\mathcal{N}' = \mathcal{N} \cup \Delta\mathcal{N}$. Although $\mathcal{N}'$ can be determinized out of its context by *Subset Construction*, this may result in poor performance, especially if $\mathcal{N}$ is large and $\Delta\mathcal{N}$ relatively small. A better solution is to perform context-sensitive determinization by updating $\mathcal{D}$ into $\mathcal{D}'$ based on $\Delta\mathcal{N}$, thereby avoiding recomputing the whole transition function of $\mathcal{D}'$. To this end, several incremental determinization techniques have been proposed in recent years, both for acyclic automata [13–15] and

I. Czarnowski et al. (eds.), *Intelligent Decision Technologies 2017*,
Smart Innovation, Systems and Technologies 72, DOI 10.1007/978-3-319-59421-7_3

cyclic automata [3–5,19]. In addition, a conspicuous set of works for somewhat 'incremental processing' of FAs exists in the literature, including [6–11,23,24], although they are not designed for (and do not solve) the problem stated above.

In this paper we consider a different determinization problem, for which no specific algorithm exists in the literature, namely the *decremental determinization problem*. To solve this problem, a *Decremental Subset Construction* algorithm is then proposed.

2 Determinization of Finite Automata by *Subset Construction*

A finite automaton (FA) can be either deterministic (DFA) or nondeterministic (NFA). A DFA is a 5-tuple $(D, \Sigma, T_{\mathrm{d}}, d_0, F_{\mathrm{d}})$, where D is the set of states, Σ a finite set of symbols called the alphabet, T_{d} the transition function, $T_{\mathrm{d}} : D_\Sigma \mapsto D$, where $D_\Sigma \subseteq D \times \Sigma$, d_0 the initial state, and $F_{\mathrm{d}} \subseteq D$ the set of final states. Determinism comes from the transition function mapping a pair state-symbol into a single state. An NFA is a 5-tuple $(N, \Sigma, T_{\mathrm{n}}, n_0, F_{\mathrm{n}})$ where the fields have the same meaning as in the DFA, only that the transition function is nondeterministic, $T_{\mathrm{n}} : N_\Sigma^\varepsilon \mapsto 2^N$, where $N_\Sigma^\varepsilon \subseteq N \times (\Sigma \cup \{\varepsilon\})$, with ε being the *empty* symbol, $\varepsilon \notin \Sigma$. A transition from n to n' and marked by symbol ℓ is denoted $n \xrightarrow{\ell} n'$. Let $\mathbb{N} \subseteq N$. The ε-*closure* of $\mathbb{N}$ is the union of $\mathbb{N}$ and the set of states that are reached by a path of transitions exiting a state in $\mathbb{N}$ and marked by ε. Let ℓ be a symbol in the alphabet of the NFA. The ℓ-*closure* of $\mathbb{N}$ is the ε-*closure* of the set of states reached by the transitions leaving states in $\mathbb{N}$ and marked by ℓ. Each FA is associated with a regular language, namely the set of strings on the alphabet Σ generated by a path from the initial state to a final state. Two FAs are equivalent iff they share the same regular language. For each NFA there exists an equivalent DFA that can be generated by *Subset Construction* [22].

Example 1. Displayed on the left of Fig. 1 is the representation of an NFA, where 0 is the initial state, while 3 is the (only) final state (double circled). Next to the NFA are the intermediate representations of the equivalent DFA generated by *Subset Construction*. Each state of the DFA is identified by a subset of states of the NFA. Specifically, the initial state is the ε-*closure* of the initial state of

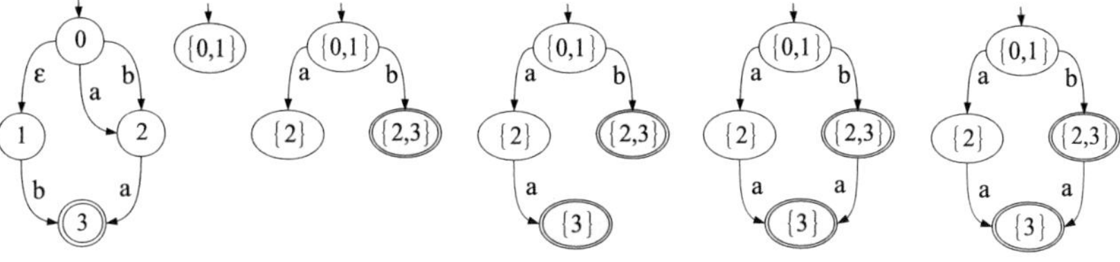

Fig. 1. Steps taken by *Subset Construction* for determinizing the NFA outlined on the left.

the NFA, namely $\{0, 1\}$. The DFA is generated with the support of a stack of DFA states (those to be processed, displayed in gray in Fig. 1). To this end, one state d at a time is popped from the stack, while the transitions exiting d are created, thereby possibly generating new states that are pushed into the stack. Transitions are created by considering each symbol $\ell \in \Sigma$ marking a transition exiting an NFA state in d. A transition $d \xrightarrow{\ell} d'$ is created where $d' = \ell\text{-}closure(d)$. A state of the DFA is final iff it includes a final state of the NFA. This process continues until the stack is empty, thereby obtaining the DFA on the right in Fig. 1. The resulting DFA share the same regular language of the NFA, namely $\{aa, b, ba\}$.

3 Decremental Subset Construction

The *Decremental Subset Construction* algorithm is designed to solve a decremental determinization problem, whose definition is formalized below (Definition 3).

Definition 1 (Decrement). *Let $\mathcal{N}$ be an NFA. A* decrement *$\Delta\mathcal{N}$ of $\mathcal{N}$ is a pair $(\Delta N, \Delta T)$, where ΔN is the set of removed states and ΔT the set of removed transitions. The decremented NFA is denoted $\mathcal{N} \setminus \Delta\mathcal{N}$.*

Definition 2 (*SC*-equivalence). *The DFA generated by* Subset Construction *by determinization of an NFA $\mathcal{N}$ is said to be SC-*equivalent *to $\mathcal{N}$.*

Definition 3 (Decremental Determinization Problem). *Let $\mathcal{N}$ be an NFA, $\mathcal{D}$ the DFA* SC*-equivalent to $\mathcal{N}$, and $\Delta\mathcal{N}$ a decrement of $\mathcal{N}$. Generate the DFA $\mathcal{D}'$* SC*-equivalent to the decremented NFA $\mathcal{N}' = \mathcal{N} \setminus \Delta\mathcal{N}$.*

In order to solve a decremental determinization problem efficiently, rather than applying *Subset Construction* to $\mathcal{N}'$, thereby disregarding both $\mathcal{D}$ and $\Delta\mathcal{N}$, $\mathcal{D}'$ is determined by updating $\mathcal{D}$ based on $\mathcal{N}$ and $\Delta\mathcal{N}$. To grasp such *Decremental Subset Construction* algorithm, we give a preliminary example, based on the notion of a *bud*, defined below.

Definition 4 (Bud). *Let d be the identifier of a state of the automaton $\mathcal{D}$ being processed by* Decremental Subset Construction*, ℓ either a symbol of the alphabet or ε, and $\mathbb{N}$ the ℓ-closure of $\|d\|$ in $\mathcal{N} \setminus \Delta\mathcal{N}$. The triple $(d, \ell, \mathbb{N})$ is a* bud *for $\mathcal{D}$.*

Intuitively, a bud $(d, \ell, \mathbb{N})$ indicates that the transition function of state d needs some processing. Specifically, when $\ell \neq \varepsilon$, the pair (d, ℓ) maps to a state identified by $\mathbb{N}$.

Example 2. Pictured on the left of Fig. 2 is an NFA $\mathcal{N}$, whose equivalent DFA is $\mathcal{D}_0$, with states d_0, d_1, and d_2. We distinguish the identifier d of a state from its *extension*, this being the set of NFA states included in d, denoted $\|d\|$. The dashed part of $\mathcal{N}$ indicates $\Delta\mathcal{N}$. Starting from $\mathcal{D}_0$, the algorithm performs a sequence of updates up to $\mathcal{D}_4$, with the latter being the DFA equivalent to $\mathcal{N} \setminus \Delta\mathcal{N}$. A stack $\mathcal{B}$ of buds is exploited. For reasons detailed in Algorithm 1, $\mathcal{B}$ is initialized

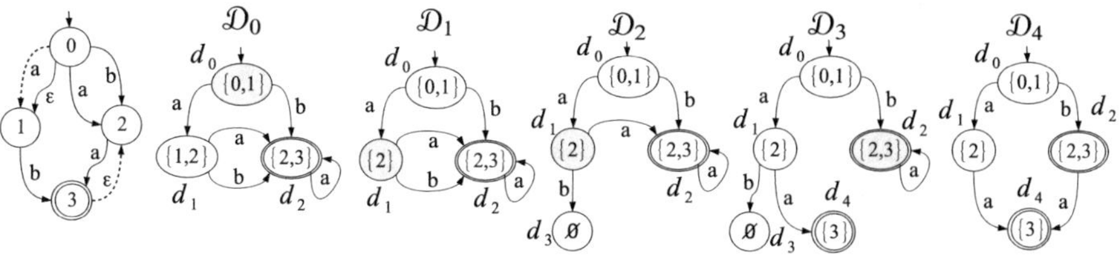

Fig. 2. Determinization by *Decremental Subset Construction* of the decremented NFA on the left.

with buds $(d_0, a, \{2\})$, $(d_1, b, \{3\})$, $(d_1, a, \{3\})$, $(d_2, a, \{3\})$. One bud at a time is processed. Processing $(d_0, a, \{2\})$ brings $\mathcal{D}_0$ to $\mathcal{D}_1$, where $\|d_1\|$ is reduced from $\{1, 2\}$ to $\{2\}$. This creates a new bud $(d_1, b, \emptyset)$, which replaces $(d_1, b, \{3\})$ in $\mathcal{B}$. Processing $(d_1, b, \emptyset)$ brings $\mathcal{D}_1$ to $\mathcal{D}_2$, with the additional transition $d_1 \xrightarrow{b} d_3$, where $\|d_3\| = \emptyset$. Processing $(d_1, a, \{3\})$ brings $\mathcal{D}_2$ to $\mathcal{D}_3$, with the new transition $d_1 \xrightarrow{a} d_4$. Processing $(d_2, a, \{3\})$ creates transition $d_2 \xrightarrow{a} d_4$. Since $\mathcal{B}$ is empty, the empty state d_3 and its entering transition are removed, thereby yielding the final DFA $\mathcal{D}_4$, which is equivalent to $\mathcal{N} \setminus \Delta\mathcal{N}$.

To avoid waste of processing, *Decremental Subset Construction* keeps each state d in $\mathcal{D}$ still reachable from the initial state d_0. In so doing, d is marked by its *distance*, written $\delta(d)$, this being the minimum number of transitions connecting d_0 with d. In particular, $\delta(d_0) = 0$. Based on the notion of distance, the set of states in $\mathcal{D}$ is partitioned into a finite set of *strata*, namely $D_0, D_1, \ldots, D_k$, with each stratum D_i including the set of states d such that $\delta(d) = i$. In particular, $D_0 = \{d_0\}$. Buds $(d, \ell, \mathbb{N})$ in the bud stack $\mathcal{B}$ are partially ordered based on $\delta(d)$. Consequently, *Decremental Subset Construction* updates $\mathcal{D}$ top-down, stratum by stratum. The stratum under processing, called the *front stratum*, is denoted $D_{\hat{\delta}}$, where $\hat{\delta}$ is the *front distance*.

While processing, $\mathcal{D}$ is partitioned into three regions: (1) the *prefix* of $\mathcal{D}$ (already processed), including strata D_i such that $i < \hat{\delta}$, (2) the front stratum $D_{\hat{\delta}}$ (under processing), and (3) the *suffix* of $\mathcal{D}$ (not yet processed), including strata D_j such that $j > \hat{\delta}$. States in the prefix are complete (in both extension and transition function) and cannot be changed by subsequent processing. By contrast, states in the suffix can subsequently change, in both extension and transition function. States in the front stratum are fixed, in both number and extension, while their transition function may change.

Since states in $\mathcal{D}$ are marked by their distance, updating $\mathcal{D}$ may require distance changes. *Decremental Subset Construction* performs distance relocation only as much as necessary, specifically up to stratum $D_{\hat{\delta}+1}$. Only when the bud stack becomes empty is distance relocation performed on the suffix of $\mathcal{D}$. To this end, a *relocation sequence* $\mathbb{R}$ is exploited, where states with possible changed distance are inserted.

The pseudo-code of *Decremental Subset Construction* (lines 1–53) is split into two parts, namely Algorithms 1 and 2. First, the main data structures are

Algorithm 1. Decremental Subset Construction (Part 1)

1: **procedure** DecrementalSubsetConstruction($\mathcal{N}$, $\mathcal{D}$, $\Delta\mathcal{N}$)
2: $\mathcal{N} = (N, \Sigma, T_{\mathrm{n}}, n_0, F_{\mathrm{n}})$: an NFA
3: $\mathcal{D} = (D, \Sigma, T_{\mathrm{d}}, d_0, F_{\mathrm{d}})$: the DFA SC-equivalent to $\mathcal{N}$
4: $\Delta\mathcal{N} = (\Delta N, \Delta T_{\mathrm{n}})$: a decrement of $\mathcal{N}$

5: $\bar{\mathbb{N}} \leftarrow \{n \mid n \in N, n \xrightarrow{\ell} n' \in \Delta T_{\mathrm{n}}\}$
6: Remove from $\mathcal{N}$ all states and transitions in $\Delta\mathcal{N}$
7: $\mathcal{B}_{d_0} \leftarrow \{(d_0, \varepsilon, \mathbb{N}) \mid n_0 \xrightarrow{\varepsilon} n' \in \Delta T_{\mathrm{n}}, \mathbb{N} = \varepsilon\text{-}closure(n_0)\}$
8: $\mathcal{B}_{\varepsilon} \leftarrow \{(d, \ell, \mathbb{N}) \mid d \xrightarrow{\ell} d' \in T_{\mathrm{d}}, n \in \|d'\| \cap \bar{\mathbb{N}}, n \xrightarrow{\varepsilon} n' \in \Delta T_{\mathrm{n}}, \mathbb{N} = \ell\text{-}closure(\|d\|)\}$
9: $\mathcal{B}_{\ell} \leftarrow \{(d, \ell, \mathbb{N}) \mid \ell \neq \varepsilon, d \in D, n \in \|d\| \cap \bar{\mathbb{N}}, n \xrightarrow{\ell} n' \in \Delta T_{\mathrm{n}}, \mathbb{N} = \ell\text{-}closure(\|d\|)\}$
10: $\mathcal{B} \leftarrow \mathcal{B}_{d_0} \uplus \mathcal{B}_{\varepsilon} \uplus \mathcal{B}_{\ell}$ – *The union $\uplus$ keeps buds partially ordered based on state distance*
11: $\mathbb{R} \leftarrow [\,]$, $\hat{\delta} \leftarrow 0$ – *Initialization of relocation sequence $\mathbb{R}$ and front distance $\hat{\delta}$*
12: **while** $\mathcal{B}$ is not empty, with $\bar{d}$ being the state within the first bud of $\mathcal{B}$ **do**
13: **if** $\delta(\bar{d}) > \hat{\delta}$ **then**
14: Propagate($\mathbb{R}$) – *Lazy distance propagation*
15: **end if**
16: Remove the first bud $(d, \ell, \mathbb{N})$ from $\mathcal{B}$
17: $\hat{\delta} \leftarrow \delta(d)$
18: **if** $\ell = \varepsilon$ **then** – *Rule* $\mathcal{R}_0$
19: Update($d, \mathbb{N}$)
20: **else if** no transition marked by ℓ exits d **then**
21: **if** $d' \in D, \|d'\| = \mathbb{N}$ **then** – *Rule* $\mathcal{R}_1$
22: Insert transition $d \xrightarrow{\ell} d'$ into $\mathcal{D}$
23: Relocate($d', \delta(d) + 1$)
24: **else** – *Rule* $\mathcal{R}_2$
25: Create a new state d' in D, where $\|d'\| = \mathbb{N}$, along with relevant buds
26: Insert transition $d \xrightarrow{\ell} d'$ into $\mathcal{D}$
27: $\delta(d') \leftarrow \delta(d) + 1$
28: **end if** – *Part 2 continued in Algorithm 2*

initialized (lines 5–11), this being bud sequence $\mathcal{B}$, front distance $\hat{\delta}$, and relocation sequence $\mathbb{R}$. Then, one bud at a time is removed from $\mathcal{B}$ and processed in the main loop (lines 12–48). In lines 13–15, before actually processing the bud, a check is performed on the distance of the involved state $\bar{d}$: if $\bar{d}$ belongs to the suffix of $\mathcal{D}$ then distance propagation is carried out by auxiliary procedure *Propagate* (Algorithm 7). After removing the first bud $(d, \ell, \mathbb{N})$ from $\mathcal{B}$ and setting front distance $\hat{\delta}$ (lines 16–17), the bud is processed based on ℓ and the current transitions exiting d. Seven processing rules are considered, namely $\mathcal{R}_0, \ldots, \mathcal{R}_6$, which are detailed below.

In rule $\mathcal{R}_0$ (lines 18–19), when $\ell = \varepsilon$, $\|d\|$ is replaced by $\mathbb{N}$. Notice how, based on line 7, d is necessarily the initial state d_0.

In rule $\mathcal{R}_1$ (lines 21–23), when d is not exited by a transition marked by ℓ and there is a state d' such that $\|d'\| = \mathbb{N}$, a transition $d \xrightarrow{\ell} d'$ is created in $\mathcal{D}$,

Algorithm 2. Decremental Subset Construction (Part 2)

29: **else** – *At least one transition marked by ℓ exits d*
30: **for all** $t \in T_{\mathrm{d}}, t = d \xrightarrow{\ell} d'$ **do**
31: **if** $d' \neq d_0$ and no other transition enters d' **then** – *Rule* $\mathcal{R}_3$
32: UPDATE($d', \mathbb{N}$)
33: **else if** $d' = d_0$ **or** $t' \in T_{\mathrm{d}}, t' \neq t, t' = d_p \xrightarrow{x} d', \delta(d_p) \leq \hat{\delta}$ **then**
34: **if** $d'' \in D, \|d''\| = \mathbb{N}$ **then** – *Rule* $\mathcal{R}_4$
35: Redirect t toward d''
36: RELOCATE($d'', \delta(d) + 1$)
37: **else** – *Rule* $\mathcal{R}_5$
38: Create a new state d'' in D, where $\|d''\| = \mathbb{N}$, along with relevant buds
39: Redirect t toward d''
40: $\delta(d'') \leftarrow \delta(d) + 1$
41: **end if**
42: **else** – *Rule* $\mathcal{R}_6$
43: Remove transitions entering d' other than t and surrogate them with buds
44: UPDATE($d', \mathbb{N}$)
45: **end if**
46: **end for**
47: **end if**
48: **end while**
49: **if** $d \in D, \|d\| = \emptyset$ **then**
50: Remove from $\mathcal{D}$ all transitions entering/exiting d, as well as d
51: **end if**
52: PROPAGATE($\mathbb{R}$) – *Distance propagation on the suffix of $\mathcal{D}$*
53: **end procedure**

with the distance of d' being possibly updated by auxiliary procedure *Relocate* (Algorithm 6).

In rule $\mathcal{R}_2$ (lines 25–27), when d is not exited by a transition marked by ℓ and there is no state d' such that $\|d'\| = \mathbb{N}$, both a new state d', with $\|d'\| = \mathbb{N}$, and a new transition $d \xrightarrow{\ell} d'$ are created. To subsequently generate the transition function of d', buds for d' are inserted into $\mathcal{B}$ based on the transition function of NFA states in $\|d'\|$.

Rules $\mathcal{R}_3, \ldots, \mathcal{R}_6$ (outlined in Part 2, Algorithm 2), occur when there is at least one transition exiting d and marked by ℓ.[1] Hence, each of such rules can be applied several times for the same bud, as specified by the loop in lines 30–46.

In rule $\mathcal{R}_3$ (lines 31–32), when d' is not the initial state and no other transition enters d', the extension of d' is updated by auxiliary procedure *Update* (Algorithm 3).

[1] Because of possible merging of states by auxiliary procedure *Merge* (Algorithm 5), several transitions marked by the same symbol may exit the same state in $\mathcal{D}$. However, such nondeterminism in $\mathcal{D}$ disappears before ending the processing of buds, as guaranteed by Theorem 1.

Algorithm 3. *Update* replaces $\|d\|$ with $\mathbb{N}$, possibly generating new buds and removing existing ones. If there is already a state d' with same extension as d, depending on the distance of d' relative to $\hat{\delta}$, either a pruning of $\mathcal{D}$ or a merging of d and d' is performed.

```
procedure UPDATE(d, ℕ)
  if ℕ ⊂ ‖d‖ then
    ℕ_d ← ‖d‖ \ ℕ
    𝔹 ← {(d, ℓ, ℕ') | n ∈ ℕ_d, n -ℓ-> n' ∈ 𝒩', ℕ' = ℓ-closure(ℕ)}
    for all B ∈ 𝔹, B = (d, ℓ, ℕ') do
      if B' ∈ ℬ, B' = (d, ℓ, ℕ''), ℕ'' ≠ ℕ' then
        Remove B' from ℬ
      end if
      ℬ ← ℬ ⊎ [B]
    end for
    ‖d‖ ← ℕ
    if d ∈ F_d, ℕ ∩ F_n = ∅ then
      Remove d from F_d
    end if
    if d' ∈ D, ‖d'‖ = ‖d‖ then
      if δ(d') < δ̂ then
        PRUNE(d, d')
      else
        MERGE(d, d')
      end if
    end if
  end if
end procedure
```

In rule $\mathcal{R}_4$ (lines 34–36), when d is the initial state or there is another transition entering d' from a parent state d_p such that $\delta(d_p) \leq \hat{\delta}$ and there is d'' such that $\|d''\| = \mathbb{N}$, the transition exiting d is redirected toward d'', whose distance needs relocation.

Rule $\mathcal{R}_5$ (lines 38–40) is applied based on the same conditions of $\mathcal{R}_4$, except that there is no state d'' such that $\|d''\| = \mathbb{N}$. If so, a new state d'' is created (along with relevant buds), with $\|d''\| = \mathbb{N}$, and t is redirected toward d''.

Rule $\mathcal{R}_6$ (lines 43–44) occurs when condition in line 33 is not fulfilled. When so, all other transitions entering d' are removed and surrogated by buds, while the extension of d' is updated based on $\mathbb{N}$.

Notice that, when condition in line 33 holds, the connection of d' with the initial state does not depend on t, which can then be redirected to another state (rules $\mathcal{R}_4$ and $\mathcal{R}_5$). By contrast, when such condition does not hold, redirecting t away from d' may result in the disconnection of d'. This is why in $\mathcal{R}_6$, all other transitions entering d' are removed (and surrogated by buds), while preserving $d \xrightarrow{\ell} d'$.

Algorithm 4. *Prune* is called in Line 17 of *Update* (Algorithm 3) because $\|d\| = \|d'\|$ and $\delta(d') < \hat{\delta}$. Since the transition function of d' is complete, it is more efficient to perform a cascade pruning on the transition function of d (as long as state distances are greater than $\hat{\delta}$) rather than performing a cascade merging starting with d and d'.

procedure PRUNE(d, d')
  Redirect to d' all transitions entering d
  $\mathbb{P} \leftarrow [(d, d')]$
  repeat
    Remove the first pair (d_1, d_2) from $\mathbb{P}$
    for all transition $t = d_1 \xrightarrow{\ell} d'_1$ in $\mathcal{D}$ **do**
      Remove t from $\mathcal{D}$
      if $\delta(d'_1) > \hat{\delta}$ **then**
        Remove all transitions entering d'_1 and surrogates them with buds
        Let d'_2 be the state entered by the transition exiting d_2 and marked by ℓ
        if $\delta(d'_2 < \hat{\delta})$ **then**
          $\mathbb{P} \leftarrow \mathbb{P} \cup [(d'_1, d'_2)]$
        else
          UPDATE$(d'_1, \|d'_2\|)$
        end if
      end if
    end for
    Remove d_1 from $\mathcal{D}$ and relevant buds from $\mathcal{B}$
  until $\mathbb{P} = [\,]$
end procedure

When $\mathcal{B}$ becomes empty, in lines 49–51, if $\mathcal{D}$ includes a state d with empty extension then d and all transitions entering/exiting d are removed from $\mathcal{D}$.[2] Eventually, in line 52, distance relocation is propagated on the suffix of $\mathcal{D}$ (*Propagate*, Algorithm 7).

The processing state of *Decremental Subset Construction* is called a *configuration*, namely a pair $(\bar{\mathcal{D}}, \bar{\mathcal{B}})$, where $\bar{\mathcal{D}}$ is the current instance of automaton $\mathcal{D}$ and $\bar{\mathcal{B}}$ the current instance of bud stack $\mathcal{B}$. As such, the algorithm performs a *trajectory*, namely a finite sequence $[\alpha_0, \alpha_1, \ldots, \alpha_q]$ of configurations, where $\alpha_0 = (\mathcal{D}_0, \mathcal{B}_0)$ is the initial configuration, with $\mathcal{D}_0$ being the DFA equivalent to $\mathcal{N}$ and $\mathcal{B}_0$ the initial instance of $\mathcal{B}$ (line 10, Algorithm 1), while $\alpha_q = (\mathcal{D}_q, \mathcal{B}_q)$ is the final configuration, with $\mathcal{D}_q$ being the DFA equivalent to the decremented NFA $\mathcal{N}' = \mathcal{N} \setminus \Delta\mathcal{N}$ and $\mathcal{B}_q = \emptyset$.

Example 3. Depicted on the left of Fig. 3 is an NFA $\mathcal{N}' = \mathcal{N} \setminus \Delta\mathcal{N}$, with $\Delta\mathcal{N}$ being dashed. On the center is the DFA $\mathcal{D}$ equivalent to $\mathcal{N}$. On the right is the DFA $\mathcal{D}'$ equivalent to $\mathcal{N}'$. The trajectory of *Decremental Subset Construction* in determinizing $\mathcal{N}'$ is outlined in Fig. 4. For each configuration $\alpha_i = (\mathcal{D}_i, \mathcal{B}_i), i \in [0\,..\,9]$,

[2] Retaining the empty state until $\mathcal{B} = \emptyset$ is essential in order to avoid the disconnection of $\mathcal{D}$.

Algorithm 5. *Merge* merges two states d and d' with same extension into a single state. It also performs distance relocation.

```
procedure MERGE(d, d')
  if δ(d) > δ(d') then
    Swap d and d'
  end if
  Redirect to/from d all transitions entering/exiting d', while removing duplicated transitions
  Remove d' from 𝒟
  Convert to d and recompute all buds in ℬ relevant to d'
  for all additional child state d_c of d do
    RELOCATE(d_c, δ(d) + 1)
  end for
end procedure
```

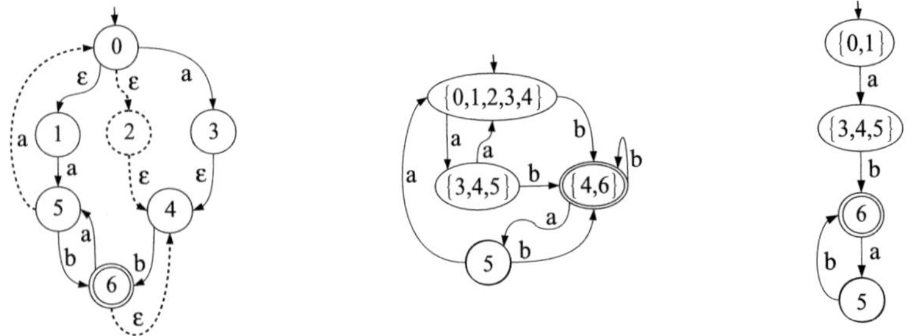

Fig. 3. An NFA $\mathcal{N}' = \mathcal{N} \setminus \Delta\mathcal{N}$, with $\Delta\mathcal{N}$ being dashed (left), the DFA $\mathcal{D}$ equivalent to $\mathcal{N}$ (center), and the DFA $\mathcal{D}'$ equivalent to $\mathcal{N}'$ (right).

the processed bud and the activated rule are highlighted. In configuration α_9, bud stack $\mathcal{B}$ is empty. Thus, according to Line 50, the dashed part relevant to empty state d_2 is removed. As expected, the resulting automaton equals the DFA $\mathcal{D}'$ displayed on the right of Fig. 3.

4 Soundness and Completeness

Decremental Subset Construction is functionally equivalent to *Subset Construction*, as supported by Theorem 1, whose proof is based on Definitions 5–6, and Lemmas 1–10.

Definition 5 (Balanced Transition). *Let* $(\bar{\mathcal{D}}, \bar{\mathcal{B}})$ *be a configuration of the algorithm. A transition* $d \xrightarrow{\ell} d'$ *in* $\bar{\mathcal{D}}$ *is* balanced *iff* $\|d'\| = \ell\text{-}closure(\|d\|)$. *Otherwise, the transition is* unbalanced.

Definition 6 (Viable Configuration). *A configuration* $\alpha_i = (\mathcal{D}_i, \mathcal{B}_i)$ *is* viable *iff, for each unbalanced transition* $d \xrightarrow{\ell} d'$ *in* $\mathcal{D}_i$, *bud sequence* $\mathcal{B}_i$ *includes a set of buds whose processing makes the transition balanced.*

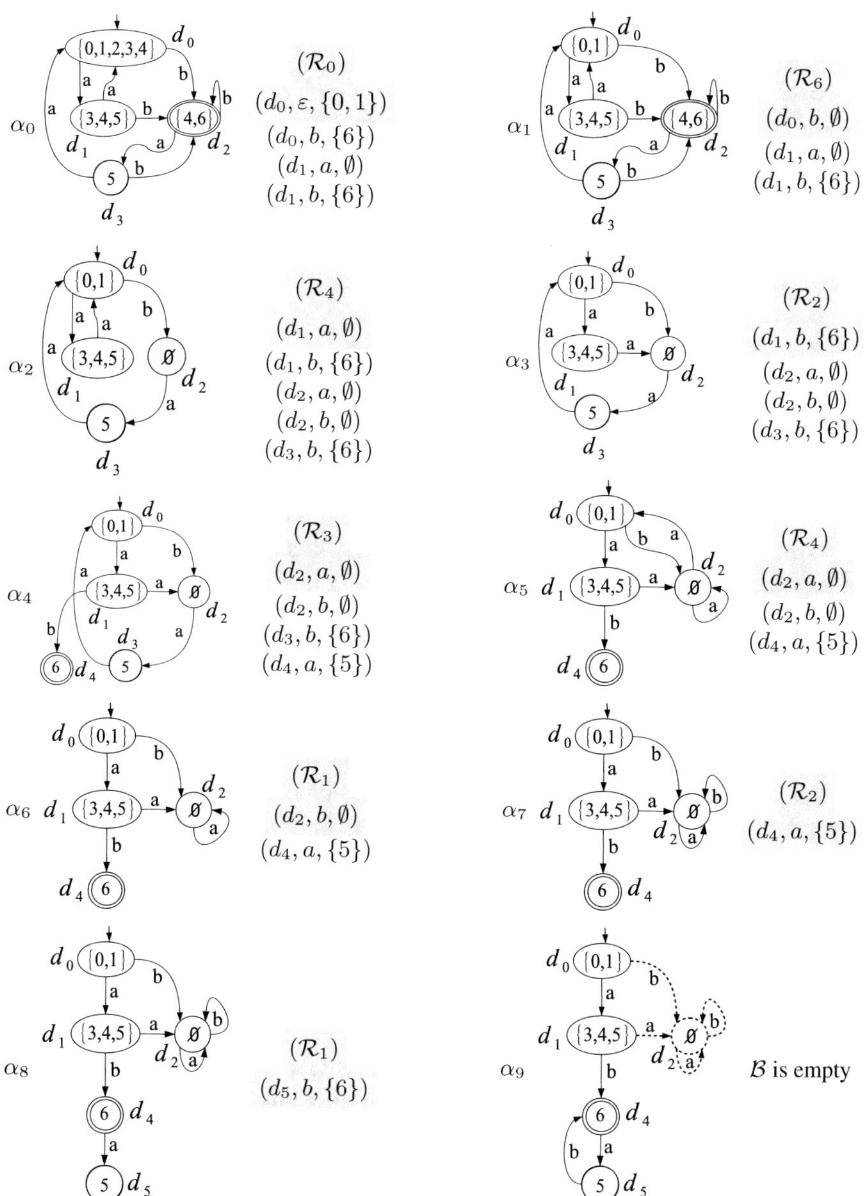

Fig. 4. Trajectory of *Decremental Subset Construction* for the decremented NFA $\mathcal{N}'$ in Fig. 3. For each configuration $\alpha_i = (\mathcal{D}_i, \mathcal{B}_i), i \in [0 \mathinner{..} 9]$, both the processed bud and the activated rule are highlighted.

Algorithm 6. *Relocate* performs distance relocation of state d based on distance $\bar{\delta}$. It also rearranges buds of d within $\mathcal{B}$ based on the new distance of d, and inserts d into relocation distance $\mathbb{R}$ for successive distance propagation by *Propagate* (Algorithm 7).

procedure RELOCATE$(d, \bar{\delta})$
  if $\delta(d) > \bar{\delta}$ **then**
    $\delta(d) \leftarrow \bar{\delta}$
    Rearrange buds of d within $\mathcal{B}$ based on distance $\delta(d)$
    $\mathbb{R} \leftarrow \mathbb{R} \cup [d]$
  end if
end procedure

Theorem 1. *Let $\mathcal{N}$ be an NFA, $\mathcal{D}$ the DFA* SC*-equivalent to $\mathcal{N}$, $\Delta\mathcal{N} = (\Delta N, \Delta T)$ a decrement of $\mathcal{N}$, $\mathcal{N}' = \mathcal{N} \setminus \Delta\mathcal{N}$ the decremented NFA, $\mathcal{D}'$ the DFA* SC*-equivalent to $\mathcal{N}'$, and $\mathcal{D}'_\Delta$ the DFA generated by* Decremental Subset Construction *by updating $\mathcal{D}$ based on $\mathcal{N}$ and $\Delta\mathcal{N}$. We have that $\mathcal{D}'_\Delta$ is identical to $\mathcal{D}'$.*

The proof of Theorem 1 is grounded on Lemmas 1–10. Each lemma is followed by a short rationale.

Lemma 1. *Let $(\mathcal{D}_i, \mathcal{B}_i)$ be a configuration of* Decremental Subset Construction, *$i \geq 0$, d a state in $\mathcal{D}_i$, $\delta(d)$ the distance marking d, and $\delta^*(d)$ the actual distance of d. If $\delta^*(d) \leq \hat{\delta} + 1$ then $\delta(d)$ equals $\delta^*(d)$ else $\delta(d) > \hat{\delta}$.*

This property is essential to distinguish between states with correct extension (up to distance $\hat{\delta}$) from states with possibly incorrect extension (at distance greater than $\hat{\delta}$). Lemma 1 can be proven by induction on the trajectory $[\alpha_0, \alpha_1, \ldots]$ of the algorithm. Specifically, each rule, involving a possible change in state distance, performs distance relocation correctly up to $\hat{\delta} + 1$.

Lemma 2. *Algorithm* Decremental Subset Construction *terminates.*

Nontermination might be caused by two reasons: either nontermination of auxiliary procedures or nontermination of the main loop (lines 12–48). It can be shown that both scenarios are impossible. In particular, nontermination of the main loop implies the consumption of an infinite number of buds. Being the number of possible buds $(d, \ell, \mathbb{N})$ finite, the same bud shall be generated (and consumed) an infinite number of times when processing the front stratum $D_{\hat{\delta}}$. Since the number of transitions exiting states in $D_{\hat{\delta}}$ is finite and since such transitions cannot be removed by rule $\mathcal{R}_6$, at a certain point, when all such transitions have been generated, the creation of a bud $(d, \ell, \mathbb{N})$ for a state $d \in D_{\hat{\delta}}$ is processed by no actions as $\|d\| = \mathbb{N}$, thus terminating bud generation.

Lemma 3. *Each transition in $\mathcal{D}'_\Delta$ is balanced* (Definition 5).

Algorithm 7. In Line 14 of Algorithm 1, *Propagate* performs distance propagation up to stratum $\delta(d)+1$, where d is the state of the first bud in $\mathcal{B}$. Instead, in Line 52 of Algorithm 2, since $\mathcal{B}$ is empty, distance propagation is performed on the suffix of $\mathcal{D}$.

```
procedure PROPAGATE(ℝ)
  if ℝ ≠ [] then
    repeat
      Remove from ℝ the first state d
      for all child state d_c of d do
        if δ(d_c) > δ(d) + 1 then
          δ(d_c) ← δ(d) + 1
          Rearrange buds of d_c within 𝓑 based on distance δ(d_c)
          ℝ ← ℝ ∪ [d_c]
        end if
      end for
      if 𝓑 = [(d, ℓ, ℕ), ...] then  – 𝓑 is not empty, with (d, ℓ, ℕ) being the first bud
        δ̄ ← δ(d) + 1
      end if
    until ℝ = [] or ℝ = [d', ...], δ(d') > δ̄
  end if
end procedure
```

This can be proven by induction on the trajectory of the algorithm, by showing that each configuration is viable (Definition 6). Specifically, the proof comes from the fact that $\mathcal{D}'_{\Delta}$ is the automaton in the last configuration, with the bud sequence being empty.

Lemma 4. *Let $n \xrightarrow{\ell} n'$ be a transition in $\mathcal{N}'$. Each state d in $\mathcal{D}'_{\Delta}$, such that $n \in \|d\|$, is exited by an ℓ-transition.*

If $n \in \|d\|$ in $\mathcal{D}$ already and n is not removed from $\|d\|$, then the property holds (even if the exiting ℓ-transition is temporarily removed and surrogated by buds). If n is not initially in $\|d\|$ (possibly because d is created later) then the insertion of n into $\|d\|$ is accompanied by the creation of a bud $(d, \ell, \mathbb{N})$, whose processing will create the ℓ-transition exiting d.

Lemma 5. *The initial states of $\mathcal{D}'$ and $\mathcal{D}'_{\Delta}$ have equal extension.*

Since the extension of the initial state d_0 of $\mathcal{D}$ can be changed only by rule $\mathcal{R}_0$, if a bud $(d_0, \varepsilon, \mathbb{N})$ is created then $\|d_0\|$ will become equal to that of the initial state of $\mathcal{D}'$.

Lemma 6. *$\mathcal{D}'_{\Delta}$ is deterministic.*

Since $\mathcal{D}'_{\Delta}$ cannot include ε-transitions, nondeterminism in $\mathcal{D}'_{\Delta}$ can only be caused by two ℓ-transitions exiting d and entering two states, d' and d'', respectively, where $\|d'\| \neq \|d''\|$ (otherwise, d' and d'' would be merged). However, based on Lemma 3, the two transition are balanced, hence, $\|d'\| = \|d''\|$, a contradiction.

Lemma 7. *Each state in $\mathcal{D}'_{\Delta}$ is reachable from the initial state.*

Since the property initially holds for $\mathcal{D}$, a disconnection may be caused by the processing of buds, either in auxiliary procedures *Merge* and *Prune* or in rules $\mathcal{R}_4$, $\mathcal{R}_5$, and $\mathcal{R}_6$. However, by exploiting state distance, states keep being connected.

Lemma 8. *The transition function of $\mathcal{D}'_{\Delta}$ equals the transition function of $\mathcal{D}'$.*

This property is grounded on three facts: equality of initial states (Lemma 5), balanced transitions (Lemma 3), and property of exiting transitions (Lemma 4).

Lemma 9. *The set of states of $\mathcal{D}'_{\Delta}$ equals the set of states of $\mathcal{D}'$.*

This property is a consequence of Lemmas 5 and 8.

Lemma 10. *The set of final states of $\mathcal{D}'_{\Delta}$ equals the set of final states of $\mathcal{D}'$.*

Grounded on the implicit maintenance of final states in $\mathcal{D}$: d is final iff $\|d\| \cap F_{\mathrm{n}} \neq \emptyset$.

5 Is *Decremental Subset Construction* Worthwhile?

Decremental Subset Construction is meant to deal with decremental determinization of large NFAs. However, this algorithm is not an absolute panacea. To compare the performances of *Subset Construction* versus *Decremental Subset Construction*, we consider the worst-case time complexity, with $\mathcal{C}_{SC}$ and $\mathcal{C}_{DSC}$ denoting the complexity of the two algorithms, respectively. By approximation, we assume that time complexity is proportional to the number of processed states. Let $\mathcal{N}$ be an NFA, $\Delta\mathcal{N}$ a decrement of $\mathcal{N}$, and $\mathcal{N}' = \mathcal{N} \setminus \Delta\mathcal{N}$. Let n and n' be the numbers of states in $\mathcal{N}$ and $\mathcal{N}'$, respectively. Let $\mathcal{D}$ and $\mathcal{D}'$ be the DFAs *SC*-equivalent to $\mathcal{N}$ and $\mathcal{N}'$, respectively, and D and D' the sets of states in $\mathcal{D}$ and $\mathcal{D}'$, respectively. Starting from $\mathcal{N}'$, the generation of $\mathcal{D}'$ by *Subset Construction* requires the creation of $2^{n'}$ states. Instead, the generation of $\mathcal{D}'$ via *Decremental Subset Construction* requires the removal of all states in $D \setminus D'$, hence:

$$\mathcal{C}_{DSC} = 2^n - 2^{n'} = 2^n \left(1 - \frac{1}{2^{n-n'}}\right) \approx 2^n. \tag{1}$$

Surprisingly, $\mathcal{C}_{DSC}$ is exponential with n, the number of states in $\mathcal{N}$. Hence, in a worst-case scenario, *Decremental Subset Construction* is not better than *Subset Construction*. However, the point is, worst-case scenario is very pessimistic in nature. For instance, if $n' = 100$ then $\mathcal{D}'$ will include $2^{100} \approx 10^{30}$ states, a number that goes well beyond the computational power of any computer. If these figures compare with those in real applications, then there will be no point in using either determinization algorithm, be it decremental or not. In fact, evidence from massive experimentation conducted in [5] shows that, more often than not, the size of $\mathcal{D}$ resulting from determinization of $\mathcal{N}$ compares with the size of $\mathcal{N}$ (rather than being exponential with the size of $\mathcal{N}$). This is why it is possible determinizing an NFA with several tens of thousands of states without memory explosion. Besides, as shown in [5], there are some classes of expanding automata which result in a small variation of $\mathcal{D}$. The same applies in reverse, when the automaton shrinks by a decrement. Thus, assuming that $\mathcal{N}$ is large and $\Delta\mathcal{N}$ relatively small, so that $n \approx n'$, the actual number of states processed by *Decremental Subset Construction*, namely n_d, may be realistically small. By contrast (and this is unquestionable), *Subset Construction* always creates all states in $\mathcal{D}'$, possibly much more than n_d. In these circumstances, *Decremental Subset Construction* may outperform *Subset Construction* to a large degree.

6 Conclusion

Worst-case complexity analysis indicates that there is no benefit in *Decremental Subset Construction*. However, in real contexts, when $\mathcal{N}$ is large and $\Delta\mathcal{N}$ relatively small, chances are that the difference between $\mathcal{D}$ and $\mathcal{D}'$ is small. If so, *Decremental Subset Construction* may outperform *Subset Construction* to a large degree. *Decremental Subset Construction* opens the way to solve more general determinization problems, where an NFA changes its topology over time either by augmentation or decrement. Solving such general *dynamical determinization problems* is a topic for future research.

Acknowledgment. This work was supported in part by Zhejiang Provincial Natural Science Foundation of China (No. LY16F020004); National Natural Science Foundation of China (No. 61472369).

References

1. Aichernig, B., Jöbstl, E., Kegele, M.: Incremental refinement checking for test case generation. In: Veanes, M., Viganò, L. (eds.) TAP 2013. LNCS, vol. 7942, pp. 1–19. Springer, Heidelberg (2013)
2. Aichernig, B., Jöbstl, E., Tiran, S.: Model-based mutation testing via symbolic refinement checking. Sci. Comput. Program. **97**(4), 383–404 (2015)
3. Balan, S., Lamperti, G., Scandale, M.: Incremental subset construction revisited. In: Neves-Silva, R., Tshirintzis, G., Uskov, V., Howlett, R., Jain, L. (eds.) Smart Digital Futures. Frontiers in Artificial Intelligence and Applications, vol. 262, pp. 25–37. IOS Press, Amsterdam (2014)
4. Balan, S., Lamperti, G., Scandale, M.: Metrics-based incremental determinization of finite automata. In: Teufel, S., Tjoa, A.M., You, I., Weippl, E. (eds.) CD-ARES 2014. LNCS, vol. 8708, pp. 29–44. Springer, Heidelberg (2014)
5. Brognoli, S., Lamperti, G., Scandale, M.: Incremental determinization of expanding automata. Comput. J. **59**(12), 1872–1899 (2016)
6. Carrasco, R., Daciuk, J., Forcada, M.: Incremental construction of minimal tree automata. Algorithmica **55**(1), 95–110 (2009)
7. Carrasco, R., Forcada, M.: Incremental construction and maintenance of minimal finite-state automata. Comput. Linguist. **28**(2), 207–216 (2002)

8. Daciuk, J.: Incremental construction of finite-state automata and transducers, and their use in the natural language processing. Ph.D. thesis, University of Gdansk, Poland (1998)
9. Daciuk, J.: Semi-incremental addition of strings to a cyclic finite automaton. In: Klopotek, M., Wierzchon, S., Trojanowski, K. (eds.) Intelligent Information Processing and Web Mining. Advances in Soft Computing, vol. 25, pp. 201–207. Springer, Heidelberg (2004)
10. Daciuk, J., Mihov, S., Watson, B., Watson, R.: Incremental construction of minimal acyclic finite state automata. Comput. Linguist. **26**(1), 3–16 (2000)
11. Daciuk, J., Watson, B., Watson, R.: Incremental construction of minimal acyclic finite state automata and transducers. In: Proceedings of FSMNLP 1998, pp. 48–56. Bilkent University, Ankara, June 30 – July 1 1998
12. Hopcroft, J., Motwani, R., Ullman, J.: Introduction to Automata Theory, Languages, and Computation, 3rd edn. Addison-Wesley, Reading (2006)
13. Lamperti, G., Scandale, M.: From diagnosis of active systems to incremental determinization of finite acyclic automata. AI Commun. **26**(4), 373–393 (2013)
14. Lamperti, G., Scandale, M.: Incremental determinization and minimization of finite acyclic automata. In: IEEE International Conference on Systems, Man, and Cybernetics – SMC 2013, Manchester, United Kingdom, pp. 2250–2257 (2013)
15. Lamperti, G., Scandale, M., Zanella, M.: Determinization and minimization of finite acyclic automata by incremental techniques. Softw. Pract. Exp. **46**(4), 513–549 (2016)
16. Lamperti, G., Zanella, M.: Diagnosis of discrete-event systems from uncertain temporal observations. Artif. Intell. **137**(1–2), 91–163 (2002)
17. Lamperti, G., Zanella, M.: Diagnosis of Active Systems – Principles and Techniques. Engineering and Computer Science, vol. 741. Springer, Dordrecht (2003)
18. Lamperti, G., Zanella, M.: Monitoring of active systems with stratified uncertain observations. IEEE Trans. Syst. Man Cybern. Part A Syst. Hum. **41**(2), 356–369 (2011)
19. Lamperti, G., Zanella, M., Chiodi, G., Chiodi, L.: Incremental determinization of finite automata in model-based diagnosis of active systems. In: Lovrek, I., Howlett, R., Jain, L. (eds.) Knowledge-Based Intelligent Information and Engineering Systems. LNAI, vol. 5177, pp. 362–374. Springer, Heidelberg (2008)
20. Lamperti, G., Zanella, M., Zanni, D.: Incremental processing of temporal observations in model-based reasoning. AI Commun. **20**(1), 27–37 (2007)
21. Lamperti, G., Zhao, X.: Diagnosis of active systems by semantic patterns. IEEE Trans. Syst. Man Cybern. Syst. **44**(8), 1028–1043 (2014)
22. Rabin, M., Scott, D.: Finite automata and their decision problems. IBM J. Res. Dev. **3**(2), 114–125 (1959)
23. Watson, B.: A fast new semi-incremental algorithm for the construction of minimal acyclic DFAs. In: Champarnaud, J.M., Maurel, D., Ziadi, D. (eds.) WIA 1998. LNCS, vol. 1660, pp. 121–132. Springer, Heidelberg (1999)
24. Watson, B., Daciuk, J.: An efficient incremental DFA minimization algorithm. Nat. Lang. Eng. **9**(1), 49–64 (2003)

Using Alloy for Verifying the Integration of OLAP Preferences in a Hybrid What-If Scenario Application

Mariana Carvalho and Orlando Belo(✉)

School of Engineering, ALGORITMI R&D Center, University of Minho, 4710-057 Braga, Portugal
mqvcarvalho@gmail.com, obelo@di.uminho.pt

Abstract. Owning the right and high quality set of information is a crucial factor for developing business activities and consequently gaining competitive advantages. However, retrieving information is not enough. The possibility to simulate hypothetical scenarios without harming the business using What-If analysis tools and to retrieve highly refined information is an interesting way of achieving such advantages. Based on this, we designed and developed a specific piece of software especially oriented for discovering the best recommendations for What-If analysis scenarios' parameters, using OLAP usage preferences. In this paper, we propose a formal description and verification of one of the phases of the hybridization model we developed related to the extraction of OLAP usage preferences. We used Alloy to specify and verify the viability of the process, and discover possible ambiguity and inconsistencies cases.

Keywords: Decision support systems · What-If analysis · On-Line Analytical Processing · Usage preferences · Analysis systems specification · Alloy

1 Introduction

Within a company, the presence of analytical information systems and the availability of techniques and models for multidimensional data exploration and analysis is no longer a novelty in enterprise business environments. To compete in a knowledge-based society and to be a part of a rapidly changing global economy, companies must try gaining some competitive advantage from a better use of information and knowledge they have. Based on previous experiences, we know that is possible to improve the effectiveness of What-If analysis scenarios with *On-line Analytical Processing* (OLAP) usage preferences [7, 11]. The use of OLAP usage preferences may help filtering business information, meeting the users' needs and business requirements without losing data quality. In this paper we present a formal description and verification of the phase of extracting OLAP usage preferences, applying a specific hybridization process we designed. The enrichment of What-If scenarios with OLAP usage preferences improves the quality of decision models from the user point of view. This allows for avoiding the lack of expertise of users in the implementation of What-If scenarios and models. The system we developed has the ability to suggest OLAP preferences, providing to the user the most adequate scenario parameters according to its needs and making What-If scenarios more valuable. Given the complexity

I. Czarnowski et al. (eds.), *Intelligent Decision Technologies 2017*,
Smart Innovation, Systems and Technologies 72, DOI 10.1007/978-3-319-59421-7_4

of such hybridization process, it is imperative to run a formal verification of the process to check for defects or inconsistencies in the process using formal methods [5]. Thus, this paper is organized as follows. In Sect. 2 we present some related work about the process of extracting OLAP preferences in analytical environments. Then, in Sect. 3, we describe briefly the hybridization model we designed and implemented for extracting OLAP preferences, how we extracted them from OLAP sessions, and how we used Alloy for verifying the viability of the extraction process. Finally, in Sect. 4, we present the conclusions about the work we done.

2 Related Work

OLAP usage preferences reflect the most interesting data that decision-making agents selected and analyzed in past OLAP sessions [9]. More recently, preferences capture the attention of many researchers in the OLAP domain, approaching the extraction of preferences using data mining over MDX queries logs [2], or specifying OLAP preferences algebras [7]. Meanwhile, in [6] it was presented a recommender system for OLAP users having the ability to recommend to the user discoveries detected on former sessions. Later, in [18] it was proposed another framework with the ability for assisting users in the automation of their activities in the context of the next generation of business intelligence systems using query recommendation support. More recently, Kozmina [13] provided a method for generating report recommendations taking into consideration user preferences, while Bimonte and Negre [4] showed the usefulness of OLAP recommender systems on decision-making activities. However, only few papers have focused on formal verification in OLAP systems. Most of the formal verification done so far was in the verification process of data warehouses construction. For instance, Zhao and Ma [19] started developing an *Abstract State Machine* (ASM) ground model for data warehouses and OLAP system to overcome the data warehouse design complexity. Later, Sen, et al. [16] proposed an optimal aggregation methodology on the multidimensional data cube, using Galois connection formal analysis to guarantee the storage space and time complexity optimizations. In the same direction, Salem, et al. [15] and Stefanov [17] proposed, respectively, a UML profile to design data warehouses, and some formal algorithms (methods) for the detection of conflicts and heterogeneities that can arise in the integration of data marts. Finally, Mezzanzanica, et al. [14] proposed a model-based approach quite useful to identify poor quality data, using formal methods to perform sensitivity analysis.

3 Integrating OLAP Preferences in What-If Scenarios

OLAP preferences can be used as recommendations for enriching What-If application scenarios. This makes possible to simulate the behavior of a system based on past data extracted from OLAP sessions. Preferences have the ability to recommend to business users the axes of analysis that are strongly related to each other, helping them to introduce valuable information in the application scenario under construction. To do that, we designed a specific process (Fig. 1) for enhancing the effectiveness of a What-If application. We start the process using an OLAP data cube as input, we define an application scenario based on historical data extract from previous OLAP sessions and stored in a MDX querying log file. Then, the scenario settings are defined, delineating the axis of analysis, the set of values for analyzing, and the set of values to change according to previous defined goals - usually this differs among distinct analysis tools. Finishing this, the What-If process proceeds with choosing a tool, and returning an OLAP data cube for prediction support.

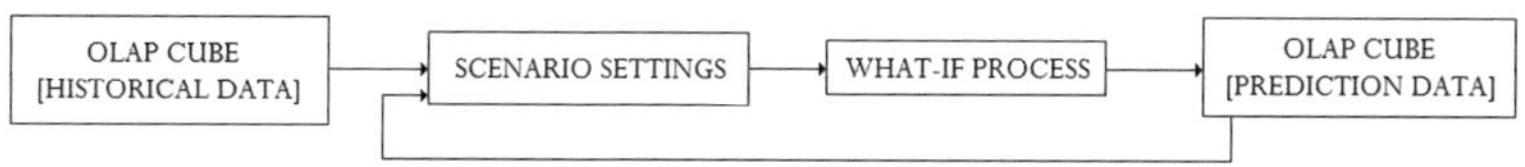

Fig. 1. The hybridization process.

3.1 Dealing with Preferences

According to [12], given a set of attributes A, a preference P is a strict partial order defined as P (A, <P), where <P is an irreflexive, transitive and asymmetric binary relation, <P ⊆ dom(A) × dom(A). If X <P Y, then 'Y is preferred to X'. A preference P = (A, <P) is an irreflexive, transitive and asymmetric binary relation <P on the domain of values of attributes set A. Consider the following example. If we consider to analyze, in some retail application, how sales vary with the number of customers having children living at home, to set its preferences a user need to choose one of the elements included in the set of the frequent item sets. Given the set of attributes "Marital Status", "Gender", "Yearly Income", "Number Cars Owned", "Birth Date", "English Education", "Total Children" and "Number Children At Home" and assuming that the user chooses "Number Children At Home", using the previous defined semantics, we got something like:

"Marital Status" ≡ "Gender"
"Gender" ≡ "Yearly Income"
"Yearly Income" ≡ "Number Cars Owned"
"Number Cars Owned" ≡ "Birth Date"
"Birth Date" ≡ "English Education"
"English Education" ≡ "Total Children"
"Total Children" <P "Number Children At Home"

In other words, this can be interpreted as the attribute "Number Children At Home" is preferred to the attribute "Total Children", "Marital Status", "Gender", "Yearly Income", "Number Cars Owned", "Birth Date", and so on. Thus, "Marital Status" is equivalent to "Gender"; "Gender" is equivalent to "Yearly Income", "Yearly Income" is equivalent to "Number Cars Owned", and so on. Based on this set of previous preferences, it is possible to select a set of association rules that contains the attribute "Number Children At Home".

3.2 Extracting OLAP Preferences

The identification and extraction of OLAP preferences provide us strong arguments for improving the simulation of a given system's behavior based on the preferences of its users. This approach gave us the ability to suggest recommendations, providing the user with the most adequate scenario parameters and thus supporting better a potential development of a What-If scenario application. OLAP preferences allow for discovering the strongly related axes of analysis based on the usage of a certain user – predicting future demands valuable information of provable new scenarios. The main differences between our hybrid approach and a standard What-If analysis method is the introduction of a process of extraction of usage preferences on a business multidimensional database and their use in the simulation of the model, allowing for setting the basis to predict the behavior of a given application scenario. The process of extracting OLAP usage preferences and using them to enrich What-If scenarios considers five distinct phases, namely: (1) data warehouse's view selection; (2) OLAP cube construction; (3) association rules extraction; (4) OLAP usage preferences extraction; and (5) What-If analysis using suggested preferences. In this paper, we focus exclusively on the description of the process of extraction of usage preferences of the association rules – a filter process. But first we need to explain how to get the mining model.

The data cube was created using data from the database example "AdventureWorksDW2014" [3], integrating a fact table - "Internet Sales" - and some related dimension tables – "Dim Currency", "Dim Customer", "Dim Product", "Dim Promotion", "Dim Sales Territory", and "Dim Date". Next, we applied a mining association process over the data of the cube, using an OLAP mining process [8]. To do that, we selected the Microsoft Association Rules algorithm [1] that comes with Microsoft Analysis Services. We created the mining structure and the mining model. The mining structure defines the data from which mining models are built and the mining model is created by applying an algorithm to data. All the rules and item sets extracted were stored in the mining model. Item set information nodes include the definition of the item set, the number of cases that it contains, and other diverse information. In turn, a rule information node describes a general pattern for the association of items. The filter process starts with the list of association rules and item sets extracted and stored in the mining model and with the choice of a user preference item from a list of frequent item sets of the mining model, ordered by decreasing values of probability. The list of association rules is filtered, returning only a list of association rules that contains the item set chosen by the user. This item set corresponds to the user's main research focus the user intends to analyze. The returned list of association rules is organized by importance, using the performance measures of each rule. This part helps to get first the strong association rules, which help us to discover the item sets strongly related to the chosen attribute. This list is suggested to the user as its business preferences that are taken into consideration in the What-If scenario as configuration parameters. In order to discover the list of preferences, we use the performance measures of the rules obtained in the previous phase.

Lets consider the following example. Initially the process starts with the definition of a what-if question, which corresponds to an analysis goal: "What-if we want to increase the profit sales by 10%, focusing mainly on customers who have children living at home?" This corresponds to increase the sales profit and to focus on people who have children. To do this, we intend to discover which attributes in the data are related to the goal attribute ("Number of Children"). Next, we apply the association mining algorithm over the data of the cube and the returned list of association rules is filtered, returning only rules containing "Number of Children" (Fig. 2).

Number Children At Home >= 4, Yearly Income = 37386.1278744576 - 74999.9999942656 -> Total Children >= 4
Birth Date < 9/14/1946 7:00:54 AM, Yearly Income < 37386.1278744576 -> Number Children At Home < 1
English Education = Graduate Degree, Yearly Income < 37386.1278744576 -> Number Children At Home < 1
Total Children < 1, Yearly Income = 37386.1278744576 - 74999.9999942656 -> Number Children At Home < 1
Total Children < 1, Yearly Income < 37386.1278744576 -> Number Children At Home < 1
Birth Date < 9/14/1946 7:00:54 AM, Yearly Income = 37386.1278744576 - 74999.9999942656 -> Number Children At Home < 1
Birth Date < 9/14/1946 7:00:54 AM, Number Cars Owned = 2 - 3 -> Number Children At Home < 1
Number Cars Owned < 1, Total Children < 1 -> Number Children At Home < 1
Number Cars Owned = 1 - 2, Total Children < 1 -> Number Children At Home < 1
Total Children < 1, English Education = Bachelors -> Number Children At Home < 1
538.English Product Name = LL Road Tire, Total Children < 1 -> Number Children At Home < 1
Total Children < 1, Number Cars Owned = 2 - 3 -> Number Children At Home < 1

Fig. 2. The association rules for a given preference - 'Number Children At Home'.

The returned list of association rules is then organized by importance, using the performance measures of each rule, showing the strong association rules. With these rules, we discover the attributes' set strongly related to "Number Children At Home", for example the top 3 strong association rules (Fig. 3) of the previous set (Fig. 2), which will be used later to define his OLAP preferences. For example, if the returned list of association rules is the list presented in Fig. 3, the recommendations to the user will are "Number Children At Home", and, obviously, "Birth Date", "Yearly Income", "English Education" and "Total Children". After this step, the user chooses the item sets of his preference that will be used in the What-If scenario as configuration parameters.

Number Children At Home >= 4, Yearly Income = 37386.1278744576 - 74999.9999942656 -> Total Children >= 4
Birth Date < 9/14/1946 7:00:54 AM, Yearly Income < 37386.1278744576 -> Number Children At Home < 1
English Education = Graduate Degree, Yearly Income < 37386.1278744576 -> Number Children At Home < 1

Fig. 3. A list of some filtered association rules.

3.3 A Formal Approach Using Alloy

The model we designed was formalized using Alloy [10]. It is a formal object-oriented modeling language based on first-order logic. Alloy is designed for performing automatic analysis and includes friendly tool-support (Alloy Analyzer), which is based on a bounded to a SAT (boolean satisfiability) technology. It works by translating all the constraints in the model into Boolean constraints and calls a SAT solver to answer them. Alloy was chosen due to its ability to generate an abstract initial model that becomes more complex, as the project evolves. We can use Alloy as a modeling language for specification, and the Alloy Analyzer, which provides graphic instant feedback for verification and validation. We can run the specification model and get an instance of the abstract model or we can check an assertion by looking for counter-examples and validate the model. In the first case, the Alloy Analyzer finds an instance of the model that satisfies the specification within the specified scope, which means that the Alloy Analyzer returns an example of an instance with signatures and respective relations that make facts in the specification always true. In the latter case, the Alloy Analyzer finds counter-examples that violate the assertion. When a counter-example is found, this means that the model contains errors or inconsistencies. We can use Alloy Analyzer to visualize and analyze the counter-example to understand where it failed, and consequently correct the specification of the model. First we start by specify the whole process and define the objects using signature and the relation between them. A signature declaration represents an object. For example, if we want to represent an OLAP cube in the model, we specify sig OLAPCube{ }. Signature declarations can introduce fields, which represent a relation among objects, for example: sig OLAPCube{getMiningStructure : MiningStructure}. In the Alloy specification, we start with the definition of all objects and the relation among them:

```
sig OLAPCube {
     getMiningStructure :  MiningStructure,
     disj a1,a2,a3 : one Attribute,
     m : one Measure }
sig MiningStructure {getMiningModel :  MiningModel}
sig MiningModel { getRules: some Rules }
```

Other object that needs attention in the specification phase is the object Rule. We need to specify the format of the association rules, which can only be one of two types: each rule can be either ruleType1 (A → A) or ruleType2 (A × A → A). And each rule must have associated with it a pair of PerformanceMeasures (Support and Confidence).

```
sig Rule extends Rules {ruleType1 : Attribute ->  Attribute,
ruleType2 : Attribute -> Attribute ->  Attribute, getPerformance :
one PerformanceMeasures }
sig PerformanceMeasures {getS : Support, getC : Confidence }
```

In this phase, there are some restrictions that need some attention in the Alloy specification, in order to keep the consistency between the relations between the objects of the model. To do this, we need to define some facts in the Alloy specification. These facts describe invariants or constraints that are always true, being represented by the fact X { }. In our case, we need to guarantee the coherence of the relations between the objects. Without the following fact miningModel{…}, some objects appear isolated in the instance. For example, the object MiningModel needs to be related to the object MiningStructure, through a mining model algorithm and cannot appear isolated in the Alloy instance. Other facts like: the set of Rules must be related with a mining model through a getRules relation. We can specify all these restrictions with:

```
fact miningModel{
     all ms : MiningStructure , o : OLAPCube | ms in
o.getMiningStructure
all mm : MiningModel, ms : MiningStructure | mm in ms.getMiningModel
     all mm : MiningModel, r : Rules | r in mm.getRules getRules }
```

Next, we specify the extraction of preferences from the mining rules discovered in the previous phase (Fig. 4), as follows. We want to discover the strongly related attributes with the attribute (item set) of the main research focus. It starts with displaying to the user the set of returned association rules by the mining algorithm. The user chooses an attribute, which usually it is the main focus of analysis. In other words, we want to know how sales vary considering the parameter A1. The user chooses attribute A1, and it is preferred to the attribute A2, A3, A4, A5, A6, and so on. All the association rules returned by the previous phase are filtered and a list of rules is created.

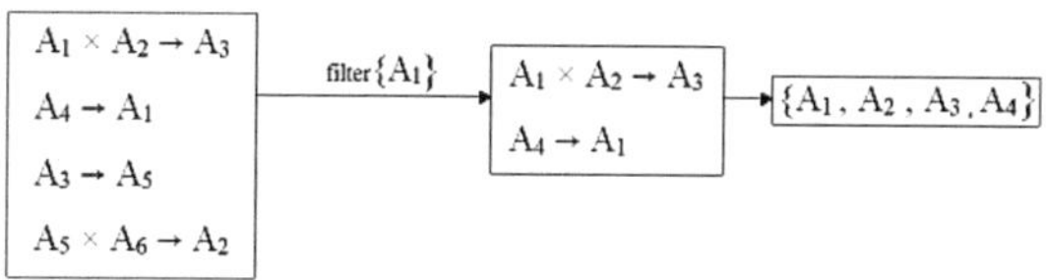

Fig. 4. OLAP preferences extraction – an example.

All of the rules of this created list contain the chosen Attribute. With this list, we can analyze which attributes are strongly related with the chosen attribute, since this list only contains the “best/higher” performance measures’ rules. Therefore, we can conclude that the remaining list contain the attributes strongly related with the chosen attribute. This list is used then to form the set of OLAP preferences for the user. This means that the attributes of the association rules filtered and returned are suggested to the user as its business preferences and then they will be taken into consideration in the What-If scenario as configuration parameters. To specify this filter process in Alloy, we need to add a field to the object Rules (a set of Rules, not a single Rule): the relation getStrongRules, which is going to include an Attribute and PerformanceMeasures; and return a set of Rules denoted by StrongRules. This set of rules (StrongRules) defines the SourceVariables that will be used in the What-If analysis scenario. We can specify the filter process using the following Alloy statement:

```
sig Rules {getStrongRules : Attribute -> PerformanceMeasures ->
StrongRules}
sig StrongRules {getPreferences : SourceVariables}
```

After the Alloy specification, we define a predicate (pred showInstance(){ }), in order to get an instance of the specified model. A predicate is similar to a fact: it always describes something true but is only verified when invoked (unlike facts which are always true in returned instances). To invoke a predicate, Alloy uses a run (run showInstance). With these commands, the Alloy Analyzer returns an example of an instance consistent with the defined specifications. We start by defining an empty predicate, which is often a useful starting point to determine whether the model is consistent or not, in other words, if the Alloy Analyzer can find an instance of the model that satisfies the specified facts. The specification model can be resumed in the following instance found by the Alloy Analyzer represented in Fig. 5. We get a set of Rules by apply mining to the OLAP Cube. The objects Mining Structure and Mining Model were omitted in the figure in order to keep it legible. A set of rules (Rule represented by the top circle) is filtered based on a specific attribute and performance measures (PerformanceMeasures) using the getStrongRules and the returned set of strong rules (StrongRules) is used to create the preferences (SourceVariables) that are used as recommendations to the What-If scenario (represented by the object OLAPCubePrediction). The new prediction is obtained using getPrediction with the data from an OLAPCube. We specify our model using signatures and facts. To validate the abstract model, we need to define assertions.

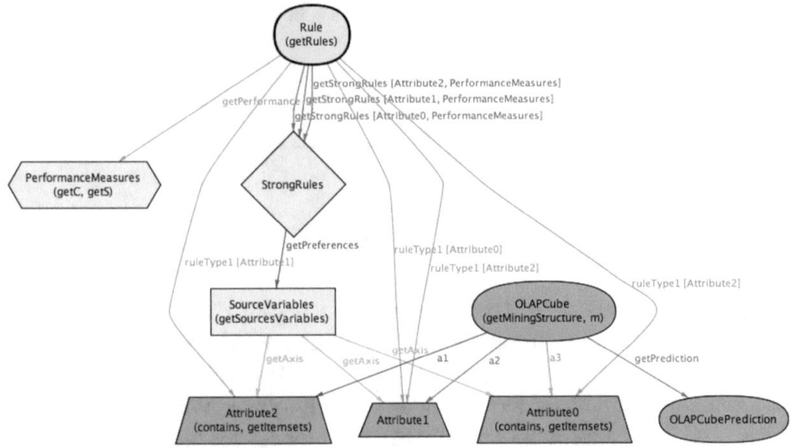

Fig. 5. An example of an instance of the alloy analyzer.

The assert instructs the Alloy Analyzer to search for counter-example within the scope. Assertions allow for expressing properties that are expected to hold as consequence of specified facts. Alloy Analyzer checks an assertion through the command check. This command verifies an assertion by searching for a counter-example, which violates the assertion property. If no counter-example is found, the model may be valid, but it is not guarantee. There are several properties that we could verified (some very obvious): all the data contained in the association rules must be contained in the OLAPCube data; in an association rule, the antecedent and the consequent must be different, in X -> Y, X must be different of Y. Considering the latter property, we already have the definition of association rules:

```
sig Rule extends Rules
{ruleType1: Attribute -> Attribute}
```

We want to ensure that there is no association rule (attribute1 -> attribute2) that has the antecedent (attribute1) equal to the consequent (attribute2). To verify this property we define an assertion. If the Alloy Analyzer creates a counter-example, it means that our assertion failed: there is an instance of our model where exists a rule that could have the antecedent attribute equal to the consequent attribute (which cannot happen). Considering this property we define the assertion:

```
assert rule{ no a1, a2 : Attribute,   r:Rule
     | a1 -> a2 in r.ruleType1 implies (a1=a2)}
check rule for 2
```

Using check rule for 2, a counter-example was not found, meaning that our assertion is valid. Figure 6 illustrates a counter-example that was found by Alloy Analyzer by using the same defined assertion (assert rule) and check rule for 3, which is a different scope from the previous case (3 Attributes instead of 2 Attributes). The counter-example shows up because in the defined assertion, we only take into account the existence of two attributes (ruleType1 : Attribute -> Attribute) and the Alloy Analyzer creates an counter-example adding a third attribute (Attribute0 represented by the red boxes), which invalidates the previous assertion. This particular case does not invalidate the syntax of our model. We can check that the syntax is still correct in Fig. 6, there is no rule that has the same antecedent and consequent in ruleType1.

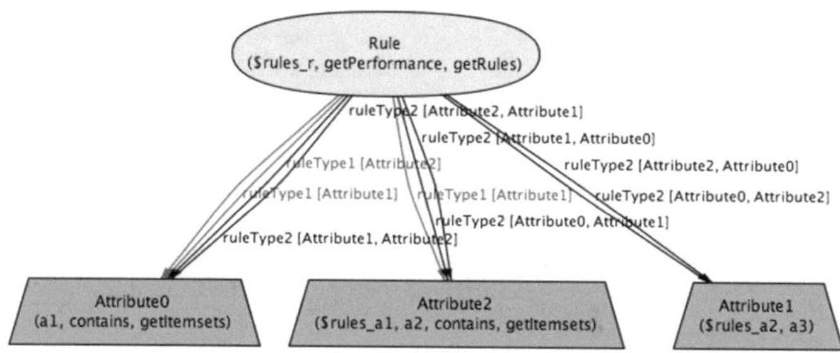

Fig. 6. An example of a counter-example of the alloy analyzer.

4 Conclusions

In this paper we presented and discussed the formal verification of the process of extracting usage preferences in a hybrid model for enhancing What-If scenarios, in order to check for inconsistencies and prove the validity of the model we developed. Alloy was chosen to support such task. The preferences extraction process gave us the ability for enriching What-If analysis scenarios using OLAP usage preferences. Usage preferences can be defined based on historical data provided by a simple business application or from a more sophisticated piece of software like a data mining system. Often, a preference reflects also hidden patterns that were detected in the data set. Using association rules based on preferences has the advantage that the user does not need to know the business domain. Preferences can also help to control over the returned information, providing access to relevant information and eliminating the irrelevant one. One may not know the proportions of the outcome: it may be an empty result, or an information flooding. Due to this, query runtime can be enhanced against cases without preferences. Consequently, in our process, we get more focused and refined results, which helps both a user who is not familiar with the business analysis and an analyst who is familiar with the business modeling data. We have shown that we were able use Alloy to successfully specify the model and validate some properties of the syntax of our model. The model produced was validated using formal verification and no inconsistencies or flaws were found.

Acknowledgments. This work has been supported by COMPETE: POCI-01-0145-FEDER-007043 and FCT - Fundação para a Ciência e Tecnologia within the Project Scope: UID/CEC/00319/2013. Our thanks to Nuno Macedo, from HASLab R&D Centre, for the comments and suggestions he did during the specification of this work.

References

1. Agrawal, R., Srikant, R.: Fast algorithms for mining association rules. In: Proceedings of the 20th International Conference on Very Large Data Bases, VLDB, vol. 1215, pp. 487–499 (1994)
2. Aligon, J., Golfarelli, M., Marcel, P., Rizzi, S., Turricchia, E.: Mining Preferences from OLAP Query Logs for Proactive Personalization. Advances in Databases and Information Systems. Springer, Heidelberg (2011)
3. AWC: Microsoft Adventure Works Database Example (2017). https://msftdbprodsamples.codeplex.com. Accessed 17 Mar 2017
4. Bimonte, S., Negre, E.: Evaluation of user satisfaction with OLAP recommender systems: an application to RecoOLAP on an agricultural energetic consumption datawarehouse. Int. J. Bus. Inf. Syst. **21**(1), 117–136 (2016)
5. Clarke, E.M., Wing, J.: Formal methods: state of the art and future directions. ACM Comput. Surv. (CSUR) **28**(4), 626–643 (1996)
6. Giacometti, A., Marcel, P., Negre, E., Soulet, A.: Query recommendations for OLAP discovery driven analysis. In: Proceedings of the ACM Twelfth International Workshop on Data Warehousing and OLAP (2009)
7. Golfarelli, M., Rizzi S.: Expressing OLAP preferences. In: Scientific and Statistical Database Management. Springer, Heidelberg (2009)
8. Han, J.: OLAP mining: an integration of OLAP with data mining. In: Proceedings of the 7th IFIP, pp. 1–9 (1997)
9. Harinarayan, V., Rajaraman, A., Ullman, J.: Implementing data cubes efficiently. ACM SIGMOD Rec. **25**(2), 205–216 (1996)
10. Jackson, D.: Alloy, a lightweight object modelling notation. ACM Trans. Softw. Eng. Methodol. (TOSEM) **11**(2), 256–290 (2002)
11. Jerbi, H., Ravat, F., Teste, O., Zurfluh, G.: Preference-Based Recommendations for OLAP Analysis. Springer, Heidelberg (2009)
12. Kießling, W.: Foundations of preferences in database systems. In: Proceedings of the 28th International Conference on Very Large Data Bases, pp. 311–322. VLDB Endowment (2002)
13. Kozmina, N.: Producing report recommendations from explicitly stated user preferences. Baltic J. Mod. Comput. **3**(2), 110 (2015)
14. Mezzanzanica, M., Boselli, R., Cesarini, M., Mercorio, F.: A model-based approach for developing data cleansing solutions. J. Data Inf. Qual. (JDIQ) **5**(4), 13 (2015)
15. Salem, A., Triki, S., Ben-Abdallah, H., Harbi, N., Boussaid, O.: Verification of security coherence in data warehouse designs. In: Trust, Privacy and Security in Digital Business, pp. 207–213. Springer, Heidelberg (2012)
16. Sen, S., Chaki, N., Cortesi, A.: Optimal space and time complexity analysis on the lattice of cuboids using galois connections for data warehousing. In: 4th International Conference Computer Sciences and Convergence Information Technology (ICCIT 2009), pp. 1271–1275. IEEE (2009)

17. Stefanov, G.: Formal methods for conflict detection during multi-dimensional data mart integration. In: International Conference on Application of Information and Communication Technology and Statistics in Economy and Education (ICAICTSEE), p. 232 (2013)
18. Varga, J., Romero, O., Pedersen, T. B., Thomsen, C.: Towards next generation BI systems: the analytical metadata challenge. In: Data Warehousing and Knowledge Discovery, pp. 89–101 (2014)
19. Zhao, J., Ma, H.: Quality-assured design of on-line analytical processing systems using abstract state machines. In: Proceedings of the Fourth International Conference on Quality Software (QSIC 2004), Braunschweig, Germany, pp. 224–231 (2004)

Electrohydrodynamic Effect Simulation and Method of Its Optimization

Jolanta Wojtowicz and Hubert Wojtowicz(✉)

Faculty of Mathematics and Nature, The University of Rzeszów,
16c Al. Rejtana, 35-959 Rzeszów, Poland
hubert.wojtowicz@gmail.com

Abstract. The paper presents an optimization method for electrohydrodynamic effect basing on results of its simulation. The EHD effect is simulated using neural networks and differential equations describing a mathematical model of the phenomenon, which are solved using Runge - Kutta algorithm. The optimization of this effect allows finding for particular input parameters of the generator an optimal diameter of the wire, which burned in a thermo-physical process gives a maximal energy release in the form of an acoustic pressure wave.

Keywords: Electrohydrodynamic effect · Neural networks · Optimization

1 Introduction

The electrohydrodynamic effect finds its application in a geophysical prospecting method, which involves excitation of a seismic wave with the EHD generator [8] or with the controlled spark generator [7]. The generator through the release of electrical energy stored in its capacitors to the material of the conductor initiates an explosion of the conductor in a liquid or a gaseous medium [3]. The phenomenon is modelled as a regression problem using neural networks [9]. The input parameters for the neural network are temperature, voltage and current intensity in the generator's circuit. The values of input parameters are derived through solving a set of nonlinear differential equations using Runge-Kutta numerical method. The output parameter is current in the release circuit registered by the oscilloscope. The amount of energy converted into the seismic wave is dependent on the electrical parameters of the generator's circuit as well as on the mechanical parameters such as material and geometric dimension of the conductor placed in the sonde [5,6]. The diameter of the conductor has a decisive influence on the process of the electrohydrodynamic phenomenon. If the diameter is too small a premature burning of the conductor occurs, whereas in the case of choosing too large a parameter the conductor's material won't even reach boiling temperature [1].

I. Czarnowski et al. (eds.), *Intelligent Decision Technologies 2017*,
Smart Innovation, Systems and Technologies 72, DOI 10.1007/978-3-319-59421-7_5

2 Mathematical Model of the EHD Explosion

The mathematical model of the electrohydrodynamic phenomenon is comprised of three major phases. The first of them is a heating phase of the conductor from the initial temperature T_0 to the flow temperature T_{pl}. The second phase encompasses the process of melting of exploding conductor. The third phase covers the process of heating the melted conductor from the temperature T_{pl} to the boiling temperature denoted as T_{wrz}. The system of differential equations describing EHD phenomenon can be presented as:

$$\frac{di}{d\tau} = \frac{1}{l_k + l_p}\left\{\left\{u - i\alpha_0\left\{a_k + z_{pr} + \frac{\Delta l_{pr}}{R}k_\nu M i^2\right\}\right\}\right\}, \tag{1}$$

$$\frac{du}{d\tau} = -i, \tag{2}$$

$$\frac{d\Theta}{d\tau} = 2B\alpha_0 M i^2 \tag{3}$$

$$\frac{dv}{d\tau} = \frac{K_\nu \alpha_0}{\sqrt{R^3}} M i^2 - A\frac{\alpha_0 \alpha_p k_{pl} P}{\nu N h_{\Theta pl}} i^2, \tag{4}$$

where:

$$\left.\begin{aligned}
&P = 1 + c_1(n_\Theta \Theta - 1),\\
&Q = 1 + c_2(n_\Theta \Theta - 1),\\
&N = [(1 + k_{pl})\nu^2 - 2\nu]D + 1,\\
&l_{pr} = l_p + \tfrac{\Delta l_{pr}}{2} lnR,\\
&R = 1 - k_\nu(n_\Theta \Theta - 1),\\
&z_{pr} = a_p l_{pl}\tfrac{PQR}{N},\\
&M = \tfrac{z_{pr}}{q_0[1 + b_1(2n_\Theta \Theta - 1)]}.
\end{aligned}\right\} \tag{5}$$

τ time, Θ temperature, i current intensity in the discharge circuit, u voltage on plates of the capacitor, v wire diameter, l_k, l_p, α_0, a_k, a_p, q_0, c_1, b_1 dimensionless parameters.

Following equations and initial conditions must be satisfied:

- for the first phase ($0 \le \tau \le \tau_1$) of heating the exploding conductor from temperature T_0 to the flow temperature T_{pl}:

$$\tau = 0 \begin{cases} u = 1, \Theta = 1, \\ i = 0, \nu = 1. \end{cases} \tag{6}$$

$A = 0$, $B = 1$, $D = 1$, $n_\Theta = 1$, it is also assumed that $\Delta l_{pl} = 0$, $k_\gamma = 0$, $c_2 = 0$, $k_{pr} = 1$;

– for the second phase $\tau_1 \leq \tau \leq \tau_2$ flow phase of the exploding conductor:

$$\tau = \tau_1 \begin{cases} u = u_{\tau_1}, \Theta = \Theta_{pl}, \\ i = i_{\tau_1}, \nu = 1, \xi = 1. \end{cases} \tag{7}$$

$A = 1$, $B = 0$, $D = 1$, $n_\Theta = \frac{\Theta_{pl}}{\Theta}$, it is formally assumed that $\Delta l_{pl} = 0$, $k_\gamma = 0$, $c_2 = 0$;

– for the third phase ($\tau_2 \leq \tau \leq \tau_3$) heating phase of the melted conductor from T_{pl} to the boiling temperature T_{wrz}:

$$\tau = \tau_2 \begin{cases} u = u_{\tau_2}, \Theta = \Theta_{pl}, \\ i = i_{\tau_2}, v = 1, \end{cases} \tag{8}$$

$A = 0$, $B = 1$, $D = 0$, $n_\Theta = \frac{\Theta_{pl}}{\Theta}$.

The differential equations, under proper initial conditions, can be simulated using Runge-Kutta method. Figure 1 shows results of solving Eqs. (1)–(5) under initial conditions (6)–(8) and under following values of dimensionless coefficients:

Particular curves in Fig. 1 correspond to: dimensionless voltage on plates of the capacitor u, current intensity in the discharge circuit i, temperature t.

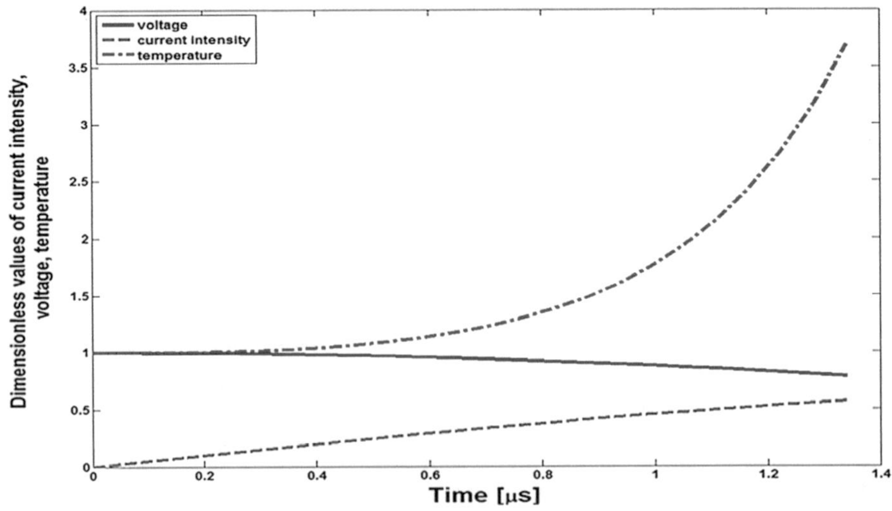

Fig. 1. Numerically calculated, using Runge-Kutta method, values of dimensionless voltage on plates of the capacitor u, current intensity in the discharge circuit i and temperature t, of the electro-explosion of a 0.20 mm diameter and 55 mm lenght Cu wire.

3 Data for the Training of Neural Networks

In the research two types of data were utilized. The first type is data resulting from solving differential equations, which describe the EHD phenomenon. A set of these results is used as an input vector for neural networks training. The second type is data obtained during practical experiments i.e. oscillographs containing current intensity and pressure values. These data after appropriate processing are used as input signals for neural networks. The complete information describing a case of electro-explosion is constructed from pairs of data, which are comprised of a vector fed to the input layer of a neural network and a vector containing output reference values, which the neural network should learn to predict. A prepared set of input data as well as knowledge about the content of vectors, which are considered as output signals for a neural network allows for building of a tool with which a simulation of the phenomenon is carried out.

4 Training of Neural Networks with Data Describing EHD Phenomenon

For the approximation of current intensity variation registered by the oscilloscope the following algorithms were used: a three-layered MLP neural network trained with Levenberg-Marquardt algorithm, a three-layered neural network trained with Bayesian regularization algorithm and a general regression neural network. For the training of neural networks sets of data representing electro-explosions of copper wires of 0.15 mm, 0.3 mm and 0.45 mm diameters were used. The set of data for the copper wire of 0.22 mm diameter was used to test prediction accuracy of the trained neural networks' models. Additionally in order to improve training and testing results for the three-layered MLP neural network the data was standardized. The generalization capabilities of trained neural networks were tested with leave-one-out cross validation method. The training error goal was ascertained by the early stopping method. The optimal architecture for three-layered neural networks was obtained by gradual change of network configuration by adding neurons to the hidden layer and computing values of criterion functions for each of the trained models. The best architecture was selected considering a trade-off between error values in validation and testing phases. The three-layered MLP neural network showed high accuracy of training results (Fig. 2a, b, c and d) and capabilities to accurately predict outcomes of electro-explosions for the sets of input data not shown in the training phase (Fig. 3). Ordinate axes of the graphs correspond to the time measured in microseconds. Abscissa axes of the graphs correspond to the values of current intensity registered by the oscilloscope. Circles denote data on the graph representing real values of current intensity; points on the graphs denoted with crosses represent values of training data simulated by neural network and values of neural network output in the testing phase.

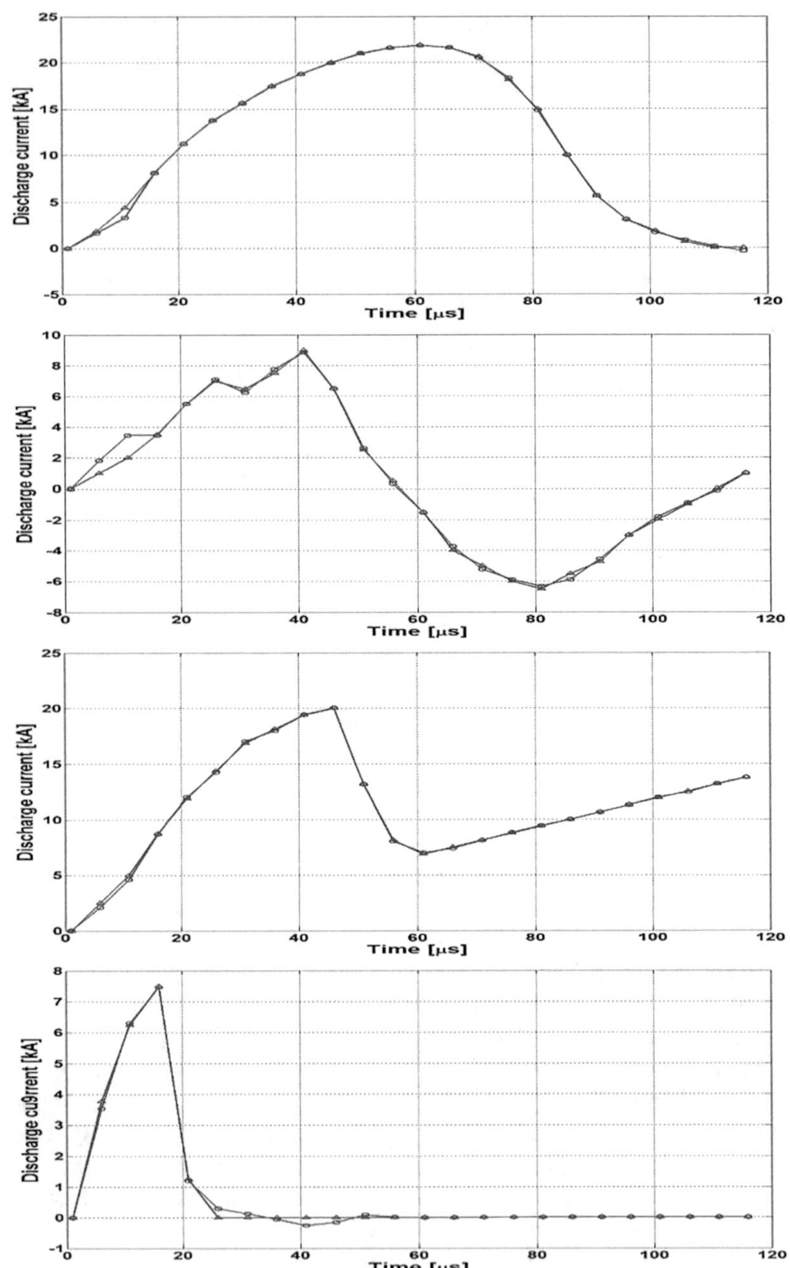

Fig. 2. MLP training results

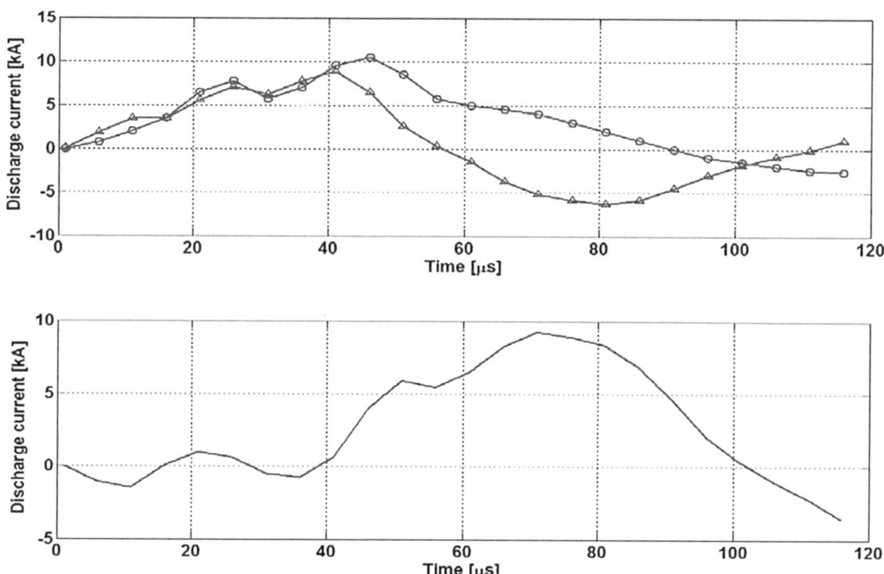

Fig. 3. Test results of the simulation of an electro-explosion of a Cu wire of 0.3 mm diameter and 143 mm length (a), error calculated between real values and neural network output values (b).

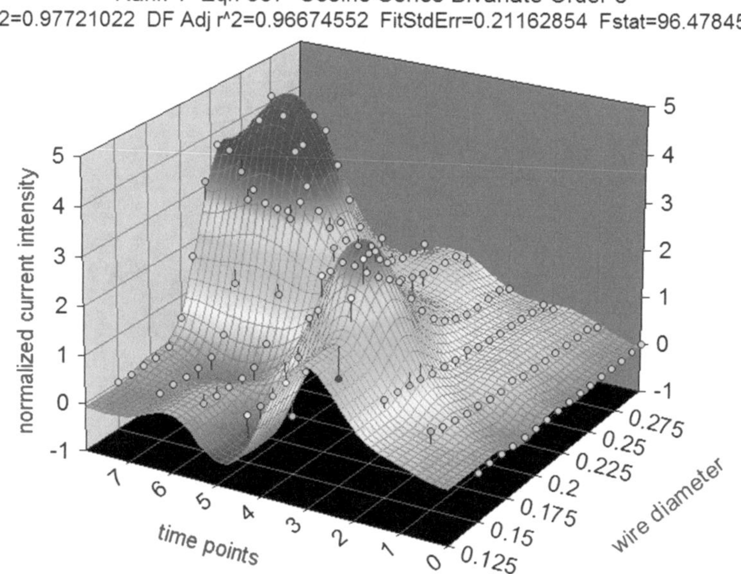

Fig. 4. Graph of the function of current intensities in the release circuit occurring in time for the continuous interval of copper wire diameters from 0.15 mm to 0.3 mm approximated basing on outputs of three-layered neural network model.

5 Approximation of the Function Describing Changes in the Release Circuit of the EHD Generator

The application of algorithmic procedures for finding optimal geometrical parameters of the exploding wire is possible after approximating the function describing variations in current intensity values in the release circuit for the particular continuous interval of wire diameters. For the approximation of this function a TableCurve3D was used taking advantage of 'Selective Subset' fitting algorithm. This algorithm uses single pass Gauss elimination method to calculate coefficients matrix and single pass Gauss-Jordan algorithm for matrix inversion in order to calculate standard errors and confidence bounds. The result of the algorithm application for the empirical data from electro-explosions is a function, which general equation and values of constant parameters are shown in Fig. 4. Figure 5 shows deviations between real values and values of the function found in the approximation process.

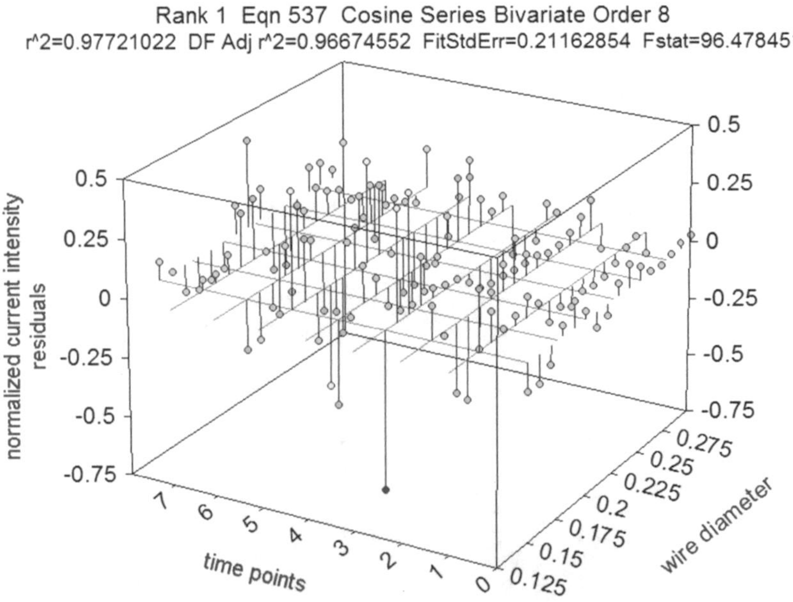

Fig. 5. Graph of errors between values of the approximated function of current intensities in the release circuit occurring in time for the continuous interval of copper wire diameters from 0.15 mm to 0.3 mm and values of outputs of three-layered neural network model.

6 Optimization of EHD Phenomenon

In order to create an accurate optimization model the target function was approximated using first nine values of sets obtained from a trained three-layer feed-forward neural network model consisting of five neurons in an input layer, nine neurons in a first hidden layer and three neurons in a second hidden layer. Optimization is limited to the first nine values for the reason that at this point in time the first local maximum occurs on the plot of current intensities registered by the oscilloscope for the explosion of the 0.22 mm diameter copper wire. The vertex at this point coincides with the end of the third phase of electro-hydrodynamic effect, and marks the end of the usable range for the secondary discharge. For this reason it is desirable to achieve at this graph point the highest possible value of current intensity. The smooth shape of the plot without the first discernible increase of current intensity vale on the oscillograph is interpreted as disadvantageous and corresponds to the too small amount of energy supplied to the wire of too large diameter. The comparison of plots for electro-explosions of copper wires of 0.22 mm and 0.45 mm diameters shows that the area under the plot of current intensity for the 0.45 mm diameter is larger but its plot is a completely smooth curve, which means that the ratio of the amount of energy supplied to the wire diameter is worse than the ratio for the diameter of 0.22 mm. The value of energy registered during the electro-explosion at this point marking the end of the third phase of EHD phenomenon is the highest in the range of empiric data for the interval of diameters between 0.15 mm and 0.30 mm. The responses of three-layer MLP network of the structure consisting of five neurons in an input layer, nine neurons in a first hidden layer and three neurons in a second hidden layer are approximated with the function described by cosine bivariate series of the eight order and their plot is shown in Fig. 4. The analysis of the surface constructed through merging of real data with data from the neural network model shows, that the responses are accurate both in the interval of smaller diameters 0.15 mm–0.18 mm as well as for the more typical diameters in the interval of 0.20 mm–0.25 mm. The obtained target function accurately maps variances of plots representing electro-explosions of conductors. The optimal diameter of the copper wire calculated on the basis of this function is 0.24 mm. In result of the comparison of the quality of obtained models, the model of three-layer MLP of the structure consisting of five neurons in the input layer, nine neurons in the first hidden layer and three neurons in the second hidden layer was considered to be the best, and the optimal diameter of copper wire determined by this model on the basis of quadratic programming calculations is equal to 0.24 mm. This diameter is close to the diameter of 0.22 mm for which field explosion was carried out and, which was evaluated after a series of field experiments by the human expert to be very close to the optimal diameter. The plot of the electro-explosion of copper wire of 0.24 mm diameter obtained from the approximated surface and the plot of neural network responses for the same diameter of copper wire are shown in Fig. 6a. Deviations between values of these plots are shown in Fig. 6b. Due to the fact that the conductor's diameter has a far greater influence on the course of the electro-explosion than its length it

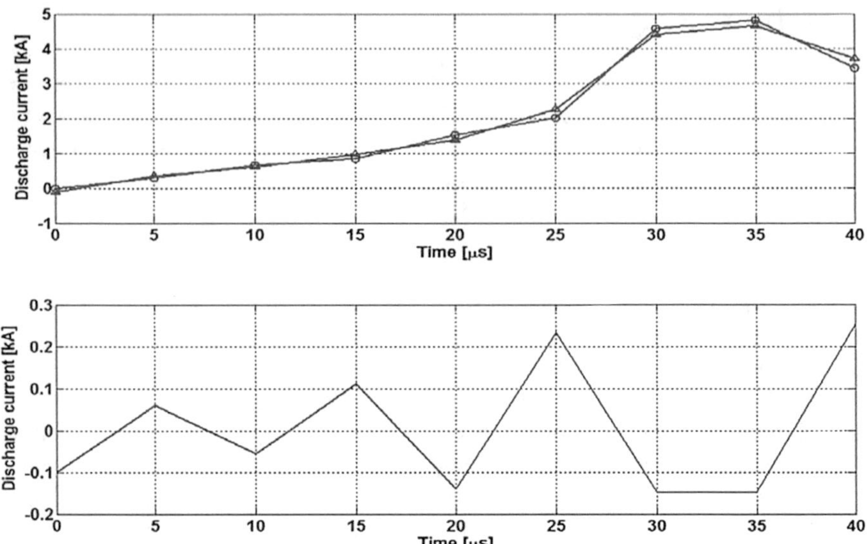

Fig. 6. Plot of neural network responses for the Cu conductor of diameter 0.24 mm and plot of values obtained from the target function for the conductor of the same diameter (a), deviations between values of neural network responses and values calculated from the target function for the Cu wire of 0.24 mm diameter (b).

can be predicted that in case of copper wires of different lengths the optimum can be found in the interval of diameters between 0.21 mm and 0.25 mm. The analysis carried out of the electro-explosion phenomenon of conducting material submerged in water allows for optimization of the electrical energy conversion process into the energy of seismic impulse. The energy of impulse pressure in water and the remaining parameters such as: steepness of the impulse rise, amplitude of the impulse front, time duration of impulse and its orientation, for the constant values of voltage and discharge capacity are dependent on the electro-mechanical properties and geometry of the conducting material subjected to the process of electro-explosion.

7 Conclusions

The analysis of trained neural network models' responses allowed for finding in the optimization process the best networks' response, which corresponds to the highest amount of energy passed in the form of pressure wave to a copper conductor of a particular diameter. Through the approximation of real data sets and combining them with outputs from trained neural networks an optimization function was created for all considered neural network models of different architectures used for the prediction of EHD phenomenon. Combining in the optimization function of real data with data obtained from neural network model made it possible to precisely determine the best conductor's diameter for which the efficiency of EHD process is the highest. Additionally plotting outputs of neural network models in the considered interval of conductors' diameters in the three dimensional presentation space, which is easily understandable and convenient for human interpretation, allowed for better estimation of their quality. The appraisal of the shape of obtained surfaces by the human expert enables selection of the best model without relying solely on mathematical criteria for which calculated values for some of the models were very similar despite the varying quality of responses in the most interesting local areas of the generated surfaces. The results achieved in the course of research fully justify purposefulness of using neural network algorithms for the analysis of real data sets describing EHD phenomenon. The trained neural network models allow for generation of new sets of data without the need for carrying out series of time consuming and costly field experiments. The optimization procedures applied to the optimization function created on the basis of real data sets and data generated by neural network allow for the selection of wire and circuit parameters, for which the highest working capacity is achieved, manifested as the magnitude of the pressure wave front being a result of a wire explosion.

References

1. Krasik, Y.E., Grinenko, A., Sayapin, A., Efimov, S., Fedotov, A., Gurovich, V.Z., Oreshkin, V.I.: Underwater electrical wire explosion and its applications. IEEE Trans. Plasma Sci. **36**(2), 423–434 (2008)
2. Gurovich, V.T., Grinenko, A., Krasik, Y.E., Felsteiner, J.: Simplified model of underwater electrical discharge. Phys. Rev. **E69**, 036402 (2004). The American Physical Society
3. Grinenko, A., Sayapin, A., Gurovich, V.T., Efimov, S., Felsteiner, J., Krasik, Y.E.: Underwater electrical explosion of a Cu wire. J. Appl. Phys. **97**, 023303-1–023303-6 (2005). American Institute of Physics
4. Grinenko, A., Gurovich, V.T., Sayapin, A., Efimov, S., Krasik, Y.E.: Strongly coupled copper plasma generated by underwater electrical wire explosion. Phys. Rev. E **72**, 066401-1–066401-7 (2005). The American Physical Society
5. Lisitsyn, I.V., Muraki, T., Akiyama, H.: Wire induced flashover as a source of shock waves for destruction of solid materials. Japan. J. Appl. Phys. Part I **36**, 1258 (1997)
6. Lisitsyn, I.V., Muraki, T., Akiyama, H.: Mechanism of shock waves generation and material destruction in wire induced surface flashover. J. Phys. B **239**, 6 (1997)
7. Staszczak, J., Wojtowicz, J.: Controlled spark generator. Patent application no P-355 248, application date 29 July 2002
8. Staszczak, J., Wojtowicz J.: Electrohydrodynamic device for initiation of seismic waves. Patent application no P-351 418, application date 28 December 2001
9. Wojtowicz, J., Wojtowicz, H., Wajs, W.: Simulation of electrohydrodynamic phenomenon using computational intelligence methods. In: Proceedings of the 19th International Conference on Knowledge-Based and Intelligent Information and Engineering Systems, Elsevier Procedia Computer Science, vol. 60, pp. 188–196 (2015)

Specialized Decision Techniques for Data Mining, Transportation and Project Management

Incremetal GEP-Based Ensemble Classifier

Joanna Jedrzejowicz[1(✉)] and Piotr Jedrzejowicz[2]

[1] Faculty of Mathematics, Physics and Informatics, Institute of Informatics, University of Gdansk, 80-308 Gdansk, Poland
jj@inf.ug.edu.pl

[2] Department of Information Systems, Gdynia Maritime University, Morska 83, 81-225 Gdynia, Poland
pj@am.gdynia.pl

Abstract. In this paper we propose a new incremental Gene Expression Programming (GEP) ensemble classifier. Our base classifiers are induced from a chunk of data instances using GEP. Size of the chunk controls the number of instances with known class labels used to induce base classifiers iteratively. Instances with unknown class label are classified in sequence, one by one. It is assumed that after a decision as to the class label of the new instance has been taken its true class label is revealed. From a set of base classifier a metagene is induced and used to predict class label of instances with unknown class labels. To validate the approach an extensive computational experiment has been carried-out.

Keywords: Incremental classification · Gene expression programming · Metagene

1 Introduction

Incremental learners allow an existing model to be updated using only newly available individual data instances, without having to re-process all of the past instances. Incremental learners date back to eighties with paper of [17]. Later on [21] proposed an efficient method for incrementally inducing decision trees. Very fast incremental decision tree learner was suggested by [7]. Another important step in constructing effective incremental learners was info-fuzzy approach of [3]. More recently several Extreme Learning Machine (ELM) approaches to incremental learning have been discussed. For example [14] proposed a forgetting parameters concept named FP-ELM.

One of the effective approaches to incremental learning is constructing and applying an ensemble of classifiers. Ensembles are constructed from base classifiers. Combining base classifiers requires some kind of voting (including bagging and boosting) or building a meta-classifier. The approach is known as stacked generalization or stacking [25]. Example application of stacked generalization to batch learning can be found in [18]. Combining base classifiers for incremental learning was discussed in [4]. The approach of [26] utilizes ELMs as base classifiers and adaptively decides the number of the neurons in hidden layer.

I. Czarnowski et al. (eds.), *Intelligent Decision Technologies 2017*,
Smart Innovation, Systems and Technologies 72, DOI 10.1007/978-3-319-59421-7_6

Finally, the algorithm trains a series of classifiers and the decision results for unlabeled data are made by weighted voting strategy. Another useful approach to ensemble incremental learning was suggested by [12] where several different base classifiers form input to the voting scheme. In this paper we also propose to use the power of ensemble classifiers. Our base classifiers are induced using gene expression programming. Gene expression programming (GEP) introduced by Ferreira [5] is an automatic programming approach. In GEP computer programs are represented as linear character strings of fixed-length called chromosomes which, in the subsequent fitness evaluation, can be expressed as expression trees of different sizes and shapes. The approach has flexibility and power to explore the entire search space, which comes from the separation of genotype and phenotype. As it has been observed by Ferreira [6] GEP can be used to design decision trees, with the advantage that all the decisions concerning the growth of the tree are made by the algorithm itself without any human input, that is the growth of the tree is totally determined and refined by evolution. The ability of GEP to generate decision trees makes it a natural tool for solving classification problems. Ferreira [6] showed several example applications of GEP including classification.

In the proposed approach there are two crucial parameters controlling process of classification. The first one is the number of ensemble members. The second is the chunk size. Chunk of instances has, in our case, a special meaning. It is used to control the number of instances with known class labels used to induce genes, which are base classifiers. Independently from the chunk size, instances with the unknown class labels are classified sequentially, one by one. It is assumed that after decision as to the predicted class label of an instance has been taken, its true class label is revealed. As soon as the number of instances that have been already classified (correctly or wrongly) reaches the chunk size a set of base classifiers can be induced from such chunk. It is also assumed that true class labels of instances forming the first chunk in the sequence of all instances are known. Size of the chunk determines how often the internal hypotheses of our classifier is adopted to changes in the data distribution. It can be observed that setting chunk size to 1 means adopting the internal hypotheses each time a new instance is processed.

After the set of base classifiers has been induced we use the concept of stacked generalization to predict a class label of the considered instance. To this end we propose a procedure to induce from the set of base classifiers a metagene representing classifier internal hypotheses and directly providing prediction of the required class label. The remainder of this paper is organized as follows. Section 2 explains using GEP to induce classifiers. Section 3 contains a detailed description of the proposed incremental ensemble classifier. Section 4 presents results of the validating computational experiment. Finally, Section 5 summarizes conclusions and future work.

2 Using Gene Expression Programming to Induce Classifiers

Consider the following model of data classification. The learning algorithm is provided with the training set $TR = \{< d, c > \mid d \in D, c \in \{0,1\}\}$, where D is the space of attribute vectors $d = (w_1^d, \ldots, w_n^d)$ with w_i^d being symbolic or numeric values of attribute i. The learning algorithm is used to find the best possible approximation $\bar{f}$ of the unknown function f such that $f(d) = c$.

As usual when applying GEP methodology, the algorithm uses a population of chromosomes, selects them according to fitness and introduces genetic variation using several genetic operators. Each chromosome is composed of a single gene (chromosome and gene mean the same in what follows) divided into two parts as in the original head-tail method [5]. The size of the head (h) is determined by the user with the suggested size not less than the number of attributes in the dataset. The size of the tail (t) is computed as $t = h(m-1)+1$ where m is the largest arity found in the function set. In the computational experiments the functions are: logical AND, OR, XOR, NOR and NOT. Thus $m = 2$ and the size of the gene is $h + t = 2h + 1$. The terminal set contains triples $(op, attrib, const)$ where op is one of relational operators $<$, $\leq$, $>$, $\geq$, $=$, $\neq$, *attrib* is the attribute number, and finally *const* is a value belonging to the domain of the attribute *attrib*. As usual in GEP, the tail part of a gene always contains terminals and head can have both, terminals and functions. Observe that in this model each gene is syntactically correct and corresponds to a valid expression. Each attribute can appear once, many times or not at all. This allows to define flexible characteristics like for example ($attribute1 > 0.57$) AND ($attribute1 < 0.80$). On the other hand, it can also introduce inconsistencies like for example ($attribute1 > 0.57$) AND ($attribute1 < 0.40$). This does not cause problems since a decision subtree corresponding to such a subexpression would evaluate it to $false$. Besides, when studying the structure of the best classifiers in our experiments the above inconsistencies did not appear.

Attaching an expression tree to a gene is done in exactly the same manner as in all GEP systems. Consider the gene g with head $= 6$, defined below.

0	1	2	3	4	5	6	7	8
OR	AND	AND	$(>, 1, 0)$	$(=, 2, 5)$	$(>, 3, 0)$	$(<, 1, 10)$	$(>, 3, 0)$	$\cdots$

The start position (position 0) in the chromosome corresponds to the root of the expression tree (OR, in the example). Then, below each function branches are attached and there are as many of them as the arity of the function -2 in our case. The following symbols in the chromosome are attached to the branches on a given level. The process is complete when each branch is completed with a terminal. The number of symbols from the chromosome to form the expression tree is denoted as the termination point. For the discussed example, the termination point is 7. For the attribute vector $rw = (8.0, 5.0, 2.5, \cdots)$ the value of the above gene g is

$$g(rw) = true.$$

The algorithm for learning the best classifier using GEP works as follows. Suppose that a training dataset is given and each vector in the dataset has a correct label representing the class. In the initial step the minimal and maximal value of each attribute is calculated and a random population of chromosomes is generated. For each gene the symbols in the head part are randomly selected from the set of functions AND, OR, NOT, XOR, NOR and the set of terminals of type $(op, attrib, const)$, where the value of *const* is in the range of *attrib*. The symbols in the tail part are all terminals. To introduce variation in the population the following genetic operators are used:

- mutation,
- transposition of insertion sequence elements (IS transposition),
- root transposition (RIS transposition),
- one-point recombination,
- two-point recombination.

Details on GEP operators can be found in [8–10].

For a fixed training set TR and fixed gene g the fitness function counts the proportion of vectors from TR classified correctly

$$fit_{TR}(g) = \frac{\sum_{rw \in TR,\ g(rw)\ is\ true} sg(\text{rw is from class 1})}{|TR|} \tag{1}$$

where

$$sg(\varphi) = \begin{cases} 1 & \text{if } \varphi \text{ is true} \\ 0 & \text{otherwise} \end{cases}$$

2.1 Creating Metagenes

Having generated a population of genes it is possible to create a population of metagens which corresponds to creating an ensemble classifier. The idea is as follows. Let *pop* be a population of genes, with each gene identified by its *id*. To create metagens from *pop* define the set of functions again as boolean ones as above and set terminals equal to identifiers of genes. For example the metagene mg makes use of two genes $g1 = g$ and $g2$ defined below:

mg :

0	1	2	3	4
AND	NOT	$g1$	$g2$	$\cdots$

$g2$:

0	1	2
NOT	$(=, 2, 7)$	$\cdots$

For a fixed attribute vector rw each terminal (i.e. gene) has a boolean value and thus the value of metagene can be computed. For example for the above metagene mg and $rw = (8.0, 5.0, 2.5, \cdots)$ we have

$$g1(rw) = true,\ g2(rw) = false,\ mg(rw) = true$$

Similarly as in (1), for a fixed training set TR and fixed metagene mg the fitness function counts the proportion of vectors from the testing set classified correctly:

$$FIT_{TR}(mg) = \frac{\sum_{rw \in TR,\ mg(rw)\ is\ true} sg(\text{rw is from class 1})}{|TR|} \tag{2}$$

3 Incremental Classification

The incremental algorithm works in rounds. In each round a chunk of training data $c1$ is used to create a population of genes (as in Algorithm 1), next chunk of data $c2$ is used to create the population of metagenes and to choose one best fitted mg (as in Algorithm 2), and the following chunk $c3$ is tested by metagene mg. In the next round $c1 = c2$, $c2 = c3$ and next chunk is used as $c3$. Thus the number of rounds is $\frac{|TR|}{|chunk|} - 2$. The first K steps of the Algorithm 3 (lines $3-9$) are used to start the population of genes, K is a parameter.

Algorithm 1. Generating the population of genes

Input: training data TR, nG - number of iterations, $popS$ - population size, $eliteS$ - size of elite
Output: population of best fitted genes.

generate random $popS$ genes as an initial population pop
calculate fitness of all genes using (1)
for $i = 0$ *to* nG **do**
    select genes for the new population using tournament selection method
    apply genetic operations : mutation, transposition of insertion sequence elements (IS transposition), root transposition (RIS transposition), one-point recombination, two-point recombination,
    keep $eliteS$ best fitted genes in the population
return pop

Algorithm 2. Creating metagenes

Input: training data TR, population of genes $popG$, nG - number of iterations, $popS$ - population size, $eliteS$ - size of elite,
Output: best fitted metagene mg.

generate random $popS$ metagens using population of genes $popG$
calculate fitness of all metagenes using (2)
for $i = 0$ *to* nG **do**
    select metagenes for the new population using tournament selection method
    apply genetic operations : mutation, transposition of insertion sequence elements (IS transposition), root transposition (RIS transposition), one-point recombination, two-point recombination,
    keep $eliteS$ best fitted metagenes in the population
return *mg=best fitted metagene from the population*

Algorithm 3. Incremental algorithm

```
Input: data D, chunk size ch, K-number of genes in population
Output: qc - quality of the incremental classification
initialize correctClsf ← 0
dataTrain ← first ch vectors from D
for i = 1 to K do
    apply Algorithm 1 to generate gene g
    dataTest ← next ch vectors from D
    corr ← number of vectors from dataTest correctly classified by g
    correctClsf ← correctClsf + corr
    add g to population of genes pop
    dataTrain ← dataTest
dataTrainMeta ← next ch vectors from D
while vectors in D not considered yet do
    apply Algorithm 1 to dataTrain to generate gene g
    add g to population of genes pop
    delete from pop the gene with the lowest fitness
    apply Algorithm 2 to dataTrainMeta and pop to generate metagene mg
    dataTest ← next ch vectors from D
    corr ← number of vectors from dataTest correctly classified by mg
    correctClsf ← correctClsf + corr
    dataTrain ← dataTrainM
    dataTrainM ← dataTest
qc ← correctClfs / (|D| − 2·ch)
return qc
```

4 Computational Experiment Results

To evaluate performance of the proposed approach we tested it over a set of publicly available benchmark datasets including data often used to test incremental learning algorithms. Datasets used in the experiment are shown in Table 1.

In the experiment the following settings have been used: number of classifiers - 5; population size and number of iterations - both 100 for dataset with 2000 or less instances and both 50 for the remaining datasets, probability of mutation - 0,5; probability of RIS transposition, IS transposition, 1-point and 2-point recombination - each 0,2. For selection we used tournament-from-the-pair rule. Chunk size have been set to 1 in case of a small dataset. For a bigger dataset it has been set to assure time feasibility during computation. Average accuracy and standard deviation achieved by the proposed approach in 20 runs is shown in Table 2. In Table 3 we compare accuracy of the incremetal GEP-based ensemble classifier with results of some state-of-the-art incremental classifiers. There are several factors affecting performance of the proposed incremental GEP-based ensemble classifier. The most important are dataset size, number of attributes, chunk size and set of parameters controlling evolution of genes.

Example interrelations between chunk size, classifier accuracy and computation time needed to classify single instance (Intel Core i7 processor) for Thyroid and Sonar datasets are shown in Table 4.

Table 1. Benchmark datasets used in the experiment.

Dataset	Source	Instances	Attributes	Classes
Banana	[13]	5300	3	2
Bank M	[13]	4522	17	2
Breast cancer	[13]	263	10	2
Chess	[27]	503	9	2
CMC	[13]	1473	10	3
Diabetes	[13]	768	9	2
Electricity	[1]	44976	6	2
Heart	[13]	303	14	2
Hepatitis	[13]	155	20	2
Image	[13]	2086	19	2
Ionosphere	[13]	351	35	2
Luxembourg	[27]	1901	32	2
Magic	[13]	19020	11	2
Poker hand	[13]	141179	11	10
Sonar	[13]	208	61	2
Solar flare	[13]	1066	12	6
Spam	[13]	4601	58	2
Thyroid	[13]	7000	22	3
Twonorm	[15]	7400	21	2
Waveform	[13]	5000	41	3

Table 2. Accuracy of the incremental GEP-based ensemble classifier

Dataset	Accuracy (%)	St.dev. (%)	Chunk size
Banana	89,3	2,3	1
Bank M	93,3	1,8	1
Breast	82,1	0,7	1
Chess	85,9	1,6	1
Cmc	82,6	0,6	1
Diabetes	85,0	0,9	1
Electricity	88,1	3,1	50
Heart	88,9	1,1	1
Hepatitis	94,6	0,8	1
Image	89,6	2,3	1

Table 2. (*Continued*)

Dataset	Accuracy (%)	St.dev. (%)	Chunk size
Ionosphere	92,9	1	1
Luxembourg	100,0	0	200
Poker-hand	59,4	1,2	1000
Solar flare	81,2	0,8	20
Sonar	90,6	0,5	1
Spam	87,7	1,2	10
Thyroid	99,4	0,3	50
Twonorm	92,3	1,3	20
Waveform	90,3	0,9	10

Table 3. Comparison of accuracy incremental GEP-based ensemble classifier vs. other incremental classifiers.

Dataset	GEP ensemble	Other	Algorithm	Source
Banana	89,3	89,3	Inc. SVM	[22]
Bank M	93,3	86,9	LibSVM	[24]
Breast cancer	82,1	72,2	IncSVM	[22]
Chess	85,9	71,8	EDDM	[27]
CMC	82,6	86,3	Granular SVM	[23]
Diabetes	85,0	75,7	Inc. N-B	[12]
Electricity	88,1	90,7	KFCM	[11]
Heart	88,9	83,8	IncSVM	[22]
Hepatitis	94,6	83,8	Inc. Ensemble	[12]
Image	89,6	94,2	SVM	[16]
Ionosphere	92,9	92,4	Inc. Ensemble	[12]
Luxembourg	100,0	85,6	LCD	[20]
Magic	59,4	52,0	LWF	[19]
Poker hand	81,2	64,1	FPA	[22]
Sonar	90,6	81,8	Inc. Ensemble	[12]
Solar flare	87,7	85,5	K-a graph	[2]
Spam	99,4	95,8	LibSVM	[24]
Thyroid	92,3	97,6	FPA	[22]
Twonorm	90,3	86,7	Inc. SVM	[22]

Table 4. Accuracy, chunk size and computation time for Thyroid and Sonar dataset

Chunk size	Accuracy (%)	Time (s.)
50	99,4	0,222
100	99,2	0,097
200	99,0	0,052
300	98,9	0,027
400	98,8	0,017
500	98,8	0,013
1000	98,4	0,009

Chunk size	Accuracy (%)	Time (s.)
1	90,6	0,855
5	82,2	0,158
10	81,2	0,098
15	78,6	0,056
20	76,9	0,024

5 Conclusions

It can be observed that the proposed incremental GEP-based ensemble classifier performs very well. Comparison with the state-of-the-art incremental classifiers shows that our approach outperforms in majority of cases the existing solutions in terms of classification accuracy. In several cases the improvement has been dramatically high. A weaker side of the proposed approach is the number of parameters that have to be set by the user. This results in costs needed for fine-tuning. Another disadvantage is a rather high demand for computational resources, typical for population based algorithms. In the future work we will focus on improving efficiency of the approach especially with respect to computation time.

References

1. MOA Analysis: UCI Machine Learning Repository (2013). http://moa.cms.waikato.ac.nz/datasets/
2. Bertini, J.R.J., Zhao, L., Lopes, A.A.: An incremental learning algorithm based on the k-associated graph for non-stationary data classification. Inf. Sci. **246**, 52–68 (2013)
3. Cohen, L., Avrahami, G., Last, M., Kandel, A.: Info-fuzzy algorithms for mining dynamic data streams. Appl. Soft Comput. **8**(4), 1283–1294 (2008). http://dx.doi.org/10.1016/j.asoc.2007.11.003
4. Fern, A., Givan, R.: Online ensemble learning: an empirical study. Mach. Learn. **53**(1-2), 71–109 (2003). http://dx.doi.org/10.1023/A:1025619426553
5. Ferreira, C.: Gene expression programming: a new adaptive algorithm for solving problems. CoRR cs.AI/0102027 (2001). http://arxiv.org/abs/cs.AI/0102027
6. Ferreira, C.: Gene Expression Programming: Mathematical Modeling by an Artificial Intelligence. SCI, vol. 21. Springer, Heidelberg (2006)
7. Hulten, G., Spencer, L., Domingos, P.M.: Mining time-changing data streams. In: Lee, D., Schkolnick, M., Provost, F.J., Srikant, R. (eds.) Proceedings of the Seventh ACM SIGKDD International Conference on Knowledge Discovery and Data Mining, San Francisco, CA, USA, 26–29 August, pp. 97–106. ACM (2001). http://portal.acm.org/citation.cfm?id=502512.502529
8. Jedrzejowicz, J., Jedrzejowicz, P.: GEP-induced expression trees as weak classifiers. In: Perner, P. (ed.) ICDM 2008. LNCS, vol. 5077, pp. 129–141. Springer, Heidelberg (2008). doi:10.1007/978-3-540-70720-2_10

9. Jedrzejowicz, J., Jedrzejowicz, P.: A family of GEP-induced ensemble classifiers. In: Nguyen, N.T., Kowalczyk, R., Chen, S.-M. (eds.) ICCCI 2009. LNCS, vol. 5796, pp. 641–652. Springer, Heidelberg (2009). doi:10.1007/978-3-642-04441-0_56
10. Jedrzejowicz, J., Jedrzejowicz, P.: Experimental evaluation of two new gep-based ensemble classifiers. Expert Syst. Appl. **38**(9), 10932–10939 (2011). http://dx.doi.org/10.1016/j.eswa.2011.02.135
11. Jedrzejowicz, J., Jedrzejowicz, P.: Distance-based online classifiers. Expert Syst. Appl. **60**, 249–257 (2016). http://dx.doi.org/10.1016/j.eswa.2016.05.015
12. Kotsiantis, S.B.: An incremental ensemble of classifiers. Artif. Intell. Rev. **36**(4), 249–266 (2011). http://dx.doi.org/10.1007/s10462-011-9211-4
13. Lichman, M.: UCI Machine Learning Repository (2013). http://archive.ics.uci.edu/ml/
14. Liu, D., Wu, Y., Jiang, H.: FP-ELM: an online sequential learning algorithm for dealing with concept drift. Neurocomputing **207**, 322–334 (2016). http://www.sciencedirect.com/science/article/pii/S0925231216303125
15. Mldata.org: Machine learning data set repository (2013). http://mldata.org/repository/tags/data/
16. Moreno-Torres, J.G., Sáez, J.A., Herrera, F.: Study on the impact of partition-induced dataset shift on k -fold cross-validation. IEEE Trans. Neural Netw. Learning Syst. **23**(8), 1304–1312 (2012). http://dx.doi.org/10.1109/TNNLS.2012.2199516
17. Schlimmer, J.C., Granger, R.H.: Incremental learning from noisy data. Mach. Learn. **1**(3), 317–354 (1986). http://dx.doi.org/10.1023/A:1022810614389
18. Todorovski, L., Dzeroski, S.: Combining classifiers with meta decision trees. Mach. Learn. **50**(3), 223–249 (2003). http://dx.doi.org/10.1023/A:1021709817809
19. Torres, D.M., Aguilar-Ruiz, J.S.: A similarity-based approach for data stream classification. Expert Syst. Appl. **41**(9), 4224–4234 (2014). http://dx.doi.org/10.1016/j.eswa.2013.12.041
20. Turkov, P., Krasotkina, O., Mottl, V.: Dynamic programming for bayesian logistic regression learning under concept drift. In: Maji, P., Ghosh, A., Murty, M.N., Ghosh, K., Pal, S.K. (eds.) PReMI 2013. LNCS, vol. 8251, pp. 190–195. Springer, Heidelberg (2013). doi:10.1007/978-3-642-45062-4_26
21. Utgoff, P.E., Berkman, N.C., Clouse, J.A.: Decision tree induction based on efficient tree restructuring. Mach. Learn. **29**(1), 5–44 (1997). http://dx.doi.org/10.1023/A:1007413323501
22. Wang, L., Ji, H., Jin, Y.: Fuzzy passive-aggressive classification: a robust and efficient algorithm for online classification problems. Inf. Sci. **220**, 46–63 (2013)
23. Wei, X., Huang, H.: Granular twin support vector machines based on mixture kernel function. In: Huang, D.-S., Han, K. (eds.) ICIC 2015. LNCS, vol. 9227, pp. 43–54. Springer, Cham (2015). doi:10.1007/978-3-319-22053-6_5
24. Wisaeng, K.: A comparison of different classification techniques for bank direct marketing. Int. J. Soft Comput. Eng. **3**(4), 116–119 (2013)
25. Wolpert, D.H.: Stacked generalization. Neural Netw. **5**(2), 241–259 (1992). http://dx.doi.org/10.1016/S0893-6080(05)80023-1
26. Xu, S., Wang, J.: A fast incremental extreme learning machine algorithm for data streams classification. Expert Syst. Appl. **65**(C), 332–344 (2016). https://doi.org/10.1016/j.eswa.2016.08.052
27. Zliobaite, I.: Combining similarity in time and space for training set formation under concept drift. Intell. Data Anal. **15**(4), 589–611 (2011). http://dx.doi.org/10.3233/IDA-2011-0484

Applying the Intelligent Decision Heuristic to Solve Large Scale Technician and Task Scheduling Problems

Amy Khalfay(✉), Alan Crispin, and Keeley Crockett

Department of Computing, Mathematics and Digital Technology, Manchester Metropolitan University, Manchester, UK
{a.khalfay,a.crispin,k.crockett}@mmu.ac.uk

Abstract. Scheduling personnel to complete tasks is a complex combinatorial optimisation problem. In large organisations, finding quality solutions is of paramount importance due to the costs associated with staffing. In this paper we have generated and solved a set of novel large scale technician and task scheduling problems. The datasets include complexities such as priority levels, precedence constraints, skill requirements, teaming and outsourcing. The problems are considerably larger than those featured previously in the literature and are more representative of industrial scale problems, with up to 2500 jobs. We present our data generator and apply two heuristics, the intelligent decision heuristic and greedy heuristic, to provide a comparative analysis.

Keywords: Combinatorial optimisation · Large scale technician and task scheduling problems · Data generator · Intelligent decision heuristic

1 Introduction

The importance of finding quality solutions to scheduling problems was highlighted in [5]. There are many benefits to optimised scheduling; such as maintaining customer satisfaction and providing a balanced working schedule for employees. The importance of field service scheduling is growing due to the increasing number of machines that are used and therefore the number of specialized technicians needed to meet the demand [8]. There are many different types of scheduling problems that can be categorized by the constraints that they include. This paper focuses on generating and solving a set of large scale technician and task scheduling problems.

This problem is NP-hard, there are no known polynomial time algorithms for solving them optimally, which makes using exact methods prohibitive. In large businesses, there is also often a conflict between computational time and solution quality, whilst a solution of high quality is desired, it is desired within a reasonable computational time. The use of heuristic methods is popular for larger sized problems and has produced competitive results in many combinatorial optimization problems such as; graph colouring [17], the vehicle routing problem [3]

I. Czarnowski et al. (eds.), *Intelligent Decision Technologies 2017*,
Smart Innovation, Systems and Technologies 72, DOI 10.1007/978-3-319-59421-7_7

and nurse rostering [1]. Whilst approximate methods have no guarantee of finding a globally optimal solution, they generally produce high quality results in short computational times and are scalable and robust.

Literature in the field of technician and task scheduling problems has included solving both artificial and real world problems. Technician and task scheduling problems have been studied by [11,12,14]. In addition the ROADEF 2007 challenge was based on France Telecom's technician and task scheduling problem and has attracted much research interest; [2,6,7,9,10].

The ROADEF 2007 challenge problem used real world datasets [4]. France Telecom aimed to reduce the cost of its workforce whilst maintaining a satisfied client base and dominating the market share. This optimisation problem involves creating a set of teams over a scheduling horizon to service or outsource a set of jobs [15]. The problem includes many constraints such as technician unavailability, priority levels, outsourcing, skill requirements and precedence relationships. Instances in the ROADEF 2007 challenge ranged from 5 to 800 jobs, however, the problems that arise in industrial settings may include many more jobs to allocate.

A technician routing and scheduling problem was proposed by Pillac et al. [14]. In this work, vehicle routing problem instances proposed by Solomon [16] were extended to create a technician routing and scheduling problem. To do this, random skill requirements were created for each customer and each vehicle (technician) had intrinsic skill levels. This problem included location information and tools and spare parts constraints but did not include the complexity of teaming or precedence relationships. Instances in these datasets contained at most 100 jobs. A service technician routing and scheduling problem was created by Kovacs et al. [11] which also extended instances from [16]. This work, concatenated skill domain information from the ROADEF 2007 challenge problem on to the customer information from the vehicle routing instances. Again these instances contained at most 100 jobs.

A personnel task scheduling problem was also studied by [12] who procedurally generated the datasets used. In this work, the objective was to minimize the make span of the scheduling horizon (similar to ROADEF 2007 challenge but without priority levels). A heuristic approach was compared against mixed integer programming using data instances with up to 2105 jobs. This work considered a heterogeneous workforce, however not in the same way as the ROADEF 2007 challenge, which used skill domain areas and multiple levels of skill within those domains. This research concluded that mixed integer programming was not an appropriate solution technique for large scale scheduling problems. Also the heuristic approaches tested produced quality solutions in short computational times and, most importantly, were scalable.

It appears that the ROADEF 2007 challenge problem includes the most relevant features of the problems studied in the literature. The problem includes the complexity of outsourcing which itself is an NP-hard problem, as well as skill compatibility, unavailability, precedence and priority etc. However, the problems featured range from 5 to 800 jobs. To our knowledge, there is no current literature that includes solving large scale (1000 + jobs) scheduling problems that have precedence relationships, skill requirements and teaming. In this paper, we have generated large scale technician and task scheduling problems (created under the ROADEF 2007 challenge problem definition) and evaluated our intelligent decision heuristic and greedy heuristic on the data.

This paper is structured as follows; Sect. 2 presents the mathematical formulation of the ROADEF 2007 challenge problem. Section 3 describes the large scale technician and task scheduling problem datasets that have been generated. Section 4 describes the two heuristics, an intelligent decision heuristic and a greedy randomized heuristic, that have been used to test the large scale data instances. Section 5 presents the experimental results and Sect. 6 discusses the performance of the intelligent decision and greedy heuristic. Lastly, Sect. 7 identifies areas for further research.

2 ROADEF 2007 Challenge Mathematical Formulation

The aim of the ROADEF 2007 challenge problem is to construct a set of teams to service a set of jobs over a scheduling horizon $K = [1...k]$. Each job i belonging to set N has certain properties, a priority level p where $p \in [1...4]$, an execution time d_i, a domain skill requirement matrix $s^i_{\delta\alpha}$ (where δ is the domain and α is the skill level), an outsourcing cost c_i and a set of successor jobs σ_i. The set of teams is denoted by $M = [1...m]$, which are made up of technicians $T = [1...t]$. The objective function set in the challenge is shown in Eq. (1). The objective function is a weighted sum of the latest ending times, e_p, of each priority group where $w_p = [28, 14, 4, 1]$ for $p = [1, 2, 3, 4]$.

$$Minimize \sum_{p=1}^{4} w_p * e_p \tag{1}$$

The start times of jobs are denoted as b_i. Equation (2) ensures that the latest ending time for each priority group, $p \in [1...3]$, must be greater than, or equal to, the start time of every job plus the duration of the job.

$$e_p \geq b_i + d_i \qquad \forall p \in 1, 2, 3, i \in N_p \tag{2}$$

In addition, Eq. (3) ensures the latest ending time overall e_4, is greater than, or equal to, the start time of every job plus the duration of every job belonging to the entire set of jobs.

$$e_4 \geq b_i + d_i \quad \forall, i \in N \tag{3}$$

Let $x_{t,k,m} = 1$ if technician t belongs to team m on day k. Equation (4) guarantees that if a technician is available to work i.e. belongs to the set T_k, then the technician may only be a member of one team that day.

$$\sum_{m \in M} x_{t,k,m} \leq 1 \quad \forall k \in K, t \in T_k \tag{4}$$

Conversely, Eq. (5) confirms if a technician may not work i.e. does not belong to the set T_k, then the technician is not a member of any team on that day.

$$\sum_{m \in M} x_{t,k,m} = 0 \quad \forall k \in K, t \notin T_k \tag{5}$$

Let $y_{i,k,m} = 1$ if job i is assigned to team m on day k. Equation (6) states that every job belonging to the set of jobs N, must be either outsourced, $z_i = 1$, or scheduled during the scheduling horizon.

$$z_i + \sum_{k \in K} \sum_{m \in M} y_{i,k,m} = 1 \quad \forall i \in N \tag{6}$$

Equation (7) ensures that if a team is assigned a job i.e. $y_{i,k,m} = 1$, then the collective skill levels of the team are greater than or equal to the skill requirements needed to complete the job.

$$y_{i,k,m} * s^{i}_{\delta\alpha} \le \sum_{t \in T_k} v^{t}_{\delta\alpha} * x_{t,k,m} \quad \forall i \in N, k \in K, m \in M, \alpha \in A, \delta \in D \tag{7}$$

Equation (8) deals with the precedence relationships between jobs, so that if job i' is a successor of job i, i.e. belongs to the set σ_i, i' may not begin until i has been completed.

$$b_i + d_i \le b'_i \quad \forall i \in N, i' \in \sigma_i \tag{8}$$

Equations (9) and (10) deal with the working hours of the day. Equation (9) ensures that if a job is scheduled to begin on day k, then the start time of the job is greater than or equal to the beginning of that day. Equation (10) states that if a job is scheduled to be completed on day k then the job must be completed before the working day ends.

$$120(k-1) * \sum_{m \in M} y_{i,k,m} \le b_i \quad \forall i \in N, k \in K \tag{9}$$

$$120(k) * \sum_{m \in M} y_{i,k,m} \ge b_i + d_i \quad \forall i \in N, k \in K \tag{10}$$

Let $u_{i,i'} = 1$ if jobs i and i' are assigned to the same team on the same day and i' begins after i is completed. Equation (11) ensures time continuity, if two jobs happen sequentially then the end time of job i is less than or equal to the start time of the job i'. Here, G is a large number to satisfy the constraint when jobs do not happen sequentially.

$$b_i + d_i - G(1 - u_{i,i'}) \le b'_i \quad \forall i, i' \in N, i \ne i' \tag{11}$$

Equation (12) helps with the ordering of jobs. If two jobs happen sequentially then they must both be allocated to the same team and one must be scheduled before the other.

$$y_{i,k,m} + y_{i',k,m} - u_{i,i'} - u_{i',i} \le 1 \quad \forall i, i' \in N i \ne i', k \in K, m \in M \tag{12}$$

In some problem instances of the ROADEF 2007 challenge problem there is an outsourcing budget available, C. Jobs that are outsourced do not contribute to the objective function, therefore utilization of this budget is important. Let $z_i = 1$ if job i is outsourced. Equation (13) ensures that the outsourcing budget is not exceeded.

$$\sum z_i * c_i \leq C \quad \forall i \in N \tag{13}$$

The set of jobs that are outsourced must adhere to precedence constraints, so if a job is outsourced then so are all successor tasks, Eq. (14).

$$|\sigma_i| * z_i \leq \sum_{i \in \sigma_i} z_i' \quad \forall i \in N^\sigma \tag{14}$$

Equations (15)–(18) show that variables; $x_{t,k,m}$, $y_{i,k,m}$, $u_{i,i'}$ and z_i are binary.

$$x_{t,k,m} = [0,1] \quad \forall k \in K, m \in M, t \in T \tag{15}$$

$$y_{i,k,m} = [0,1] \quad \forall k \in K, m \in M, i \in N \tag{16}$$

$$u_{i,i'} = [0,1] \quad \forall i, i' \in N, i \neq i' \tag{17}$$

$$z_i = [0,1] \quad \forall i \in N \tag{18}$$

Lastly, Equations (19) and (20) show that the start and end times of jobs are non-negative.

$$e_p \geq 0 \quad \forall i \in N_p \tag{19}$$

$$b_i \geq 0 \quad \forall i \in N \tag{20}$$

3 Large Scale Technician and Task Scheduling Problem Instances

Table 1 shows the large scale technician and task scheduling problems that have been designed for this research. To our knowledge, there is only one piece of research that generated technician and task scheduling problem datasets independently, this was [12]. Two other works we are aware of extended existing vehicle routing datasets, [11] and [13], by concatenating skill requirements from other scheduling problems or generating them randomly.

The datasets created in this research are novel, they involve solving a multi-period scheduling problem, with an outsourcing budget, respecting unavailability of resources and teaming. Column one shows the name of each dataset created, column two (Jobs) shows the number of jobs to be scheduled, column three (Techs) displays the number of available technicians and column four (Budget) displays the outsourcing budget available. Lastly, columns five and six (Domains and Levels) show the number of domains and levels.

There are twelve data instances that range from 1000 to 2500 jobs. These new datasets can be split into four groups; L1–L3, L4–L6, L7–L9 and L10–L12. Each group of data contains the same set of jobs to be scheduled, but contains a varying number of available technicians.

Table 1. Large scale technician and task scheduling problem instances

Dataset	Jobs	Techs	Budget	Domains	Levels
L1	1000	25	500	3	3
L2	1000	50	500	3	3
L3	1000	100	500	3	3
L4	1500	25	1000	4	4
L5	1500	50	1000	4	4
L6	1500	100	1000	4	4
L7	2000	25	1500	3	3
L8	2000	50	1500	3	3
L9	2000	100	1500	3	3
L10	2500	25	2000	4	4
L11	2500	50	2000	4	4
L12	2500	100	2000	4	4

3.1 Generating Large Scale Instances

Each data instance was made up of three files, an instance file, a technician file, and a job file. Firstly, the instance file was generated which contains the number of jobs, technicians, domains, levels and outsourcing budget. Next the set of technicians can be created. Each technician is randomly given a level of expertise in each of the domains and assigned days off within the scheduling horizon. Lastly, the job file is created. Each job is randomly assigned a duration, an outsourcing cost and a priority level. However, the jobs also have two other important attributes, domain skill requirements and precedence and successor relationships.

Generating Job Durations. In the ROADEF 2007 challenge problem the length of a working day is limited to 120 time units. In the original problem instances, the job durations ranged from 15 time units to 120 time units, in 15 time unit intervals. Therefore, the job durations in the new datasets have been randomly assigned to be of a length that is a multiple of 15 time units and not greater than 120 time units.

Generating Skill Domain Requirements. The total number of technicians skilled in each domain skill level is recorded. When it comes to generating the skill domain requirements of a job, for the first level in each domain a random number is selected from 0 to the maximum number of technicians who possess this area of expertise. For each subsequent level of expertise in a domain a random number is selected between 0 and the previous required level of expertise. This is to ensure that skill levels are hierarchical i.e. if four technicians are required to be skilled in domain 2 to level 3, then the next level, i.e. domain 2 level 4 must require four or fewer technicians.

Generating Precedence and Successor Relationships. Generating precedence and successor relationships between jobs was a complex task. In the ROADEF 2007 challenge problem, there were multi layered precedence and successor relationships that contained many layers and many jobs. In order to generate these types of constraints an algorithm had to be designed.

This algorithm randomly selected a set of jobs, and ordered them in terms of their priority levels in descending order of importance. This is to ensure that jobs of priority group p are dependent on jobs that are priority p or higher, as this is the way the objective function is calculated, a weighted sum of priority end times.

The algorithm then iterates through the list of priority ordered jobs selecting a random number of jobs for each layer of the relationship tree. Next, each job is assigned to its layer. The algorithm then iterates through each job in the tree ensuring each job has at least one connection to another layer; either upwards (successor) or downwards (precedence).

As this algorithm has a random nature the following relationship trees, as shown in Fig. 1, have been created using the same set of jobs. In Fig. 1a the relationship tree has four layers. On the first layer are jobs 1 and 2, on the second, job 3 (which is a successor of jobs 1 and 2), the third layer contains jobs 4 and 5 (which are successors of job 3), and lastly, on the fourth layer, jobs 6 and 7 (both dependent on job 4, and one dependent on job 5).

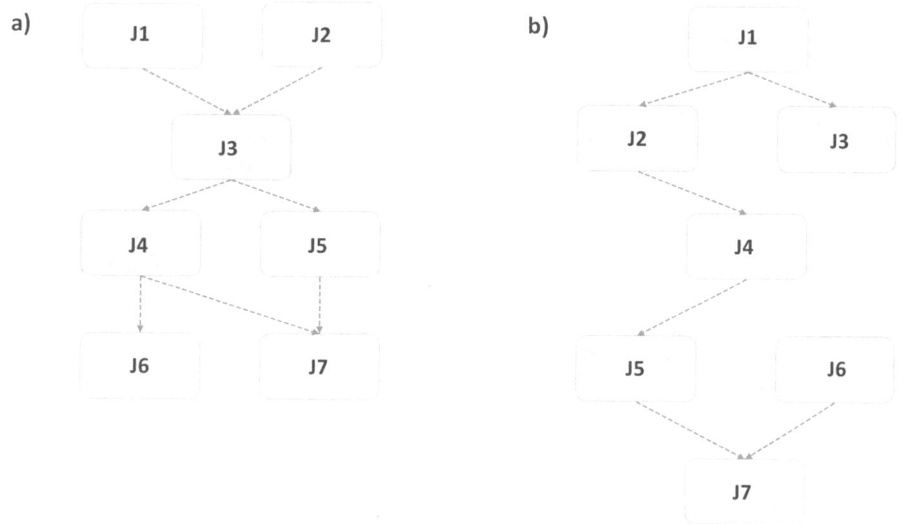

Fig. 1. Example of the dynamic precedence and successor relationship trees

In Fig. 1b, the relationship tree has 5 layers, with one initial job node and one end job node. This figure depicts that jobs can be a member of the relationship tree without having to have both a successor and predecessor (jobs 3 and 6), reiterating the complexity of job relationships within the ROADEF 2007 challenge problem framework.

4 Heuristic Approaches

4.1 Intelligent Decision Heuristic

The intelligent decision heuristic considers multiple scenarios before making a job allocation decision [10]. Given a set of jobs, the heuristic checks to see if a dummy team could be made for each job. Each dummy team is then checked to see which further job allocations could be made. Each scenario is then scored for skill utilization of the team and the utilisation of available time. The highest scoring scenario is selected and the job allocations are made.

4.2 Greedy Heuristic

In order to benchmark the intelligent decision heuristic we will also implement a greedy heuristic on the large scale technician and task scheduling problem instances. The greedy heuristic does not include the intelligent step that checks the implications of an allocation decision. The greedy heuristic selects a single job randomly, creates a team and then makes further allocation decisions based on the skill waste of the team.

5 Experimental Results

Under the competition rules of the ROADEF 2007 challenge, each run of the heuristic is allowed a 20 min computational time limit and so in this work, we have used a 20 min run time. The heuristics were programmed in Java and tested on an HP Z210 Workstation, with an i7-2600 CPU with 3.4 GHZ with 12GB of RAM. Table 2 presents the best result obtained over five runs for each heuristic. In column two the results of the intelligent decision heuristic are displayed, in column three the best result obtained for the greedy heuristic is shown and in column four (% Gap) the percentage gap between the results of the intelligent decision and greedy heuristic are displayed.

6 Discussion

The previous section shows the results obtained for each heuristic approach on the large scale technician and task scheduling problem instances. It is shown that overall the ID heuristic outperforms the greedy heuristic in all problem instances.

Table 2. Experimental results for the large scale problem instances

Dataset	ID	Greedy	% Gap
L1	192810	203850	5.7
L2	97725	103440	5.8
L3	48330	50700	4.9
L4	296940	315210	6.2
L5	147480	156960	6.4
L6	76110	80880	6.3
L7	420660	445335	5.8
L8	198900	207405	4.3
L9	97080	102870	6
L10	574465	607890	5.8
L11	280260	290745	3.7
L12	140970	144840	2.7
Average	214311	225844	5.3

Also as the number of technicians available increases, the gap in solution quality generally decreases. This can be expected, because as the number of available technicians on each day increases, there are fewer shortages of skills, and therefore less consideration can be made for team configurations.

Overall the intelligent decision heuristic finds a solution that is on average 5% better than the quality of solution found by the greedy heuristic. A saving of 5% in personnel costs has the potential to save significant amounts of money in large businesses.

This paper has demonstrated that although the ID heuristic is far more computationally expensive than the greedy heuristic, it can produce better quality results in the same computational time limit. The research has shown that the ID heuristic is both a robust and scalable approach to solving technician and task scheduling problems within strict computational time limits.

7 Conclusion

This paper has shown the benefits of finding efficient ways to solve large scale technician and task scheduling problems in time constrained conditions. In these large scale problems, finding a better quality solution of even 1% can result in large financial savings.

Our contributions to the field are; (i) a methodology for creating technician and task scheduling problem instances, (ii) twelve new large scale technician and task scheduling problems available at https://akhalfay.wordpress.com/large-scale-ttsps/, and (iii) a comparative analysis of the greedy heuristic and the intelligent decision heuristic. It is hoped that other researchers will also evaluate their algorithms on these datasets.

Future work will explore related large scale technician and task scheduling problems, that contain the complexity of routing or time windows in which a customer must be visited.

Acknowledgements. This research is sponsored by ServicePower Technologies PLC, a worldwide leader in providing innovative mobile workforce management solutions, in cooperation with Manchester Metropolitan University and KTP.

References

1. Burke, E., De Causmaecker, P., Berghe, G.V.: A hybrid tabu search algorithm for the nurse rostering problem. In: Asia-Pacific Conference on Simulated Evolution and Learning, pp. 187–194. Springer (1998)
2. Cordeau, J.F., Laporte, G., Pasin, F., Ropke, S.: Scheduling technicians and tasks in a telecommunications company. J. Sched. **13**(4), 393–409 (2010)
3. Crispin, A., Syrichas, A.: Quantum annealing algorithm for vehicle scheduling. In: 2013 IEEE International Conference on Systems, Man, and Cybernetics, pp. 3523–3528. IEEE (2013)
4. Dutot, P.F., Laugier, A., Bustos, A.M.: Technicians and Interventions Scheduling for Telecommunications. France Telecom R&D, Lannion (2006)
5. Ernst, A.T., Jiang, H., Krishnamoorthy, M., Sier, D.: Staff scheduling and rostering: a review of applications, methods and models. Eur. J. Oper. Res. **153**(1), 3–27 (2004)
6. Estellon, B., Gardi, F., Nouioua, K.: High-performance local search for task scheduling with human resource allocation. In: Engineering Stochastic Local Search Algorithms, Designing, Implementing and Analyzing Effective Heuristics, pp. 1–15. Springer (2009)
7. Hashimoto, H., Boussier, S., Vasquez, M., Wilbaut, C.: A grasp-based approach for technicians and interventions scheduling for telecommunications. Ann. Oper. Res. **183**(1), 143–161 (2011)
8. Haugen, D.L., Hill, A.V.: Scheduling to improve field service quality*. Decis. Sci. **30**(3), 783–804 (1999)
9. Hurkens, C.A.: Incorporating the strength of MIP modeling in schedule construction. RAIRO-Oper. Res. **43**(04), 409–420 (2009)
10. Khalfay, A., Crispin, A., Crockett, K.: Solving technician and task scheduling problems with an intelligent decision heuristic. In: Intelligent Decision Technologies 2016, pp. 63–75. Springer (2016)
11. Kovacs, A.A., Parragh, S.N., Doerner, K.F., Hartl, R.F.: Adaptive large neighborhood search for service technician routing and scheduling problems. J. Sched. **15**(5), 579–600 (2012)
12. Krishnamoorthy, M., Ernst, A.T., Baatar, D.: Algorithms for large scale shift minimisation personnel task scheduling problems. Eur. J. Oper. Research **219**(1), 34–48 (2012)
13. Pillac, V., Guéret, C., Medaglia, A.: On the dynamic technician routing and scheduling problem. In: Proceedings of the 5th International Workshop on Freight Transportation and Logistics (ODYSSEUS) (2012)
14. Pillac, V., Gueret, C., Medaglia, A.L.: A parallel matheuristic for the technician routing and scheduling problem. Optim. Lett. **7**(7), 1525–1535 (2013)

15. Society, F.O.R.: What is the roadef 2007 challenge (2016). http://challenge.roadef.org/2007/en/
16. Solomon, M.M.: Algorithms for the vehicle routing and scheduling problems with time window constraints. Oper. Res. **35**(2), 254–265 (1987)
17. Titiloye, O., Crispin, A.: Quantum annealing of the graph coloring problem. Discrete Optim. **8**(2), 376–384 (2011)

Manipulability of Majority Relation-Based Collective Decision Rules

Fuad Aleskerov[1,2], Alexander Ivanov[1,3], Daniel Karabekyan[1(✉)], and Vyacheslav Yakuba[1,2]

[1] National Research University Higher School of Economics, Moscow, Russia
{alesk,dkarabekyan}@hse.ru, ivanovalexalex@gmail.com, yakuba@ipu.ru

[2] Institute of Control Sciences of Russian Academy of Sciences, Moscow, Russia

[3] Skoltech, Moscow, Russia

Abstract. In the problem of aggregation of rankings or preferences of several agents, there is a well-known result that reasonable social ranking is not strategy-proof. In other words, there are some situations when at least one agent can submit insincere ranking and change the final result in a way beneficial to him. We call this situation manipulable and using computer modelling we study 10 majority relation-based collective decision rules and compare them by their degree of manipulability, i.e. by the share of the situation in which manipulation is possible. We found that there is no rule that is best for all possible cases but some rules like Fishburn rule, Minimal undominated set and Uncovered set II are among the least manipulable ones.

Keywords: Manipulation · Voting rules · Extended preferences

1 Introduction

Constructing an intelligent decision system in almost any area assumes that we need to take into account preferences of individual agents. As a result, any agent may influence the final choice by submitting insincere preferences in the system. For the situation when there are at least three alternatives and we need to choose exactly one, every non-dictatorial collective decision rule is individually manipulable [7, 15]. Then an interesting question arises: if we know that every rule is manipulable, how can we find the least manipulable one? This question initiated a lot of papers studying to which extent known rules are manipulable. The non-exahustive list of papers is [3, 4, 6, 11, 13, 16]. Similar question was considered in [5, 8, 9, 12, 14, 17].

The main problem with the analysis of the real collective decision rules is the possibility of ties: situations when the rule cannot decide between several alternatives. All the above papers use so called alphabetical tie-breaking rule, that is easy to implement but not likely to be seen in real life. Aleskerov et al. [1, 2] studied general framework for positional rules and showed that alphabetical tie-breaking crucially influences the estimated results as well. In this paper we use the similar model as in [1, 2] but look at the majority relation-based rules: set of rules based on pairwise comparison of the alternatives.

I. Czarnowski et al. (eds.), *Intelligent Decision Technologies 2017*,
Smart Innovation, Systems and Technologies 72, DOI 10.1007/978-3-319-59421-7_8

The paper is organized as follows. In Sect. 2 we describe the model. In Sect. 3 decision rules are defined. In Sect. 4 computation scheme is defined. Section 5 discusses the results. Section 6 concludes.

2 Model

We use a common notation - there is a set of m alternatives denoted as A, which consists hereafter of either 3, 4 or 5 alternatives. There are n voters or agents, each of them has a strict preference P_i which is a linear order (i.e. transitive, antisymmetric and connected binary relation) over the set of alternatives A. All possible choices can be represented as elements of the set $\mathcal{A} = 2^A \backslash \{\emptyset\}$, i.e. as all possible non-empty subsets of the set of alternatives.

A set of n agents each of them having a certain preference P_i represents a preference profile $\vec{P}$. An aggregation procedure C is a mapping of a profile $\vec{P}$ to $\mathcal{A}$. In other words, when we apply collective decision rule to a profile, we may confront a situation when two or more alternatives are chosen. In order to deal with this situation we should define preferences over the sets called extended preferences (EP_i).

In this paper we use two types of preferences introduced in [1] and [10]: PWorst and PBest.

PWorst is the ordering based on the probability of the worst alternatives under the assumption of an absence of information about tie-breaking. We assign equal probabilities to all alternatives in one multi-valued choice. When we compare two different multi-valued choices, the choice with lower probability of the worst alternative is more preferable. If these probabilities are the same then we look to the probability of second-worst alternative, and so on. For 3 alternatives and alphabetical preferences, $EP_i^{PWorst} = \{a\} \succ \{a,b\} \succ \{b\} \succ \{a,b,c\} \succ \{a,c\} \succ \{b,c\} \succ \{c\}$. PBest is similar, but the ordering is based on the probability of the best alternatives (the higher it is the better is the set). For 3 alternatives and alphabetical preferences, $EP_i^{PBest} = \{a\} \succ \{a,b\} \succ \{a,c\} \succ \{a,b,c\} \succ \{b\} \succ \{b,c\} \succ \{c\}$.

The case of individual manipulability can be described as follows. Let $\vec{P} = \{P_1, P_2, \ldots P_i, \ldots P_n\}$ be a profile of sincere preferences, and $\vec{P}_{-i} = \{P_1, P_2, \ldots P_i', \ldots P_n\}$ be a profile where agent i tries to manipulate by substituting her sincere preference P_i by insincere preference P_i'. Individual manipulation takes place if $C\left(\vec{P}_{-i}\right) EP_i C(\vec{P})$, i.e. the choice after misrepresenting preferences is better than the choice when manipulation does not hold.

3 The Rules

We consider 10 majority relation-based aggregation procedures. The majority relation μ is defined as a binary relation over the set of alternatives

$$x\mu y \leftrightarrow card\left\{i \epsilon N \mid xP_i y\right\} > card\left\{i \epsilon N \mid yP_i x\right\}$$

We will show how the rule works on a sample profile with 4 alternatives and 6 agents (higher alternative is more preferable).

P_1	P_2	P_3	P_4	P_5	P_6
a	a	a	d	d	b
b	d	c	b	b	c
c	c	d	c	c	d
d	b	b	a	a	a

Let us construct the majority relation for this profile. The relation μ is defined as $b\mu c$ and $d\mu b$.

We use the following aggregation procedures. In the brackets we use short names that later will be used in the results.

1. (**Mds**) Minimal dominant set. A set Q is called a dominant set, if each alternative in Q dominates each alternative outside Q via majority relation. A dominant set is called a minimal one if no its proper subsets are dominant. If there are more than one minimal dominant set, the choice is comprised of the union of such minimal dominant sets. For the example profile, the minimal dominant set will be {a,b,c,d}.
2. (**Mus**) Minimal undominated set. A set Q is called an undominated set, if no alternative outside Q dominates any alternative in Q via majority relation. An undominated set is called a minimal one if none of its proper subsets is undominated set. If there are more than one minimal undominated set, the choice is comprised of the union of such minimal undominated sets. For the sample profile there are two minimal undominated sets, {a} and {d}. Thus the choice is the union of them, i.e., {a,d}.
3. (**Us1**) Uncovered set I (version used in [3]). We define lower contour set of an alternative x for majority relation μ as $L(x) = \{y \in A \mid x\mu y\}$. For Uncovered set I we define a new binary relation δ such that $x\delta y \leftrightarrow L(x) \supset L(y)$. Undominated alternatives on the relation δ are chosen. The lower contours for the sample profile are $\mathrm{L(a)} = \emptyset, \mathrm{L(b)} = \{\mathrm{c}\}, \mathrm{L(c)} = \emptyset, \mathrm{L(d)} = \{\mathrm{b, c}\}$. Then the binary relation δ will include the following pairs: b δ a, b δ c, d δ b, d δ a, d δ c and the choice will be {b,d}.
4. (**Us2**) Uncovered set II (version used in [3]). We define the upper contour set of an alternative x for majority relation μ as $\mathrm{D}(x) = \{y \in A \mid y\mu x\}$. An alternative x is said to B-dominate an alternative y, i.e. xBy if $x\mu y$ and $D(x) \subseteq D(y)$. The result of Uncovered set II aggregation procedure is B-undominated alternatives. Since the upper contours for our example are $\mathrm{D(a)} = \emptyset, \mathrm{D(b)} = \{\mathrm{d}\}, \mathrm{D(c)} = \{\mathrm{b, d}\}\mathrm{D(d)} = \emptyset$, then the binary relation B will include a B b, a B c, d B b, d B c. Thus the choice will be {a,d}.
5. (**R**) Richelson's rule. First, we construct the lower and upper contour sets for the majority relation. Then, we construct a new binary relation σ such that Undominated

$$x\sigma y \leftrightarrow [L(x) \supseteq L(y) \wedge D(x) \subseteq D(y) \wedge \left(\left[L(x) \supset L(y)\right] \vee \left[D(x) \subset D(y)\right]\right)]$$

alternatives on the relation σ are chosen. In our example the binary relation σ consists of the only pair d σ c. So, the choice will be {a,b,d}.

6. (**Mws**) Minimal weakly stable set [3]. A set Q is called a weakly stable set if and only if it satisfies the following property: if for $x \in Q$ $y\mu x$ holds for some y, than either $y \in Q$ or $\exists z \in Q$ $s.t. z\mu y$. A weakly stable set is called a minimal one if no its proper subsets are weakly stable sets. If there are more than one minimal weakly stable set, the choice is comprised of the union of such minimal weakly stable sets. The majority relation for the sample profile is bμc and dμb. The minimal weakly stable sets are {a} and {d}. The resulting choice is the union of these sets {a,d}.
7. (**F**) Fishburn's Rule. First we construct upper contour set D(x) for the majority relation. Then, we construct a binary relation γ such that $x\gamma y \leftrightarrow D(x) \subset D(y)$. Undominated alternatives on the relation γ are chosen. For the sample profile the upper contours of the alternatives are $D(a) = \emptyset, D(b) = \{d\}$, $D(c) = \{b, d\}$, $D(d) = \emptyset$. The binary relation γ consists of the following pairs: aγb, aγc, dγb, bγc, dγc. Thus, the choice will be {a,d}, since alternatives "a" and "d" are undominated via γ.
8. (**C1**) Copeland's rule I [3]. First, we construct upper contour set D(x) and lower contour set L(x) in majority relation. Then, we define a function $u(x)$ such that $u(x) = card\{L(x)\} - card\{D(x)\}$. The social choice is comprised of alternatives with largest values of $u(x)$. For our example the function u is equal to u(a) = u(b) = 0, u(c) = −1, u(d) = 1. Thus the choice is {d}.
9. (**C2**) Copeland's rule II [3]. Function $u(x)$ is defined as a cardinality of lower contour set in the majority relation for a given alternative x. The social choice is comprised of the alternatives with the maximum values of $u(x)$. For our example the function u is equal to u(a) = u(c) = 0, u(b) = u(d) = 1. Thus the choice is {b,d}.
10. (**C3**) Copeland's rule III [3]. Function $u(x)$ is defined as a cardinality of upper contour set in the majority relation for a given alternative x. The social choice is comprised of the alternatives with the minimum values of $u(x)$. Since the function u for the sample example is u(a) = u(d) = 0, u(b) = u(c) = 1, then the choice is {a,d}.

4 Computation Scheme

We estimate the degree of manipulability of 10 aggregation procedures defined above. By the degree of manipulability we understand the share of all manipulable profiles. The question is how we define the number of all possible profiles. There are two main concepts: Impartial Culture (IC) and Impartial Anonymous Culture (IAC).

Under Impartial Culture it is assumed that all preferences and profiles have equal probabilities. In terms of computations, it means that if we would like to generate one profile, we need to generate s numbers from 1 to m!, where each number stands to a permutation of m alternatives.

In Impartial Anonymous Culture model only anonymous profiles (called voting situations) are considered. In other words, two profiles obtained one from another by permutation of agents are not considered. Thus, the total number of profiles under IAC is less than under IC. All voting situations are considered equally likely.

In order to estimate the degree of manipulability we use the following measure introduced by Nitzan [13] and Kelly [11]. For the case of Impartial Culture

$$NK_{IC} = \frac{d_0}{(m!)^n},$$

where d_0 is the number of manipulable profiles, $m!$ is the number of all possible preferences (linear orders) on the set of alternatives, $(m!)^n$ is the number of all possible profiles. In other words, NK index stands for the share of manipulable profiles out of all possible profiles.

For the case of Impartial Anonymous Culture the formula is

$$NK_{IAC} = \frac{d_0}{C^n_{(m!)^n+n-1}}$$

It can be noticed that the main difference is in the denominator which stands for the total number of profiles.

It is obvious that we cannot calculate NK-index directly. For this reason, we use Monte-Carlo approach and generate 1 million profiles for each case.

To estimate the degree of manipulability of aggregation procedures, we need to perform computer simulation. The scheme of computer simulation is the following

1. For each profile we determine whether it is manipulable. By definition, a profile is manipulable if there is at least one agent, which can manipulate. Whether a profile is manipulable is checked as follows:
   (a) Consider each (out of n) agent,
   (b) For each agent generate $m! - 1$ manipulations attempts (i.e. all possible ways of misrepresenting preferences),
   (c) For each manipulation attempt calculate social choices before and after a manipulation attempt. If the choice after manipulation attempt is better than the choice without a manipulation attempt, mark the profile as manipulable,
2. Calculate NK index by dividing the number of manipulable profiles by the total number of generated profiles (i.e. 1,000,000).

5 Results

We obtained the results for 10 aggregation procedures for 3, 4 and 5 alternatives for Impartial Culture and Impartial Anonymous Culture. At the Figs. 1 and 2 the results for five rules for 3 alternatives and even number of agents are given. At the Fig. 3 the results for odd number of agents are given.

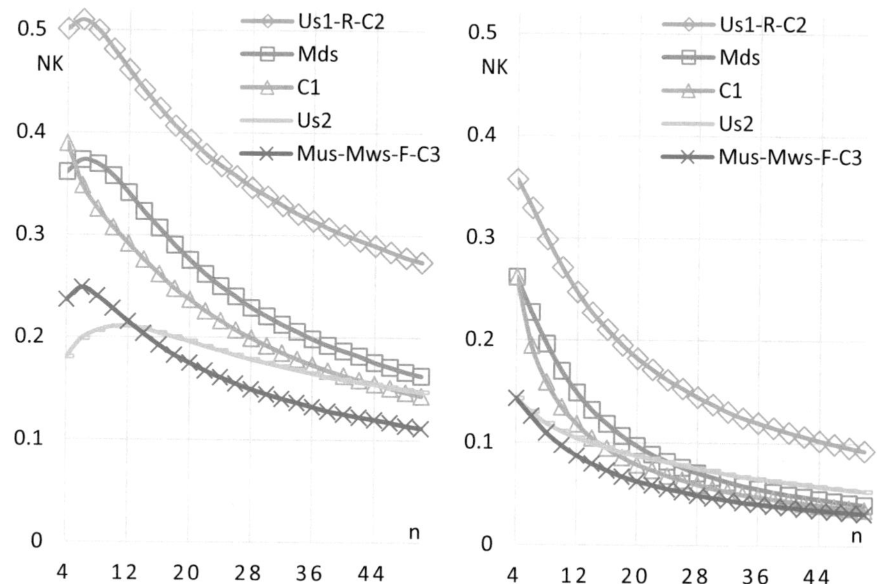

Fig. 1. NK index for PBest extension, 3 alternatives and even number of agents. Impartial Culture (left), Impartial Anonymous Culture (right)

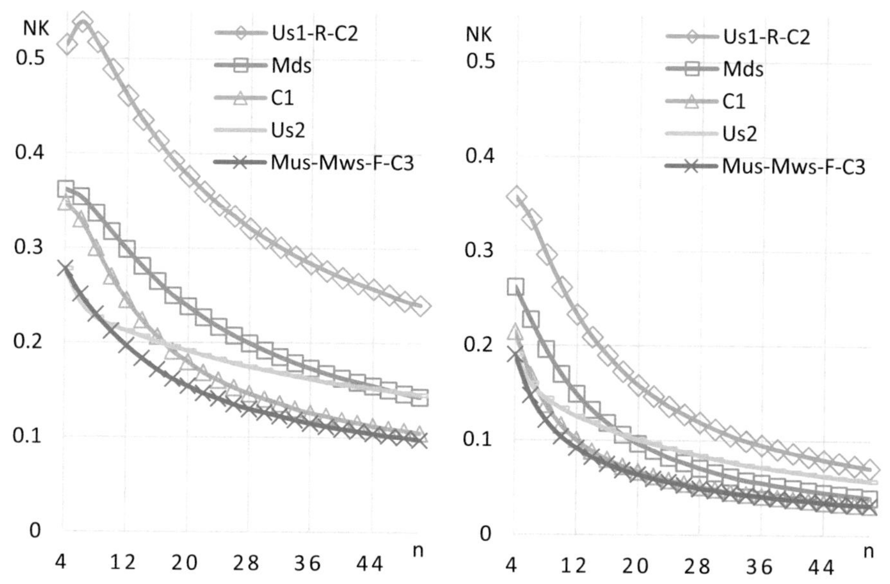

Fig. 2. NK index for PWorst extension, 3 alternatives and even number of agents. Impartial Culture (left), Impartial Anonymous Culture (right)

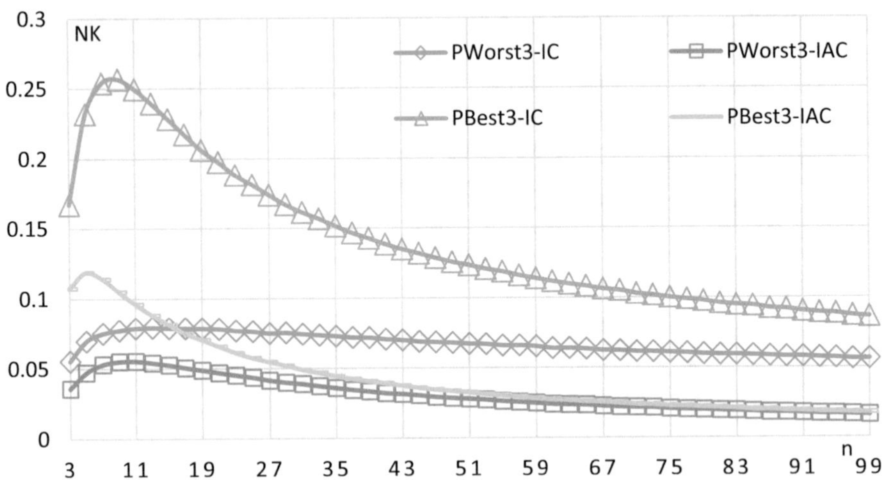

Fig. 3. NK index for 3 alternatives and odd number of agents. All rules give the same degree of manipulation for specific extension and probabilistic model

We decided to separate odd and even number of agents in order to have more representable results. For odd number of agents majority relation graph is always complete since all alternatives can be compared. Moreover for 3 alternative case all rules coincide for odd number of agents that can be seen from Fig. 3. That happens because there are just two possible different graphs of majority relation for 3 alternatives: $a\mu b, a\mu c, b\mu c$ and $a\mu b, c\mu a, b\mu c$. We mean that they are different in terms of ordering of alternatives, not their names, since all the rules under study are neutral – they are not influenced by the names of alternatives. The former type of graph has one alternative that is not dominated by any other, while latter type has cycle – all alternative dominate each other. For even number of agents we have 5 groups of rules since Us1, R and C2 coincide as well as Mus, Mws, F, C3.

From Figs. 1 and 2 we observe that the degree of manipulability decreases for all rules with the number of agents increases, but the velocity is different. Thus, if some rule is the least manipulable for small number of agents, it may not be the same for larger number (see for example, Us2). Under IAC degree of manipulability is generally lower, but the ordering of the rules is almost the same. At the same time extension axiom influence the ordering. From Fig. 3 we can see that influence of extension axiom is almost negligible for large number of agents and IAC.

We can summarize the results in the following tables where the names of the least manipulable rules are given. Since our computation scheme gives us about 0.001 error in estimation (95% confidence interval) in the table we name the rule which have the lowest Nitzan-Kelly index and all rules in 0.001 interval of it. Since as we explained above for the case of 3 alternatives all rules gives the same manipulability in Table 1 only results for even number of voters are given.

Table 1. Least manipulable rules for 3 alternatives. Note: MMF – Mus, Mws and F

Agents	4	6	8	10	even
IC-PWorst3	MMF, C3, Us2	Us2	Us2	MMF, C3	MMF, C3
IC-PBest3	Us2	Us2	Us2	Us2	MMF, C3
IAC-PWorst3	MMF, C3, Us2	MMF, C3	MMF, C3	MMF, C3	MMF, C3
IAC-PBest3	MMF, C3, Us2	MMF, C3	MMF, C3	MMF, C3	MMF, C3

While we observe from figures differences in Nitzan-Kelly indices, from tables we can see that there is almost no difference between probabilistic models if we look for names of least manipulable rules in each case. It can be seen from Tables 2 and 3, that IC-PWorst4 coincide with IAC-PWorst4, the same for PBest. For the case of 5 alternatives and IAC-PWorst5, more rules lie in the confidence interval of the least manipulable rule.

Table 2. Least manipulable rules for 4 alternatives. Note: FUUR – F, Us1, Us2 and R; MMF – Mus, Mws and F; MMM – Mds, Mus, Mws

Agents	3	4	5	6	7	8	9	10
IC-PWorst4	MMM, FUUR	MMF, C3	MMM, FUUR	MMF, C3	MMM, FUUR	MMF	MMM, FUUR	MMF
IC-PBest4	FUUR	Us2	FUUR	Us2	FUUR	Us2	FUUR	Us2
IAC-PWorst4	MMM, FUUR	MMF, C3	MMM, FUUR	MMF, C3	MMM, FUUR	MMF	MMM, FUUR	MMF
IAC-PBest4	FUUR	Us2	FUUR	Us2	FUUR	Us2	FUUR	Us2

Table 3. Least manipulable rules for 5 alternatives. Note: FUUR – F, Us1, Us2 and R; MMF – Mus, Mws and F; MMM – Mds, Mus, Mws

Agents	3	4	5	6	7	8	9	10
IC-PWorst5	MMM, FUUR	MMF, C3	MMM, FUUR	Mus, F	MMM, FUUR	MMF	MMM, FUUR	Mus, Mws
IC-PBest5	FUUR	Us2	FUUR	Us2	FUUR	Us2	FUUR	Us2
IAC-PWorst5	MMM, FUUR	MMF, C3	MMM, FUUR	MMF, C3	MMM, FUUR	MMF, C3	MMM, FUUR	MMF
IAC-PBest5	FUUR	Us2	FUUR	Us2	FUUR	Us2	FUUR	Us2

For 4 and 5 alternatives that for the case of odd number of agents some rules still coincide, but results now depend on the extension axiom and only 4 or 7 rules out of 10 are least manipulable ones. For even number of agents results are more specific: Us2, Mus and F are among the least manipulable ones, for some cases Mws and C3 are also in this list. If we take to the account even and odd number of alternatives we can see, that for PWorst extension F and Mus are the least manipulable for 3 to 10 agents, while for PBest extension Us2 is the least manipulable rule. At the same time, when number of agents is high enough situation changes. We see from figures different velocity of manipulability of rules when number of agents goes up for the case of 3 alternatives. The same situation holds for the case of 4 and 5 alternatives. For example, for 4 alternatives and 100 voters, IC and PBest extension, Us2 is no longer least manipulable rule while Mus is.

6 Conclusions

Shortly the results can be summarized as follows:

1. The values of NK index are different for even and odd numbers of agents. Main reason for that is the different nature of majority matrices for the two scenarios. For the case of odd number of agents, the majority matrix will have connection either way between each pair of alternatives. For the case of even number of agents, there can be some ties in the majority relation, i.e. incomparable pairs of alternatives.
2. For some aggregation procedures this difference can be relatively large. For example, for the Richelson rule NK-index for 23 agents is equal to 0.077, while for 24 agents it is equal to 0.347. This is more than 4 times difference.
3. The values of NK index decrease with growing number of agents. It means that for larger numbers of agents situations when one agent can influence the collective choice are less probable.
4. There is no rule that is least manipulable for any number of voters and any extension method. For example, for small number of voters and PBest extension Uncovered set II is the least manipulable rule, but for 15 and more agents this rule is not among the least manipulable ones.
5. When there are more than 20 voters the least manipulable aggregation procedures are Fishburn rule, Minimal undominated set and Minimal Weakly stable set. They show very similar results.

Acknowledgments. The article was prepared within the framework of the Basic Research Program at the National Research University Higher School of Economics (HSE) and supported within the framework of a subsidy by the Russian Academic Excellence Project '5-100'. We thank Laboratory of Decision Choice and Analysis (DeCAn) at the HSE for support.

References

1. Aleskerov, F., Karabekyan, D., Sanver, M.R., Yakuba, V.: An individual manipulability of positional voting rules. SERIEs. **2**(4), 431–446 (2011). doi:10.1007/s13209-011-0050-y
2. Aleskerov, F., Karabekyan, D., Sanver, M.R., Yakuba, V.: On the manipulability of voting rules: the case of 4 and 5 alternatives. Math. Soc. Sci. **64**(1), 67–73 (2012). doi:10.1016/j.mathsocsci.2011.10.001
3. Aleskerov, F., Kurbanov, E.: Degree of manipulability of social choice procedures. In: Alkan, P.A., et al. (eds.) Current Trends in Economics, pp. 13–27. Springer, Heidelberg (1999). doi:10.1007/978-3-662-03750-8_2

4. Chamberlin, J.R.: An investigation into the relative manipulability of four voting systems. Syst. Res. **30**(4), 195–203 (1985). doi:10.1002/bs.3830300404
5. Diss, M.: Strategic manipulability of self-selective social choice rules. Ann. Oper. Res. **229**(1), 347–376 (2015). doi:10.1007/s10479-014-1763-7
6. Favardin, P., Lepelley, D.: Some further results on the manipulability of social choice rules. Soc. Choice Welf. **26**(3), 485–509 (2006). doi:10.1007/s00355-006-0106-2
7. Gibbard, A.: Manipulation of voting schemes: a general result. Econometrica **41**(4), 587–601 (1973). doi:10.2307/1914083
8. Green-Armytage, J.: Strategic voting and nomination. Soc. Choice Welf. **42**(1), 111–138 (2014). doi:10.1007/s00355-013-0725-3
9. Green-Armytage, J., Tideman, T.N., Cosman, R.: Statistical evaluation of voting rules. Soc. Choice Welf. **46**(1), 183–212 (2016). doi:10.1007/s00355-013-0725-3
10. Karabekyan, D.: On extended preferences in a voting. Ekonomicheskij J. VShE **13**, 19–34 (2009). (in Russian)
11. Kelly, J.S.: Almost all social choice rules are highly manipulable, but a few aren't. Soc. Choice Welf. **10**(2), 161–175 (1993). doi:10.1007/BF00183344
12. Laslier, J.-F.: Heuristic voting under the alternative vote: the efficiency of "Sour Grapes" behavior. Homo Oecon. **33**(1–2), 57–76 (2016). doi:10.1007/s41412-016-0001-8
13. Nitzan, S.: The vulnerability of point-voting schemes to preference variation and strategic manipulation. Public Choice **47**(2), 349–370 (1985). doi:10.1007/BF00127531
14. Nurmi, H.: Reflections on the significance of misrepresenting preferences. In: Transactions on Computational Collective Intelligence XXIII, pp. 149–161. Springer, Heidelberg (2016). doi:10.1007/978-3-662-52886-0_10
15. Satterthwaite, M.A.: Strategy-proofness and Arrow's conditions: Existence and correspondence theorems for voting procedures and social welfare functions. J. Econ. Theory **10**(2), 187–217 (1975). doi:10.1016/0022-0531(75)90050-2
16. Smith, D.A.: Manipulability measures of common social choice functions. Soc. Choice Welf. **16**(4), 639–661 (1999). doi:10.1007/s003550050166
17. Veselova, Y.: The difference between manipulability indices in the IC and IANC models. Soc. Choice Welf. **46**(3), 609–638 (2016). doi:10.1007/s00355-015-0930-3

Stacking-Based Integrated Machine Learning with Data Reduction

Ireneusz Czarnowski(✉) and Piotr Jędrzejowicz

Department of Information Systems, Gdynia Maritime University, Morska 83, 81-225 Gdynia, Poland
{irek,pj}@am.gdynia.pl

Abstract. Integrated machine learning is understood as integration of the data reduction with the learning process. Such integration allows to introduce adaptation mechanisms within the learning process by modification of the data with a view to finding its better representation from the point of view of the learning performance criterion. Data modification can be carried out through data reduction in both dimensions, i.e. the feature and the instance ones producing the set of prototypes. Currently, data reduction has become a crucial technique for big data analysis and improvement of the machine learning process results. In this paper the stacking technique has been proposed for improving the process of the integrated machine classification and to assure diversification among prototypes. To validate the proposed approach we have carried-out computational experiment. The paper includes the description of the approach and the discussion of the validating experiment results.

Keywords: Classification · Learning from data · Data reduction · Prototype selection · Stacked generalization

1 Introduction

The paper focuses on the problem of the integrated machine classification. The integrated machine classification is understood as integration of the two main knowledge discovery stages, i.e. data pre-processing and data mining.

Classification is one of the typical data mining tasks. Its main goal is to predict unknown objects or concepts. Classification requires finding the model (or function) that describes and distinguishes data, classes or concepts. The model is called a classifier. The process of finding the classification model is called learning process or classifier learning and it can be used to predict classes of objects whose class labels are unknown [10]. So, the derived model is based on the analysis of the set of training data (i.e., data objects whose class label is known). Data pre-processing comprises data collecting, data integration, data transformation, data cleaning, and data reduction.

Integration of both important stages of the knowledge discovery process, i.e. data pre-processing with data mining, has been recognized as an important step towards improving quality of machine learning tools. It was already proposed in earlier papers of authors [10]. Such integration allows to introduce adaptation mechanisms within the

I. Czarnowski et al. (eds.), *Intelligent Decision Technologies 2017*,
Smart Innovation, Systems and Technologies 72, DOI 10.1007/978-3-319-59421-7_9

learning process by modification of the data with a view to finding its better representation from the point of view of the learning performance criterion. Introducing the proposed adaptation mechanisms exemplifies the idea of learning classifier systems introduced by [15, 24, 3].

Data modification can be carried out by data reduction. Data reduction carried-out without losing extractable information is considered as an important approach to increasing effectiveness of the learning process when the available data sets are large [17]. *Data reduction is perhaps the most critical component in retrieving information from big data (i.e., petascale-sized data) in many data-mining processes* [27]. Data reduction is carried out by selection of relevant information and construction of prototypes. It is an effective approach for analyzing complex high dimensional data and allows efficient implementation of the classification and decision algorithms.

On the modelling side, an effective approach to substantially increase accuracy of the machine learning is to use the concept of the classifier ensemble also known as the multiple model technique. Multiple model techniques have proven useful in substantially increasing quality of the results produced by the machine learning models and tools. These techniques integrate a number of learning models into a single machine learning system. In case of the classification task such system is called an ensemble of classifiers or a combined classification system. It has been shown that machine learning systems using the multiple learning models offer practical and effective solutions to different classification problems [22]. Integration of the multiple learning models aims at achieving better accuracy of the resulting system [14]. Main strategy for the classifier ensemble construction is to combine outputs of the different classification models, known as base classifiers, such that their combination outperforms each of the single classifiers.

Techniques for assuring diversification among base classifiers include using different learning algorithms or using different subsets of the original dataset. One of such techniques is stacked generalization or stacking. In stacking diversity of the ensemble is achieved by generating base learners using different training parameters, including different training sets.

The paper extends earlier research results, where the integrated data reduction with learning classifiers was considered [10], and data reduction, through instance selection, and stacked generalization for learning from examples were investigated [7, 9]. In [7, 9] two different approaches were considered for instance selection.

The proposed approach focuses on the adaptive and integrated machine classification based on data reduction, where the stacking has been applied for improving the learning from examples process. Diversified base classifiers are trained using prototypes selected from clusters produced by the kernel-based fuzzy clustering algorithm, i.e. in the process of data reduction. To validate the approach, an extensive computational experiment has been carried out using several benchmark datasets from UCI repository [1]. The paper includes the description of the approach and the discussion of the validating experiment results.

The paper is organized as follows. In Sect. 2 the problem of integrated learning from example with data reduction is described. This section also contains a short review of current instance reduction techniques. Section 3 includes brief description of the stacked generalization. Section 4 provides a detailed description of the proposed approach. Section 5 discusses computational experiment plan and results. Finally, the last section contains conclusions and suggestions for future research.

2 Integrated Machine Classification and Data Reduction

The process of finding the classification model is called the learning process or the classifier learning [10]. Formally, the learner outputs a classifier $h \in H$, called the *hypothesis*, that has been induced based on the set D, where H is the hypothesis space, i.e. a set of all possible hypotheses that a learner can draw on in order to construct the classifier. Thus, a learning algorithm outputs an element $h \in H$. The hypothesis may be viewed as a logical predicate that may be true or false with respect to the considered data, or a function that maps examples into, for example, category (data class) space [10].

The problem of learning from examples can be defined as follows: The goal of learning from examples is to find a hypothesis $h = L(D)$ assuring a good approximation of the target concept, and where the hypothesis is optimal with respect to some performance criterion F, and where D is a non-empty dataset of examples x, each x is described by a fixed set of attributes (features), $A=\{a_1,a_2,\ldots,a_m\}$, and h is a classifier belonging to a set of hypotheses H and induced by the learning algorithm L. In the end, the learning task takes the following form:

$$h = \arg \max_{h \in H} f(h). \tag{1}$$

Finally, the role of the classification model (classifier) is to predict the class of the object whose labels are unknown. If for given $x \in D$, $h(x) = d(x)$ then h makes a correct decision with respect to x (where d is a function, such that $d(x)$ is the value of decision class for the example x).

Integrated adaptive classifiers should possess two important features:

- Integration, at least partial, of the data reduction and the data mining stages.
- The existence of a positive feedback whereby more effective data reduction leads to higher classification accuracy, and in return, higher classification accuracy results in even more effective data reduction.

In Fig. 1 the idea of integrated approach to constructing machine classifiers is shown. From the technical point of view, the introduction of the integrated and adaptive learning means, that some processes within the pre-processing stage, like instance selection and/or attribute selection can be integrated with the construction of the classifier, which allows for introducing the adaptability of the approximation space. Thus, the machine learning problem can be formulated as follows: construct hypothesis performing best from the point of view of the performance criterion by finding the representation of the approximation space and, at the same time, deciding on the classifier features.

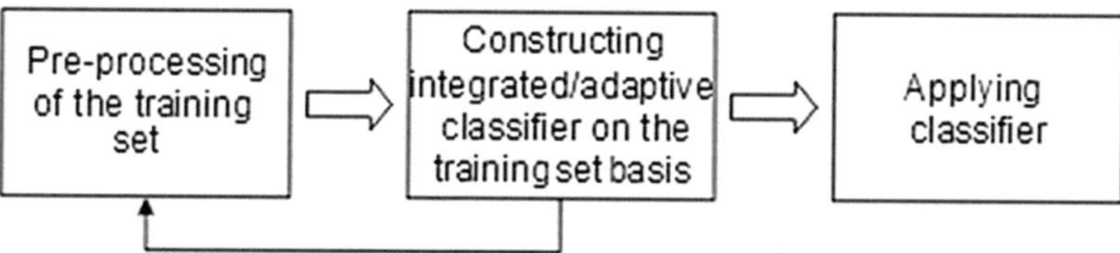

Fig. 1. Integrated and adaptive approach to constructing machine classifiers

Thus, when we assume that the goal of learning from examples is to find a hypothesis h that is a good approximation of the target concept and where the learner used to produce h requires setting of some parameters decisive from the point of view of its performance. Let the set of parameters g describe the way the training set should be transformed. Thus, it can be said that the goal of learning from examples is to find the hypothesis $h = L(D,g)$ where parameters g affect the learning process and influence the performance measure f. In such case the learning task takes the following form:

$$h = \arg \max_{h \in H, g \in G} f(h = L(D, g)), \tag{2}$$

where G is the parameter space. Thus, the learning process is realized in the hypothesis space H and the parameter space G. From the implementation point of view, it means that instances and features are selected iteratively while constructing the classifier.

Considering to apply data reduction techniques aiming at reinforcement of the learning process, one expects that classifier built on the reduced dataset is better or at least not worse than a classifier induced using the original dataset [12]. In other words, data reduction is especially useful to boost up effectiveness of the machine learning process, especially when the processed datasets are large [4, 5]. A small set of instances called prototypes, selected from an original and large dataset, can assure construction of the classification model superior to one constructed using the whole original and large dataset. Besides, data reduction is the best way to avoid working on the whole original dataset all time [29]. On the other hand, removing some instances from the training set diminishes time and memory complexity of the learning process [23, 25].

Thus, the problem of data reduction is to find the optimal prototype dataset, S_{opt}, as a subset of the original non-reduced dataset D, ensuring maximum value of the performance criterion or criteria of the learning algorithm L.

In this paper it is assumed that the data reduction is carried out in both dimensions. Thus the optimal prototype dataset, S_{opt}, is a subset of the dataset D, where each example is described by a set of $A' \subset A$. With respect to the problem of learning from data, when data reduction process is carried out, the task of learner L is to output the hypothesis $h \in H$ that optimizes performance criterion F using dataset S which is a subset of the set D, such that $|S| < |D|$ (ideally $S = S_{opt}$), where each example $x \in S$ is described by a set A', where $|A'| < |A|$. With regard to formula (2), instances and features which finally are used in the learning process are described by the set g. From mathematical point of view, it can be written that g is a vector of data transformation.

3 Stacked Generalization

Stacking, introduced by Wolpert in [26], has been proposed as a technique for increasing accuracy of the machine classification. Wolpert in [26] defines the stacked generalization as a way of combining multiple models that have been learned for a classification task. The stacking produces an ensemble of classifiers in which each base classifier is built using different training parameters, and in the next step the outputs of the base classifiers can be combined using selected schemas.

In the standard stacking algorithm q different subsets of the training data set are created using stratified sampling with replacement. The subsets are generated assuring relative proportion of the different classes as in the original dataset. Subsets of the training set, without one of them are used to induce base classifiers. Base classifiers induced on q-1 training sets are called the level-0 models. Next, the level-0 models produce vector of predictions that form the input to the level-1 model, which in turn, predicts class label for new instances with unknown class labels. At the level-1 the omitted set at the level-0 is the test for the respective iteration. The process is repeated q times following the pattern of the q-fold cross-validation procedure. In such approach a meta classifier in term of relative weights for each level-0 classifier is induced by assigning weights to classifiers, proportionally to their performance.

Based on assumptions for the standard approach, the stacking works as follows: Suppose that there are q different learners $L_1,\ldots,L_q$ and q different training sets – $D_1, \ldots, D_q$, where $D = D_1 \cup D_2 \ldots \cup D_q$,. It is also assumed, that each learner induced from training sets $D_1,\ldots,D_q$ respectively, outputs hypotheses $h_1,\ldots,h_q$, where q is the number of base classifiers and $\forall h_{i:i=1,\ldots,q} \in H$. The goal of stacking is to learn a good combined classifier h such that the final classification will be computed from $h_1(x),\ldots,h_q(x)$ as shown in the Eq. (3):

$$h = \arg \max_{h_i \in H} \sum_{i=1}^{q} w_i f(h_i), \tag{3}$$

where w represents respective weights.

Different modifications of the standard approach are possible. There are several studies where different stacking variants have been proposed and investigated. A broad overview of the stacking is presented in [26]. In this paper the stacking technique proposed by Skalak in [21] is used. We select a few prototypes per class as the level-0 classifiers, and next use their outputs at the level-1 for generating the meta-classifier.

We also suggest the procedure of selecting prototypes, through instance and feature selection carried out simultaneously in parallel, within the integrated machine classification scheme as described in Sect. 2. The stacking technique is proposed for improving of the process of the integrated machine classification and to assure diversification among prototypes. Stacking technique generates a set of different base classifiers induced using different training sets. A diversity of the ensemble is achieved by generating base learners using different training parameters, including different training sets. In our case, the final classification will be computed in the following way:

$$h = \arg \max_{h_i \in H, g_i \in G} \sum_{i=1}^{q} w_i f(h_i = L(D, g_i)), \quad (4)$$

where $g_i, \ldots, g_q$ are vectors of data transformation, respectively, for $D_1, \ldots, D_q$,

4 Integrated Machine Classification and Stacking Ensemble Learning

This paper deals with the problem of integrated machine learning using prototypes produced in the process of data reduction. Data reduction is the computationally difficult combinatorial optimization problem. Hence, the approximate algorithms seem to be a most effective way of solving it. In our case it is the metaheuristic known as the agent-based population learning algorithm [2].

The discussed approach is based on selection of prototypes from clusters of instances. That means that, at the first step (initial phase) of computations the algorithm constructs clusters of instances. Next, from the clusters the prototypes are selected using the population-based search, and the searching process is carried out in two dimensions simultaneously.

At the initial phase to produce clusters the clustering procedure must be run. In this paper we consider two procedures for clustering

- the first one is based on the similarity coefficient,
- the second is based on an applying the kernel-based C-means (KFCM) algorithm.

The procedure based on the similarity coefficient groups instances into clusters according to their similarity coefficient calculated as in [9, 11]. Finally, each produced cluster contains instances with identical similarity coefficients.

The implemented procedure for cluster initialization is based on the kernel-based C-means (KFCM) algorithm. The clustering procedure based on KFCM) algorithm has been implemented in previous works of authors (see, for example [11]) and the obtained results show that the procedure assures good clustering quality as compared with other clustering methods. Originally, the kernel-based fuzzy C-means was introduced to overcome noise and outliers sensitivity in fuzzy C-means [28] by transforming input data into a higher dimensional kernel space via a non-linear mapping, which increases the possibility of linear scalability of the instances in the kernel space and allows for fuzzy C-means clustering in the feature space [18].

Finally, both clustering procedures produce clusters (subsets of the initial data set), and next, as it has been mentioned, from obtained subsets, prototypes are selected.

In the presented approach the process of data reduction via prototype selection is carried out under umbrella of the stacked generalization. Thus the described integrated machine learning with data reduction is repeated q-times. In details, the training data set is split into q training subsets. Next the q-1 subsets are used as the input data for the agent-based population learning algorithm. The process is repeated q times and in each iteration the representative set of the prototypes is selected and used to induce decision tree implemented in this work as a base model. Each base model is evaluated using the subset omitted at the current iteration. It means that a pool of diverse decision trees is obtained, and the ensemble of base classifiers is heterogeneous. Next, the ensemble is used to predict class label of the newly coming examples. The pseudo-code of the described process is shown below as Algorithm 1.

The functionality of the above described procedure based on implementation of the agent-based population learning approach can be defined as the organized and incremental search for the best solution. The agent-based population learning algorithm, where the specialized team of agents can work asynchronously and in parallel, selects prototypes executing various improvement procedures and cooperating with a view to solve the data reduction problem. Agents working in the A-Team achieve an implicit cooperation by sharing the population of solutions, also called individuals, to the problem to be solved. Generally, A-Team can be also defined as a set of agents and a set of memories, forming a network in which every agent remains in a closed loop. Agents cooperate to construct, find and improve solutions which are read from the shared common memory. More information on the PLA and more details on implementation of A-Teams can be found in [2].

Algorithm 1: The stacking ensemble learning

Input: D - the original data set, q – predefined number of stacking folds (the user-defined parameter).
Output: $h_1,\ldots,h_q$ – base classifiers from which the meta classifier is generated.

Map randomly examples from D into q disjoint subsets $D_1,\ldots,D_q$.
For i=1 **to** q **do**
Let $D' = D - D_i$ denote q-1 subsets of D.
Map instances from D' into clusters using the clustering procedures (KFCM procedure or the similarity-based procedure).
Run the population learning algorithm for prototype selection and generate a base classifier h_i using the omitted subset D_i as the test set.
End for
Let $h_1,\ldots,h_q$ denote the obtained base classifiers forming together the meta classifier for predicting new instances.
Return $h_1,\ldots,h_q$.

The discussed approach is based on selection of prototypes from clusters of instances. The approach generates the initial population, after the cluster initialization phase, and potential solutions are constructed through randomly selecting exactly one single instance from each of the considered clusters, and through randomly selecting the number of features from the original feature set. Finally, a potential solution is formed as a string of numbers representing the selected instances and the selected features.

A selection of the representation of instances through population-based search is carried out by the team of optimizing agents. Each agent is the implementation of the local search procedure and operates on individuals. The selection in the instance dimension is carried out for each cluster and removal of the remaining instances constitute the step of the data reduction in this case. The selection in the feature dimension is carried out for all instances regardless which cluster they are belonging to. In other words, each optimizing agent tries to improve quality of the received solutions by applying the implemented improvement procedure. An agent, after being supplied with an individual to be improved, explores its neighborhood with the aim of finding a new, better solution in this neighborhood.

Each individual representing a solution from the population of individuals is evaluated through calculating value of its fitness function. In the discussed case the evaluation procedure involves estimation of the classification accuracy assuming that the classifier is induced from the reduced dataset as indicated by the individual under consideration.

To obtain the solution of the data reduction problem several kinds of the optimization agents carrying out different local search based optimization heuristics have been implemented. Among these procedures there are two simple greedy local searches, one random search and the tabu search for instance and feature selection.

Optimization agents work in cycles. In response to the signal that an agent is ready to undertake an improvement task, a solution is selected from the population and send to this agent. After an attempted improvement a solution is returned and a new cycle begins. Solutions to be forwarded for improvement are selected from the population in accordance with the user defined working strategy specifying rules for selection and replacement of individuals in the common memory.

5 Computational Experiment

This section contains results of the computational experiment carried out with a view to evaluate the performance of the discussed approach, i.e. stacking-based integrated machine learning with data reduction. The two versions of the stacking-based integrated machine learning with data reduction considered in the paper are called respectively: *ABInDRkfStE* (Agent-Based Integrated Data-Reduction based on the KFCM with the Stacking Ensemble Learning) and *ABInDRStE* (Agent-Based Integrated Data-Reduction based on the similarity coefficient with the Stacking Ensemble Learning).

In particular, the reported experiments aimed at comparing quality of the approach and investigating influence of the clustering algorithms on classifier performance. The experiment aimed also at evaluating whether the discussed approach produces, on average, better results than results produced by its earlier versions proposed in:

- *ABDRkfStE* - Agent-Based Data-Reduction based on the KFCM with Stacking Ensemble Learning and without feature selection - introduced in [9],
- *ABDRStE* - Agent-Based Data-Reduction based on the similarity coefficient with Stacking Ensemble Learning and without feature selection - introduced in [7].

The reported experiment also includes results obtained by: *ABIS* (Agent-Based Instance Selection) - proposed in [10], *ABDRE* (Agent-Based Data-Reduction with Ensemble) with *RM-RR* (Random Move and Replace Randomly strategy) and *ABDRE* with *RM-RW* (Random Move and Replaces First Worst strategy) - proposed in [6], *AdaBoost*, *Bagging* and *Random Subspace Method* - for which the results have been presented in [6].

Evaluation of the proposed approach and performance comparisons are based on solving several classification benchmark datasets obtained from the UCI Machine Learning Repository [1] (see Table 1).

Table 1. Information about the datasets

Dataset	Instances	Attributes	Classes	Best reported results classification accuracy
Heart	303	13	2	90.0% [3]
Diabetes	768	8	2	77.34% 16
Breast cancer	699	9	2	97.5% [1]
Australian credit	690	15	2	86.9% [1]
German credit	1000	20	2	77.47% [16]
Sonar	208	60	2	97.1% [1]
Shuttle	58000	9	7	95.6% [20]

The computational experiment was repeated several times. The number of stacking folds has been set, respectively, from 3 to 10. Each benchmark problem has been solved 50 times. The experiment plan has involved 10 repetitions based on the 10-cross-validation scheme. Finally, the reported values of the quality measure have been averaged over all runs.

Table 2. Classification results (%) and comparison of different classifiers

Algorithm	Heart	Diabetes	WBC	ACredit	GCredit	Sonar	Shuttle
ABInDRkfStE	**93.01**	**80.71**	98.08	92.04	78.45	90.57	98.41
ABInDRStE	92.87	79.84	**98.13**	**91.89**	**80.24**	**91.15**	98.73
ABDRkfStE	90.45	75.15	96.91	90.78	77.41	80.42	**99.66**
ABDRStE	92.12	79.12	96.91	91.45	80.21	85.63	98.75
$ABDRE_{RM\text{-}RR}$	92.84	80.4	96.4	90.8	78.2	83.4	97.51
$ABDRE_{RM\text{-}RW}$	90.84	78.07	97.6	89.45	76.28	81.75	97.74
ABIS	91.21	76.54	97.44	90.72	77.7	83.65	95.48
AdaBoost	82.23	73.55	63.09	91.05	73.01	86.09	96.13
Bagging	79.69	76.37	95.77	85.87	74.19	76.2	95.27
Random subspace method	84.44	74.81	71.08	82.14	75.4	85.18	92.81
C 4.5	77.8	73	94.7	84.5	70.5	76.09	95.6
SVM	81.5	77	97.2	84.8	72.5	90.4*	–
DROP 4	80.90	72.4	96.28	84.78	–	82.81	–

Source for $ABDRE_{RM\text{-}RR}$ and $ABDRE_{RM\text{-}RW}$ – [6]; Source for *ABIS* – [10]; C 4.5 – [6]; DROP 4 – [25]; * – [13]

The C4.5 algorithm has been applied to induce all of the base models for all ensemble classifiers [19]. In each experiment and for each investigated A-Team parameters have been set to values corresponding to those used in earlier experiments, for which results have been presented in [7–9]. Thus the population size was set to 40. The searching for the best solution of each A-Team has been stopped either after 100 iterations or after there has been no improvement of the current best solution for one minute of computation. In case of the proposed algorithm the number of instance subsets has been set to 4. In case of Bagging and Random Subspace Method the size of bags has been set to 50% of the original training set. In case of *ABDRE* with *RM-RR* and *ABDRE* with *RM-RW* the number of base models has been set to 40.

The experiment results obtained using the *ABInDRkfStE* and the *ABInDRStE*, as averages values and the best values for different values of stacking folds are shown in Table 2. Table 2 also shows results obtained by other ensemble classifiers, some example non-ensemble classifiers, like for example C 4.5 (without data reduction and without any other pre-processing actions), as well as the results of other data reduction algorithms (i.e. DROP 4).

From Table 2 it is easy to observe that the stacking-based integrated machine learning with data reduction increases accuracy of the classification as compared with cases when classifier is induced using earlier versions of the proposed algorithms, other ensemble-based algorithms and using a non-reduced dataset. The proposed approach outperforms also the DROP technique and other methods, i.e. C4.5 and SVM. Thus, it can be noted that the stacked generalization can be promising mechanism influencing the quality of the classification system and providing a way for diversification of the learning models based on the reduced data. When we compare results of the proposed approach with respect to the clustering procedure used, it can be noticed that the similarity-based clustering assures better results.

6 Conclusions

The paper contributes through proposing a novel integrated approach to machine learning from examples with data reduction and stacking, where data reduction is carried out in two dimension, by instance and feature selection. Because, the clustering procedure is a crucial for the data reduction by the proposed algorithm, the paper also investigates influence of the clustering procedure on the quality and performance of the approach. The kernel-based fuzzy clustering algorithm and the procedure based on the similarity coefficient have been used, and it can be observed, that selection of the clustering procedure can have an impact on the performance with more accurate results produced by the second of the above discussed procedures.

Computational experiment results confirm, that the stacking method and data reduction in both dimension assures generation of diversified heterogeneous members of the ensemble, which are subsequently used for forming the meta-classifier.

Future research will aim at investigating influence of several independent factors on the integrated machine learning with data reduction and based on stacking.

References

1. Asuncion, A., Newman, D.J.: UCI Machine Learning Repository. University of California, School of Information and Computer Science, Irvine, CA (2007). http://www.ics.uci.edu/~mlearn/MLRepository.html
2. Barbucha, D., Czarnowski, I., Jędrzejowicz, P., Ratajczak-Ropel, E., Wierzbowska, I.: e-JABAT – an implementation of the web-based A-Team, In: Nguyen, N.T., Jain, L.C. (eds.) Intelligence Agents in the Evolution of Web and Applications. SCI, vol. 167, pp. 57–86. Springer, Heidelberg (2009). doi:10.1007/978-3-540-88071-4_4
3. Bull, L.: Learning classifier systems: a brief introduction, applications of learning classifier systems. In: Bull, L. (ed.) STUDFUZZ. Springer (2004)
4. Cano, J.R., Herrera, F., Lozano, M.: On the combination of evolutionary algorithms and stratified strategies for training set selection in data mining. Appl. Soft Comput. **6**, 323–332 (2004)
5. Carbonera, J.L., Abel, M.: A density-based approach for instance selection. In: Proceedings of the 2015 IEEE 27th International Conference on Tool with Artificial Intelligence, pp. 768–774 (2015). doi:10.1109/ICTAI.2015.114
6. Czarnowski, I., Jędrzejowicz, P.: Agent-based data reduction using ensemble technique, In: Badica, C., Nguyen, N.T., Brezovan, M. (Eds.): Computational Collective Intelligence. Technologies and Applications, ICCCI 2013. LNAI, vol. 8083, pp. 447–456. Springer, Heidelberg (2013)
7. Czarnowski, I., Jędrzejowicz, P.: An approach to machine classification based on stacked generalization and instance selection. In: Proceedings of 2016 IEEE International Conference on Systems, Man, and Cybernetics, SMC 2016, Budapest, Hungary, 9–12 October, 2016, pp. 4836–4841. IEEE (2016)
8. Czarnowski, I., Jędrzejowicz, P.: Experimental evaluation of the agent-based population learning algorithm for the cluster-based instance selection, In: Jędrzejowicz P., Nguyen N.T., Hoang K. (eds.): Computational Collective Intelligence, Technologies and Applications, ICCCI 2011. LNAI, vol. 6923, pp. 301–310. Springer, Heidelberg (2011)
9. Czarnowski, I., Jędrzejowicz, P.: Learning from examples with data reduction and stacked generalization. J. Intell. Fuzzy Syst. **32**(2), 1401–1411 (2017)
10. Czarnowski, I.: Distributed learning with data reduction, In: Nguyen, N.T. (ed.) Transactions on CCI IV. LNCS, vol. 6660, pp. 3–121. Springer, Heidelberg (2011)
11. Czarnowski, I.: Cluster-based instance selection for machine classification. Knowl.-Based Inf. Syst. **30**(1), 113–133 (2012)
12. Dash, M., Liu, H.: Feature selection for classification. Intell. Data Anal. **1**(3), 131–156 (1997)
13. Datasets used for classification: comparison of results. Directory of Data Sets. http://www.is.umk.pl/projects/datasets.html. Accessed 1 Sep 2009
14. Ho, T.K.: Data complexity analysis for classifier combination, In: Kittler, J., Roli, F. (eds.) MCS 2001. LNCS, vol. 2096, pp. 53–67. Springer, London (2001)
15. Holland, J.H.: Adaptation, In: Rosen, R. and Snell, F.M. (eds.) Progress in Theoretical Biology 4, Plenum (1976)
16. Jędrzejowicz, J., Jędrzejowicz, P.: Cellular GEP-induced classifiers, In: Pan, J.-S., Chen, S.-M., Nguyen, N.T. (eds.) ICCCI 2010, Part I. LNAI, vol. 6421, pp. 343–352. Springer, Heidelberg (2010)
17. Kim, S.-W., Oommen, B.J.: A brief taxonomy and ranking of creative prototype reduction schemes. Pattern Anal. Appl. **6**, 232–244 (2003)

18. Li, Z., Tang, S., Xue, J., Jiang, J.: Modified FCM clustering based on kernel mapping. In: Proceedings of the International Conference on Society for Optical Engineering, vol. 4554, pp. 241–245 (2001). doi:10.1117/12.441658
19. Quinlan, J.R.: C4.5: Programs for Machine Learning. Morgan Kaufmann, SanMateo (1993)
20. Sikora, R., Al-laymoun, O.H.: A modified stacking ensemble machine learning algorithm using genetic algorithms. J. Int. Technol. Inf. Manage. **23**(1), 1–11 (2014)
21. Skalak, D.B.: Prototype selection for composite neighbor classifiers, University of Massachusetts Amherst (1997). https://web.cs.umass.edu/publication/docs/1996/UM-CS-1996-089.pdf
22. Stefanowski, J.: Multiple and hybrid classifiers. In: Polkowski, L. (ed.) Formal Methods and Intelligent Techniques in Control, Decision Making. Multimedia and Robotics, Warszawa, pp. 174–188 (2001)
23. Tsoumakas, G., Angelis, L., Vlahavas, I.: Clustering classifiers for knowledge discovery from physically distributed databases. Data Knowl. Eng. **49**, 223–242 (2004)
24. Wilson, D.R., Martinez, T.R.: An integrated instance-based learning algorithm. Comput. Intell. **16**, 1–28 (2000)
25. Wilson, D.R., Martinez, T.R.: Reduction techniques for instance-based learning algorithm. Mach. Learn. **33**(3), 257–286 (2000)
26. Wolpert, D.: Stacked Generalization. Neural Netw. **5**, 241–259 (1992)
27. Yıldırım, A.A., Özdoğan, C., Watson, D.: Parallel data reduction techniques for big datasets. In: Hu, W.-C., Kaabouch, N. (eds.) Big Data Management, Technologies, and Applications. IGI Global, pp. 72–93 (2014)
28. Zhou, S., Gan, J.Q.: Mercel kernel fuzzy c-means algorithm and prototypes of clusters. In: Proceedings of the International Conference on Data Engineering and Automated Learning. LNCS, vol. 3177, pp. 613–618 (2004). doi:10.1007/978-3-540-28651-6_90
29. Zhu, X., Wu, X.: Scalable representative instance selection and ranking. In: IEEE Proceedings of the 18th International Conference on Pattern Recognition, vol. 3, pp. 352–355 (2006)

Intelligent Data Analysis and Applications

A Hybrid Approach to Conceptual Classification and Ranking of Resumes and Their Corresponding Job Posts

Abeer Zaroor[1], Mohammed Maree[2(✉)], and Muath Sabha[2]

[1] Computer Science Department, Faculty of Engineering and Information Technology, The Arab American University, Jenin, Palestine
abeerzaroor.aauj@gmail.com

[2] Information Technology Department, Faculty of Engineering and Information Technology, The Arab American University, Jenin, Palestine
{mohammad.maree,muath.sabha}@aauj.edu

Abstract. Due to the constant growth in online recruitment, job portals are starting to receive thousands of resumes in diverse styles and formats from job seekers who have different fields of expertise and specialize in various domains. Accordingly, automatically extracting structured information from such resumes is needed not only to support the automatic matching between candidate resumes and their corresponding job offers, but also to efficiently route them to their appropriate occupational categories to minimize the effort required for managing and organizing them. As a result, instead of searching globally in the entire space of resumes and job posts, resumes that fall under a certain occupational category are only those that will be matched to their relevant job post. In this research work, we present a hybrid approach that employs conceptual-based classification of resumes and job postings and automatically ranks candidate resumes (that fall under each category) to their corresponding job offers. In this context, we exploit an integrated knowledge base for carrying out the classification task and experimentally demonstrate - using a real-world recruitment dataset- achieving promising precision results compared to conventional machine learning based resume classification approaches.

Keywords: Online recruitment · Concept-based classification · Job matching

1 Introduction

In recent years, online recruitment has expanded significantly [1, 2]. This expansion has led to a continuous growth in the number of job portals and hiring agencies on the Internet [3, 4]. It has also led to a constant increase in the number of job seekers searching for new career opportunities [5, 6]. Accordingly, online job portals are starting to receive thousands of resumes (in diverse styles and formats) from job seekers who have different fields of expertise and specialize in different domains [7]. Several approaches have been proposed to support the automatic matching between candidate resumes and their corresponding job offers [8–10]. Examples on these approaches are automatic keyword-based resume matching techniques [8] machine

I. Czarnowski et al. (eds.), *Intelligent Decision Technologies 2017*,
Smart Innovation, Systems and Technologies 72, DOI 10.1007/978-3-319-59421-7_10

leaning based approaches [9], and semantics-based techniques [10]. The main goal of these approaches is achieving high precision ratios i.e. finding the best candidates for a given job post ignoring the cost (run time complexity) of the matching process. Other systems have attempted to reduce the cost issue by segmenting the content of both resumes and job posts and finding matches between important segments in both instead of matching between the content of the whole resumes and job posts. For instance, the authors of [11, 12] propose using machine learning algorithms (Support vector machines (SVM) [11] and Hidden Markov Model (HMM) [12]) to automatically extract structured information from job posts and resumes by annotating the segments of job posts and resumes with the appropriate features and topics. While the authors of [8] use Natural Language Processing (NLP) techniques to implement the segmentation and information extraction module. Although these approaches have proved to be more efficient in carrying out the matching task [8], every newly obtained resume still needs to be matched with all job offers in the corpus. To overcome this issue, other researchers propose utilizing machine learning-based techniques to classify job posts and resumes into occupational categories prior conducting the matching task [13]. However, as pointed in [2], these techniques suffer from low precision ratios and produce high error rates. To address the issues associated with the previously highlighted techniques, we present a hybrid approach that employs conceptual-based classification of resumes and job postings and automatically ranks candidate resumes (that fall under each occupational category) to their corresponding job postings. We summarize the contributions of our work as follows:

- Automatic Occupational Category based Classification of Resumes and Job Postings.
- Employing a Section-based Segmentation heuristic by exploiting an integrated occupational categories knowledge base.

The remainder of this paper is organized as follows. In Sect. 2, we introduce the work related. Section 3 describes the overall architecture of the proposed system. Experimental validation of the effectiveness and efficiency of the proposed system is presented in Sect. 4. In Sect. 5, we discuss the conclusions and outline future work.

2 Related Work

Many techniques have been proposed to precisely match between candidate resumes and their corresponding job offers [14, 15]. However, little attention has been paid to addressing problems associated with automatic resumes and job posts classification [16]. For instance, when an employer seeks a “Web Developer” that falls under “Web Development” occupational category, the conventional systems search globally in the entire space of resumes for finding applicants that best match the offered position. In this context, each and every resume in the resumes collection will be matched to the offered job post instead of matching only those that fall under the corresponding occupational category (“Web Development” in this scenario). To address this issue, the authors of [12] have proposed resume Information Extraction (IE) with Cascaded Hybrid Model. This system employs HMM and SVM classification algorithms in order to annotate segments of resumes with the appropriate category, taking the advantage of the resume contextual structure where the related information units usually occur in the same textual segments. Accordingly, resumes pass through two layers; where in the first layer a HMM is applied to segment the entire resume into consecutive blocks and each block is annotated with a category such as Personal Information, Education, and Research Experience. After that, in the second layer SVM is applied in order to extract the detailed information from the blocks that have been labeled with Education and Personal Information respectively. However, a large fraction of the produced results by this system suffer from low precision since the information extraction process passes through two loosely-coupled stages. Another system (E-gen) [17] has been built in order to automate the recruitment process by segmenting and classifying job posts. First, job posts are transformed into vector space representations. Then, SVM classification algorithm is utilized to annotate segments of job posts with the appropriate topics and features. A correction algorithm is further applied because during the classification process some segments were incorrectly classified [17]. The main drawback of this system is the time needed to pre-process and post-process job posts in order to minimize the error and maximize the matching probability. On the other hand, JobDiSC system [18] attempts to classify job openings automatically by employing a standard classification scheme called Dictionary of Occupational Titles (DOT). The proposed system comprises three main modules: (1) Parser/Analyzer: which creates an unclassified job opening for each job listings captured from electronic forms prepared by employers. (2) Learning System to automatically generate classification rules from a set of pre-classified job openings and (3) Classifier that assigns one or more class for each job post depending on its confidence level for any potential class assigned to it. The main drawback of this system is that DOT's usefulness has waned since it doesn't cover the occupational information that is more relevant to the modern workplace [19].

3 Overview of the Architecture of the Proposed System

In this section, we present an overview of the architecture of the proposed system and discuss its main constituents.

As shown in Fig. 1, when a user submits a CV, the system directs it to the Section-based Segmentation module which is used to extract personal, education, experience information and employment history, in addition to a list of candidate matching concepts. After that, the Filtration module refines the concept lists by removing concepts that have low tf-idf [8] weights and concepts that don't contribute to the semantics of each segment. Next, the Classification module takes a set of skills from both resumes and job posts as an input in order to classify job posts and resumes to their corresponding occupational categories. At this step, we exploit an integrated knowledge base which combines two main semantic resources: DICE[1] and O*NET[2]. More details on these resources will be provided in Sect. 3.1. Then, the Category-based

[1] http://www.dice.com/skills.

[2] http://www.onetcenter.org/taxonomy/2010/list.html.

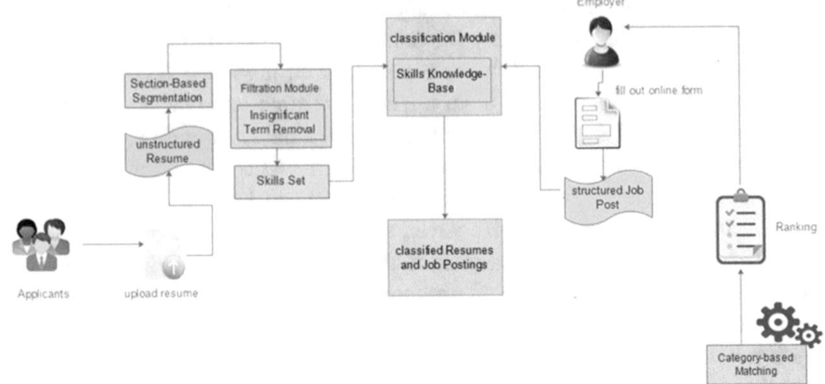

Fig. 1. Overall architecture of the proposed system

Matching module takes lists of concepts from both resumes and job posts to construct semantic networks by deriving the semantic relatedness between them using semantic resources. Finally, the matching algorithm takes the semantic networks as an input and produces the measures of semantic closeness between them as an output. The following sections detail the steps carried out in each module of the system.

3.1 Section-Based Segmentation and Conceptual Classification Modules

During this phase, an automatic extraction of segments such as Education, Experience, Loyalty and other Employment information such as Company name, Applicant Role in the company, Date of designation, Date of resignation and Loyalty is performed. In this context, the system matches segments of resumes to their relevant segments of job posts instead of matching the whole resumes and job posts. During this phase, unstructured resumes are converted into segments (semi-structured document) based on employing Natural language processing techniques (NLP) and rule-based regular expressions. As detailed in [7], the NLP steps are: document splitting, n-gram tokenization, stop word removal, part-of-Speech-Tagging (POST) and Named Entity Recognition (NER). Table 1 shows an example that illustrates the process of segmenting a sample resume.

In order to classify both resumes and job posts, we utilize an integrated knowledge-base which combines Dice skills center (henceforth stated as DICE) and a standardized hierarchy of occupation categories known as the Occupational Information Network (O*NET) (henceforth stated as O*NET). In this context, we use DICE to classify skills that belong to Information and communication technologies (ICT), and economy field because we empirically found that O*NET is not scalable enough for our classification needs. Furthermore, some skill acronyms are not classified correctly in O*NET. However, and on the contrary of Dice, O*NET is able to better classify skills that are related to the Medical and Artistic fields. Table 2 shows a comparative analysis between Dice and O*NET classification.

Table 1. Results of the Section-based Segmentation Module

Input CV	Segmentation Result	Candidate Terms	Filtered List
Input CV: I have 2 years of experience as a web designer And I have the following skills android, ios, PHP, HTML, JavaScript, CSS, JQuery, VB, .NET and c# Education: B.Sc. in CS. Employment Details I worked as web designer in ASAL Company from 2008 to 2010.	<ApplicantData> <Experience> <Years>2</Years> <Field>web designer</Field> </Experience> <Education> <Degree> B.Sc.</Degree> <Field>CS</Field> </Education> <EmploymentHistory> <role> web designer</role> <companyName>ASAL</companyName> <FromDate>2008</fromDate> <ToDate>2010</ToDate> <loyalty> 2</loyalty> </EmploymentHistory> <Skills> android, ios, PHP, HTML, JavaScript, CSS, JQuery, VB, .NET , c# </Skills> </ApplicantData>	android ios HTML Javascript Jquery designer Asal Web php skills experience company VB .NET c# CSS	android ios HTML Javascript Jquery designer web php VB .NET c# CSS

Table 2. Comparative Analysis between DICE and O*NET Classification

	O*NET		DICE	
	Skill	Classification result	Skill	Classification result
Submission of Acronyms	JPA	Accountants	JPA	Software Development
	JCA	Nursing Assistants	JCA	Software Development
	J2ME	Gem and Diamond Workers	J2ME	I.T. Administration/Technical Support
	Skill	Classification result	skill	Classification result
Correctness in Classification	xcode	Coaches and Scouts	xcode	Software Development/Mobile development
	Radiography	Radiologic Technicians	Radiography	NOT CLASSIFIED
	Medical analysis	Medical and clinical Laboratory	Medical analysis	NOT CLASSIFIED

As shown in Table 2, some skill acronyms are not recognized by O*NET, and accordingly they are not classified correctly. For instance, JPA which refers to "Java Persistence" is classified under "Accountants" category by O*NET. However, we can see that terms such as "Radiography" and "Medical analysis" are not classified in DICE, but classified correctly under "Radiologic Technicians" and "Medical and clinical Laboratory" categories in O*NET.

3.1.1 Skill-Based Resume Classification Module

In this module each skill in the skills set is submitted to the exploited skills knowledge base sequentially in order to obtain a list of candidate job categories. As shown in Fig. 2, the skill "android" is first submitted to the exploited skills knowledge base. For this skill, the knowledge base returns one occupational category, that is "Software

Development/Mobile Development". Next, the rest of the skills in the skills set are submitted to the exploited knowledge base. As a result, a list of weighted occupational categories is obtained and sorted by the highest weight (as one skill may return zero, one, or more than one occupational category). Accordingly, for the skills set that we have obtained from CV1, the occupational category "Web Development" gets the highest weight, followed by "Software Development/Application Development" and then "Soft-ware Development/Mobile Development" respectively.

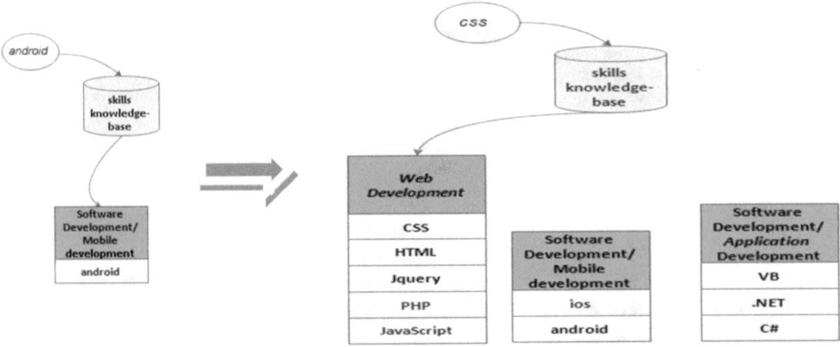

Fig. 2. List of occupational categories generated for the sample CV

To produce weights for the occupational categories, we use the following algorithm.

Algorithm 1. A weighted List of Resume category

Input: skills $[s_1, s_2, \ldots, s_n]$
Output: list of job categories sorted by the highest weight for a given resume
1: int weight=0;
2: answer $\leftarrow \langle\ \rangle$;
3: categoryList $\leftarrow \langle\ \rangle$;
4: **for each** skill $\in$ skills $[s_1, s_2, \ldots, s_n]$ **do**
5:   categoryList $\leftarrow$ **GET_FROM_KB** (skill);
6:   **while** categoryList $\neq$ NIL **do**
7:     **if** category **Not IN** answer **then**
8:       weight =1;
9:       **ADD** (answer, $\langle category, weight \rangle$);
10:     **else**
11:       weight++;
12:       **Modify** (answer, $\langle category, weight \rangle$);
13:     **end if**
14: **end for**
15: **Return** answer;

In the used algorithm, skills are submitted to the skills knowledge-base respectively. As a result, one occupational category is returned for each skill (Line 5). If the same occupational category is returned for more than one skill, the algorithm increases

Table 3. Occupational categories returned for the CV used Example 1

Job category	Skills
Software Development/Mobile Development	Android, ios
Software Development/Application Development	.Net, c#, VB
Web Development	CSS, html, php, Javascript, jquery

the weight for that particular occupational category, otherwise it sets its weight to 1. (Lines 8, 11 and 12). Finally, the algorithm returns a list of weighted occupational categories in the answer list (Line 15). Table 3 shows each occupational category assigned to its corresponding skills.

3.1.2 Job Post Classification Module

In the Job Post Categorization module, we use both the job title and the required skills from the structured job post for classification purposes. First, the job post is pre-processed and filtered through removing noisy information such as: city names, state and country acronyms that appear in the job title or job details. After that, we use the skill knowledge base to classify job posts in the same manner as we do for classifying resumes. Accordingly, we assign weights (Job Title = 70% and Required Skills = 30%) since we believe that the job title is more significant than the required skills and guides to better matching results. More examples on the results of this module are presented in Sect. 4.2.

3.2 Matching Resumes and Their Corresponding Job Postings

In the same fashion as proposed by the authors of [7], we use the same semantic resources (WordNet ontology [20] and YAGO2 ontology [21]) and statistical concept-relatedness measures to derive the semantic aspects of resumes and job posts. It is important to mention that we have considered additional weighting parameters such as: loyalty parameter (degree of devotion to the company that the candidate worked or currently working in) in order to increase the effectiveness of the matching process. It is also important to point out that we use a dynamic threshold value to fairly handle the loyalty parameter as shown in the following scoring formulae:

$$\boldsymbol{S} = \frac{|\{\boldsymbol{Sr}\}|}{|\{\boldsymbol{RSj}\}|} * \mathbf{50\%} + \frac{|\{\boldsymbol{Er}\}|}{|\{\boldsymbol{REj}\}|} * \mathbf{20\%} + \frac{|\{\boldsymbol{Xr}\}|}{|\{\boldsymbol{RXj}\}|} * \mathbf{20\%} + \frac{|\sum \boldsymbol{Yw}|}{|\sum \boldsymbol{Cw}|} * \mathbf{10\%} \quad (1)$$

Where:

- **S:** is the relevance score result.
- **Sr:** is the set of applicant's skills.
- **RSj:** the required skills in the job post.
- **Er:** is the set of concepts that describe applicant educational information.
- **REj:** is the set of concepts from the required educational information in job post.
- **Xr:** set of concepts that describe applicant experience information.
- **RXj:** concepts that represent the required experience information in the job post.

- **Yw:** the total number of employment years.
- **Cw:** number of companies that the applicant worked in.

As shown in the formula, we have set the following weighting values:

Skills weight = 50%, Educational level weight = 20%, Job experience weight = 20% and Loyalty level weight = 10%. The results of using the scoring formula are detailed in the next section (Sect. 4.3).

4 Experimental Results

This section describes the experiments that we have conducted to evaluate the efficiency and the effectiveness of the proposed system. In order to evaluate the accuracy of the proposed system, we collected a data set of 2000 resumes downloaded from:

- http://www.amrood.com/resumelisting/listallresume.htm and
- http://www.indeed.com/resumes

and used 10,000 different job posts obtained from:

- http://jobs.monster.com,
- http://www.shine.com/job-search and
- http://www.careerbuilder.com.

The collected resumes are unstructured documents in different document formats such as (.pdf) and (.doc) and we considered job posts as structured document having sections (job title, description, required skills, required years and field of experience, required education qualification and additional desired requirement). The experiments of our system prototype show that the classification process for the training data of resumes and job posts took 6 h on average on a PC with dual-core CPU (2.1 GHz) and (4 GB) RAM. The used operating system is Windows 8.1.

4.1 Execution Time for Matching Resumes with Corresponding Job Post with/Without Classification

In this section, we compare the results produced by our system with MatchingSem system [7] which is semantics-based automatic recruitment system. As shown in Fig. 3, our system Job Resume Classifier (JRC) was able to achieve better results than MatchingSem System. And this is due to the fact that, unlike MatchingSem, we only match job posts with corresponding resumes that fall under the same occupational category instead of searching globally in the entire space of resumes. For instance, "Front-End Developer" job post costs 6 h of execution time for finding the best candidate using MatchingSem, while it only took 1 h in JRC since resumes that fall under "Web Development" category were considered in the matching process.

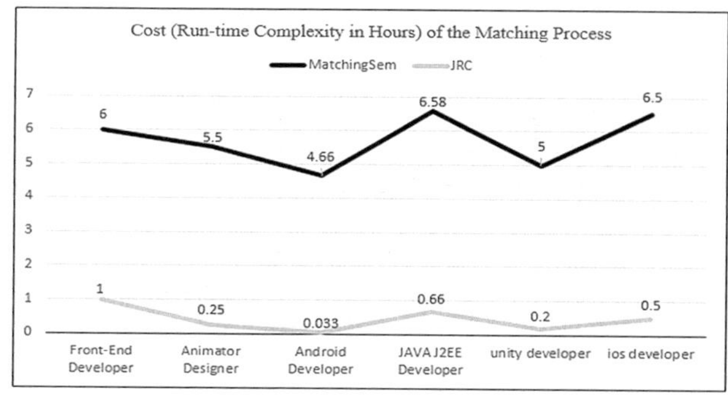

Fig. 3. Cost (Run-time complexity in hours) of the matching process

4.2 Experiments of Job Post Classification

In this section we present job post classification. As mentioned in Sect. 3.1.2 we have used job title and required skills in the classification process. In Table 4, we have used 7 job posts in order to clarify the classification process.

Table 4. Job post classification results

Job title	Required skills	Job classification	Weight
Front End Web Developer	Backbone, CSS3, HTML5, JSON, JQuery, JavaScript, Angular	Software Development/Web Development	**100%**
Unity Developer	3D, Unity	Software Development/Interactive Multimedia	**85%**
	Objective-C, Xcode	Software Development/Mobile development	15%
Animator Designer	3D, texture, compositing, lighting, animation	Software Development/Interactive Multimedia	**86.6%**
	Illustrator, After Effects, Premiere Pro, Maya	Design/Multimedia Design	13.4%
IT Support Specialist	printers, network routing protocols, TCP/IP, smartphones, VOIP, laptops, VLANs, desktops	IT Administration/Technical Support	**94%**
	SQL Express, SQL Server	Data/Databases	6%
IT Technician	DHCP, DNS, Troubleshooting, Installation, Configuration	IT Administration/Technical Support	**100%**

(continued)

Table 4. (*continued*)

Job title	Required skills	Job classification	Weight
Database Administrator	Oracle Data Guard, Oracle RMAN, RDBMS, Oracle RAC, ASM, Oracle 11 g, PeopleSoft DBA	Data/Databases	**92.5%**
	EMC Storage, Symmetrix, DMX	IT Administration/Technical Support	7.5%
Data Entry Assistant	Typing, Filing, Clerk	Industry-specific/General skills	**79%**
	Drupal, HTML, CSS	Software Development/Web Development	9%
	MS Word, Excel, PowerPoint, Outlook	Industry-specific/Microsoft Office	12%

As shown in Table 4, we can see that "Front End Web Developer" job post falls under "Software Development/Web Development" occupational category with weight equals 100%, and this is because when we submit the job title to our skills knowledge base it returns "Software Development/Web Development" category with weight 70%, then we submit the required skills and we find that all of them fall under the same space with weight 30%. However, "Unity Developer" job post falls under "Software Development/Interactive Multimedia" space with weight 85%, 70% for job title under "Software Development/Interactive Multimedia" category. When we submit "3D and unity" skills we see that they fall under the same space as job title with weight 15%, but "Objective-C, Xcode" skills fall under "Software Development/Mobile Development" space with weight 15%. And the same for "Data Entry Assistant" job post, that falls under three categories: "Industry-specific/General skills" with weight 79%, "Industry-specific/Microsoft Office" with weight 12%, and "Software Development/Web Development" with weight 9%.

4.3 Precision Results of Matching Resumes with Corresponding Job Post

In this section we evaluate our system's effectiveness using precision indicator. For each job post, we compare between the manually assigned scores and their corresponding scores that are automatically produced by the system. Table 5, shows the precision results of matching resumes with corresponding job post.

As shown in Table 5, we match job posts with their corresponding resumes that fall under the same occupational categories. For instance, "Android Developer" job post is matched only with resumes that fall under "Mobile Development" category. As such, CV1 and CV3 are matched with "Android developer" and "Web developer" job posts. And this is because these CVs exist in both "Mobile Development" and "Web Development" categories. However, the matching score differ from job post to another.

Table 5. Precision results of matching resumes with corresponding job post by the system

Occupation category	Job title	Resume index	Manual score	Auto score	Precision
Software Development\ Mobile development	Android developer	CV1	0.80	0.80	1.00
		CV2	0.80	0.76	0.95
		CV3	0.05	0.09	0.55
Software Development/Web development	Web developer	CV4	0.42	0.53	0.79
		CV3	0.70	0.75	0.93
		CV1	0.40	0.47	0.85
Software Development/Interactive Multimedia	Unity developer	CV5	0.45	0.40	0.89
		CV6	0.47	0.61	0.77
		CV7	0.57	0.66	0.86

For instance, CV3 achieved a very low matching score when matched with Android Developer job post (0.05 manual score, 0.09 automatic score), but CV1 achieved better score for the same job post (0.8 manual score, 0.8 automatic score). On the other hand, CV3 achieved better results than CV1 when it was matched with "Web developer" job post (0.70 manual score, 0.75 automatic score) and this is because CV3 falls under "Web developer" with weight 80% and falls under "Mobile Development" with weight 35%.

In order to validate our proposal and evaluate the quality of the produced results by our system, we have compared our system with one of the previously proposed systems, called MatchingSem [7]. Table 6, shows the results produced by our system compared to MatchingSem system.

Table 6. Comparison with MatchingSem System

Job title	Resume index	JRC precision	MatchingSem precision
Android Developer	**CV1**	**1.00**	0.91
	CV2	**0.95**	0.84
	CV3	**0.55**	0.50

As shown in Table 6, for the job title "Android Developer" and the three CVs, the quality of the produced results (namely, the precision) by our system is higher than MatchingSem system. The reason behind this is that – unlike MatchingSem system – we are integrating a section-based segmentation module to extract features such as educational background, years of experience and employment information from applicants' resumes. When we incorporate these features, the matching scores produced by our system are better than when using only a list of candidate concepts as proposed in MatchingSem.

5 Conclusions and Future Work

In this research work, we have introduced a hybrid approach that employs conceptual-based classification of resumes and job postings and automatically matches candidate resumes to their corresponding job postings that fall under each occupational category. The proposed system first utilizes NLP techniques and regular expressions in order to segment the resumes and extract a set of skills that are used in the classification process. Next, the system exploits an integrated skills knowledge base for carrying out the classification task. The conducted experiments using the exploited knowledge base demonstrate that using the proposed classification module assists in achieving higher precision results in a less execution time than conventional approaches. In the future work, we plan to utilize the extracted information from applicants' resumes to dynamically generate user profiles to be further used for recommending jobs to job seekers.

References

1. Faliagka, E., Iliadis, L., Karydis, I., Rigou, M., Sioutas, S., Tsakalidis, A., Tzimas, G.: On-line consistent ranking on e-recruitment: seeking the truth behind a well-formed CV. Artif. Intell. Rev. **42**(3), 515–528 (2014)
2. Kessler, R., Béchet, N., Roche, M., Torres-Moreno, J., El-Bèze, M.: A hybrid approach to managing job offers and candidates. Inf. Process. Manage. **48**(6), 1124–1135 (2012)
3. Chen, J., Niu, Z., Fu, H.: A novel knowledge extraction framework for resumes based on text classifier. In: Proceedings of the International Conference on Web-Age Information Management, pp. 540–543. Springer International Publishing (2015)
4. Schmitt, T., Philippe C., Michele, S.: Matching jobs and resumes: a deep collaborative filtering task. In: Proceedings of the 2nd Global Conference on Artificial Intelligence, pp. 1–14 (2016)
5. Hauff, C., Georgios G.: Matching GitHub developer profiles to job advertisements. In: Proceedings of the 12th Working Conference on Mining Software Repositories, pp. 362–366 (2015)
6. Pimplikar, R., Singh, A., Varshney, R., Visweswariah, K.: Efficient multifaceted screening of job applicants. In: Proceedings of the 16th International Conference on Extending Database Technology, pp. 661–671. ACM (2013)
7. Kmail, A., Maree, M., Belkhatir, M., Alhashmi, S.: An automatic online recruitment system based on exploiting multiple semantic resources and concept-relatedness measures. In: Proceedings of the IEEE 27th International Conference on Tools with Artificial Intelligence (ICTAI), pp. 620–627 (2015)
8. Kmail, A., Maree, M., Belkhatir, M.: MatchingSem: online recruitment system based on multiple semantic resources. In: Proceedings of the 12th International Conference on Fuzzy Systems and Knowledge Discovery (FSKD), pp. 2654–2659. IEEE (2015)
9. Hong, W., et al.: A job recommender system based on user clustering. J. Comput. **8**(8), 1960–1967 (2013)
10. Kumaran, V.S., Sankar, A.: Towards an automated system for intelligent screening of candidates for recruitment using ontology mapping EXPERT. Int. J. Metadata Semant. Ontol. **8**(1), 56–64 (2013)

11. Kessler, R., Béchet, N., Torres-Moreno, J.M., Roche, M., El-Bèze, M.: Job offer management: how improve the ranking of candidates. In: Rauch, J. et al.(eds.) Foundations of Intelligent Systems, pp. 431–441. Springer, Heidelberg (2009)
12. Yu, K., Guan, G., Zhou, M.: Resume information extraction with cascaded hybrid model. In: Proceedings of the 43rd Annual Meeting on Association for Computational Linguistics, pp. 499–506. Association for Computational Linguistics (2005)
13. Javed, F., et al: Carotene: a job title classification system for the online recruitment domain. In: Proceedings of the IEEE First International Conference on Big Data Computing Service and Applications (BigDataService), pp. 286–293 (2015)
14. Yi, X., Allan, J., Croft, W.B.: Matching resumes and jobs based on relevance models. In: Proceedings of the 30th Annual International ACM SIGIR Conference on Research and Development in Information Retrieval, pp. 809–810. ACM, Amsterdam (2007)
15. Faliagka, E., et al: Application of machine learning algorithms to an online recruitment system. In: The Seventh International Conference on Internet and Web Applications and Services, ICIW 2012, pp. 215–220 (2012)
16. Bekkerman, R., Gavish, M.: High-precision phrase-based document classification on a modern scale. In: Proceedings of the 17th ACM SIGKDD International Conference on Knowledge Discovery and Data Mining, pp. 231–239. ACM (2011)
17. Kessler, M., et al: E-Gen: automatic job offer processing system for human resources. In: Proceedings of the Artificial Intelligence 6th Mexican International Conference on Advances in Artificial Intelligence, pp. 985–995. Springer, Aguascalientes (2007)
18. Clyde, S., Zhang, J., Yao, C.-C.: An object-oriented implementation of an adaptive classification of job openings. In: Proceedings of the 11th Conference on Artificial Intelligence for Applications, pp. 9–16. IEEE (1995)
19. About Occupational Information Network (O*NET). https://onet.rti.org/about.cfm. Date Visited 5 Feb 2016
20. Miller, G.A.: WordNet: a lexical database for English. Comm. ACM **38**(11), 39–41 (1995)
21. Hoffart, J., et al.: YAGO2: exploring and querying world knowledge in time, space, context, and many languages. In: Proceedings of the 20th International Conference Companion on World Wide Web, pp. 229–232. ACM, Hyderabad (2011)

Chaotic Nature of Eye Movement Signal

Katarzyna Harezlak(✉) and Pawel Kasprowski

Silesian University of Technology, ul. Akademicka 16, 44-100 Gliwice, Poland
katarzyna.harezlak@polsl.pl

Abstract. The eye movement analysis undertaken in many research is conducted to better understand the biology of the brain and oculomotor system functioning. The studies presented in this paper considered eye movement signal as an output of a nonlinear dynamic system and are concentrated on determining the chaotic behaviour existence. The system nature was examined during a fixation, one of key components of eye movement signal, taking its vertical velocity into account. The results were compared with those obtained in the case of the horizontal direction. This comparison showed that both variables provide the similar representation of the underlying dynamics. In both cases, the analysis revealed the chaotic nature of eye movement for the first 200 ms, just after a stimulus position change. Subsequently, the signal characteristic tended to be the convergent one, however, in some cases, depending on a part of the fixation duration the chaotic behaviour was still observable.

Keywords: Eye movement · Fixation · Nonlinear system analysis · Chaotic behavior

1 Introduction

Eyes are admittedly one of the most important sensory systems, which facilitates people gathering information about surrounding world, developing skills and making lots of decisions accompanying our life. To improve these processes many eye tracking studies have been conducted for more than one hundred years. The first one was published even earlier - in XVIII century [23]. All these research have been undertaken to better understand the biology of the brain and oculomotor system functioning. Some of them are devoted to the analysis of the eye movement signal in terms of its features extraction and their quantification [19–22]. In others works, methods for the selection of eye movement components and mapping them to points of regard, may be found [11,25].

Two main components are distinguished in eye movement signal - a *fixation*, when eyes are almost stable, in the process of acquiring information from a scene, and a *saccade*, a very short and fast eye movement towards another fixation location, during which no information is taken [14]. There is the noteworthy group of studies, which based on selected components, try to reveal human behaviour patterns while dealing with daily tasks: reading [6,15], learning [8], information searching [9,13] or making choices [5,28].

I. Czarnowski et al. (eds.), *Intelligent Decision Technologies 2017*,
Smart Innovation, Systems and Technologies 72, DOI 10.1007/978-3-319-59421-7_11

This paper presents the studies on eye movement signal considered as an output of a nonlinear dynamic system and are concentrated on determining the chaotic behaviour existence. The analysis of this type has already been conducted in [3], yet was examined to assess an influence of a cognitive load on eyes work and were based on saccades registered during the simple, visually-guided saccade test and one with a cognitive load.

In the currently presented research the chaotic nature of eye movement signal was examined within the first of its, the above-mentioned components - the fixation - when eyes are focused on a chosen part of a scene. During the fixation eyes are not motionless but in constant micro-movements consisting of tremors (aperiodic, wave-like motion), microsaccades (small, fast movements during voluntary fixation) and drifts (slow motions of the eye that occur between microsaccades) [21]. These micro-movements' analysis may reveal dynamic patterns of eye movement signal, which has been investigated in [10], where the dynamic of eye movements during the fixation was explored taking the horizontal velocity into consideration. The studies presented here are the extension of that research and they focused on movements in the vertical direction, which is the main their contribution. The description of the performed work is organized as follows. The bases of the nonlinear dynamic system analysis are provided in the second section. Subsequently, the experiment and results obtained were discussed. The last part of the paper concludes the research.

2 Basis of the Nonlinear Dynamic Systems Analysis

Behaviour of a nonlinear dynamic system is described by a set of its states defining system's phase space, which may be determined by some differential or difference equations. The system - evolving in time - follows a path referred to as an orbit or trajectory. A subset of the phase space, towards which a system has a tendency to evolve for various starting conditions, is called an *attractor*. When a system moves in a deterministic, however not predictable way on a fractal object called 'strange attractor', it manifests chaotic behaviour, which is also revealed by the high sensitivity to initial conditions [26].

In the case of the majority of nonlinear systems, state equations are unknown and the knowledge of a system has to be built based on the system properties' observation. According to Takens' embedding theorem [27], it is feasible to reconstruct a state space spanned by a set of m variables, using N time-delayed measurements (a time series) of a single observed property. For this purpose this time series $x(t)$ has to be at first transformed into a time delay embedded one, according to the formula (1):

$$y(i) = [x(i), x(i+\tau), \cdots, x(i+(m-1)\tau)], i = \{1, 2 \cdots, M\} \tag{1}$$

where $M = N-(m-1)*\tau$ and τ is a time delay and m is an embedding dimension. Determination of these parameters is essential for the successful reconstruction of a system phase space.

2.1 Determining the Time Lag and Embedding Dimension

The process of the appropriate time lag selection should ensure that data are not correlated too much, which means that [1]:

- the time lag is large enough to separate data in the reconstructed phase space as much as possible to provide significantly different information about an underlying system in time t_i and $t_i + \tau$,
- is not larger than a system memory in order not to lose information about an initial state.

The most common approach utilised for this purpose is the usage of the averaged mutual information factor (2):

$$I(\tau) = \sum_{h=1}^{j} \sum_{k=1}^{j} P_{h,k}(\tau)\, log_2 \frac{P_{hk}}{P_h P_k} \tag{2}$$

where P_h and P_k denote the probabilities that the variable assumes a value inside the hth and k^{th} intervals, respectively, and $P_{h,k}(\tau)$ is the joint probability that x_i is taken from interval h and $x_i + \tau$ from interval k. The time lag τ is defined as the first local minimum among results returned by $I(\tau)$ [7].

The second of the aforementioned parameters - the minimal embedding dimension - is understood as a dimension guaranteeing a sufficient number of coordinates to unfold orbits from overlaps and is determined by means of the False Nearest Neighbours (FNN) method [17]. In order to find the proper value of m, it is continually increased to check whether for each point of a time series y_i and its nearest neighbours y_i^{NN}, the ratio of distances between these points seen in dimension $m + 1$ and in dimension m, is less than a predefined threshold R – Eq. 3.

$$R > \frac{\left| x_{i+(m+1)\tau} - x^{NN}_{i+(m+1)\tau} \right|}{\left\| y_i - y_i^{NN} \right\|} \tag{3}$$

According to Takens' theorem, a system attractor can be recovered properly in an embedding space whenever the embedding dimension m fulfils the requirement: $m \geq 2d + 1$, where d is the dimension of an attractor.

Given the reconstructed system phase space it is possible to examine its characteristic in terms of the chaos existence, recognizable by the sensitive dependence on initial conditions [1]. It is visible when neighbouring trajectories, which start from very close initial conditions, very rapidly move to different states. The mean divergence between such trajectories in the phase space at time t is described by the following formula [24]:

$$d(t) = Ce^{\lambda t} \tag{4}$$

where C is a normalised constant representing initial separation between neighbouring points and λ is called the Largest Lyapunov Exponent (LLE), which quantifies the strength of chaos. Negative Lyapunov exponents indicate convergence, while positive Lyapunov exponents demonstrate divergence and chaos.

3 The Experiment

Eye movement signals, used in the nonlinear time series analysis were collected during the experiment based on a 'jumping point' - one of the most popular stimulus type [12,18]. For this research purpose a set of 29 dark points of size 0.5×0.5, distributed over a white 1280×1024 (370 mm $\times$ 295 mm) screen as presented in Fig. 1, was applied. Each point was displayed for 3000 ms. The eye-screen distance was equal to 450 mm and horizontal and vertical visual angles were approximately 40 and 32 respectively. The experiment was conducted in two sessions - both of them took place in the same room at the same time during the day and on the same day in a week. Altogether 24 participants were involved in the experiment in both sessions. They were female and male students with normal vision, aged between 22 and 24, drawn from a group of unpaid volunteers, which were expected to follow the stimulus with their eyes. Before each experiment, the participants were informed about its general purpose, after which they signed the consent form.

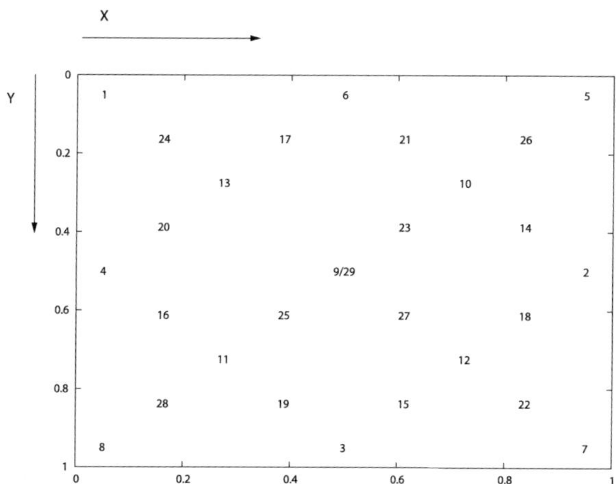

Fig. 1. Layout of stimuli applied during experiments – based on the universal scale – the top–left corner of the screen is represented by (0.0, 0.0) coordinates and the bottom–right corner by (1.0, 1.0) respectively.

Data was recorded using a head-mounted Jazz-Novo eye tracker (product by Ober-consulting [16]) - recording eye positions with sampling rate 1000 Hz. There were 48 recordings collected for 24 participants from two sessions. One recording corresponded to 29 fixations and for each of them 3000 samples were gathered, based on which $29 \times 48 = 1392$ time series were built. They were created by applying the standard procedure of the two-point signal differentiation in order to achieve the first derivative of vertical coordinates representing subsequent eyes' positions. Additionally, the ideal low-pass filter with the 50 Hz cut-off frequency (determined by means of Discrete Fourier Transform) was applied for the purpose of noise reduction.

Each of prepared time series was subjected to the processing consisting of following steps:

- estimating the time lag,
- estimation of the embedding dimension,
- system dynamics reconstruction,
- estimating the Largest Lyapunov Exponent.

4 Results Discussion

The aim of the described studies was twofold. On one hand it was revealing the characteristic of vertical eye movements and its comparison with the characteristic in the other direction on the other hand. For this reason obtained results are discussed in regard to the outcomes achieved in [10] for time series acquired during the same experiments, with the usage of the same procedure, yet built based on horizontal coordinates. Therefore distributions of studied parameters' values were collated.

The first analysed quantity was *the time lag*, which influences the selection of measurements for constructing a time delay embedded series. As it can be seen on histograms of movements registered along X and Y axes - presented in Figs. 2a and b, respectively - values of the parameter calculated for both directions are similar, yet in the case of the vertical direction are shifted toward higher numbers. It may indicate that subsequent recordings along Y-axis keep correlation longer than for X one. They are also more dispersed than in the horizontal direction, where for 83% series, one of four values - 7, 8, 9, 10 - were determined. Obtaining the same percentage along the second axis requires a set of six time lag values - 7, 8, 9, 10, 11, 12 - to be taken into account.

In the case of the second analysed parameter - *the embedding dimension* - similar parameter values were obtained as well (Figs. 3a and b). The most frequently determined values in both cases are the same (3, 4), however percentage of time series, for which three- or four-dimensional phase space was determined is different. In the case of horizontal movements it was 38%, while in the other case it is almost 71%. In the former direction, the contribution of higher dimensions are still significant almost up to the end of the second ten, whereas in latter one it is the end of the first ten.

Evaluation of both earlier-analysed parameters is the very important step of the reconstruction of the nonlinear system dynamic. Based of them it is possible to trace system's states and its trajectories and further ascertain the nature of the system, by the estimation the Largest Lyapunov Exponent (LLE) values. These calculations, in the case of the studied series, were performed in several time scopes, selected from a fixation: $<0\ldots200>$, $<0\ldots500>$, $<0\ldots700>$, $<0\ldots1500>$, $<200\ldots700>$, $<500\ldots1500>$, $<700\ldots1500>$. With the sampling rate of measurements - 1000 Hz - one point in such a segment corresponds to one millisecond. Hence, the first segment is designed to trace eye movement

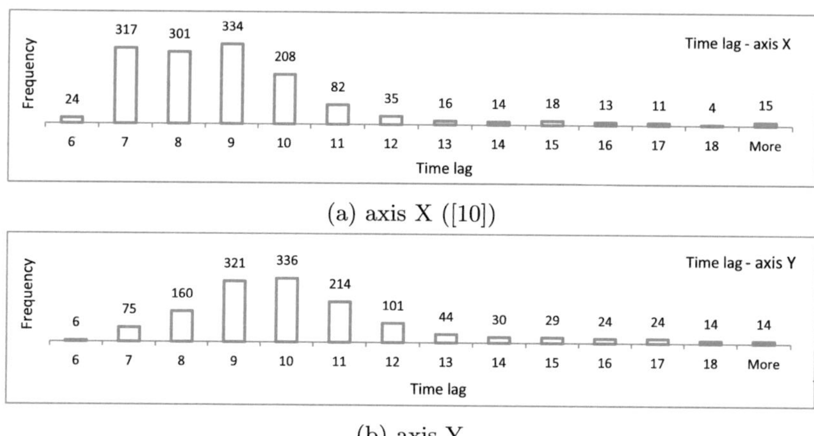

(a) axis X ([10])

(b) axis Y

Fig. 2. Histograms of *the time lag* for all tested time series along both axes

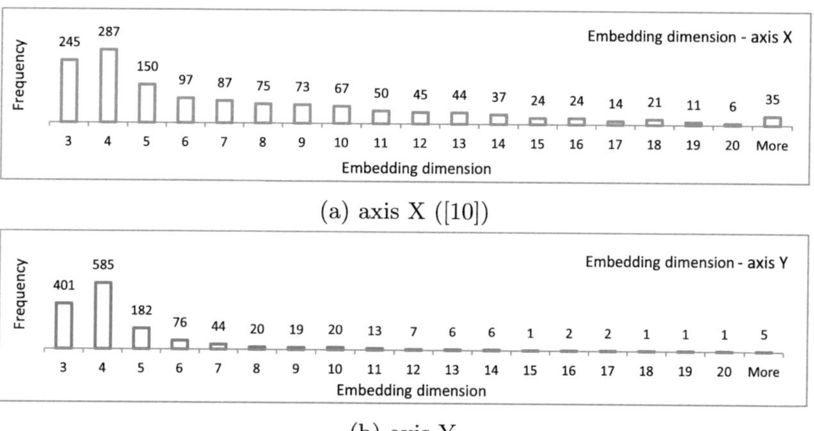

(a) axis X ([10])

(b) axis Y

Fig. 3. Histograms of *the embedding dimension* for all tested time series along both axes.

dynamics during the saccadic latency period - the time a brain needs to react to a stimulus change [4]. The second and third ones were introduced based on the research discussed in [12], which aimed at the selection of the appropriate set of measurements for defining an eye movement model. These studies revealed that rejecting data, gathered during the first 700 ms of recordings, may improve such a model adjustment. Additionally, the whole analysed scope $<0\ldots1500>$ and also those divided into three parts $<0\ldots200>$, $<200\ldots700>$, $<700\ldots1500>$ were taken into account. The last two segments in the above list – $<500\ldots1500>$, $<700\ldots1500>$ – represent time periods when a 'core fixation' takes place.

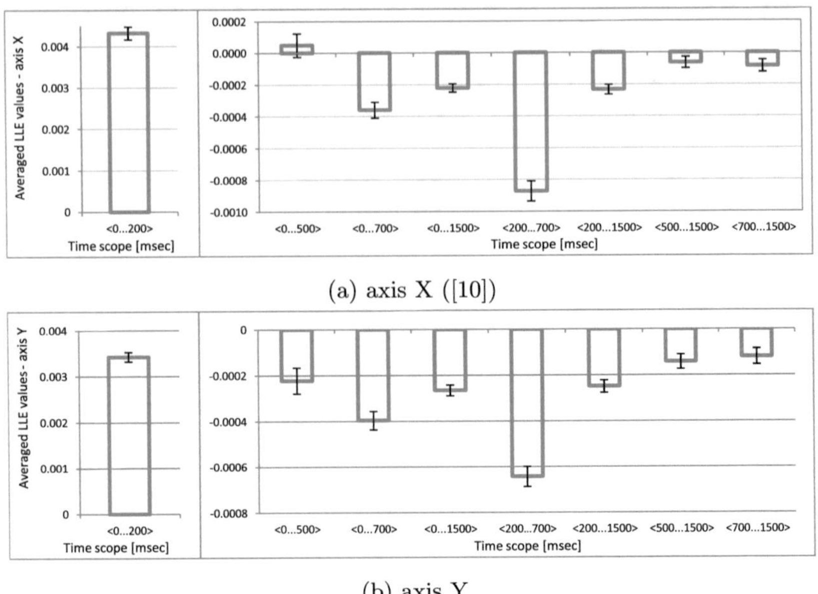

(a) axis X ([10])

(b) axis Y

Fig. 4. Averaged values of *LLE* with 95% confidence intervals within various time scopes of time series - both axes

The *LLE* values, similarly to the other quantities, were presented in regard to both horizontal and vertical movements independently in Figs. 4a and b, respectively. All of them were subjected to the t-test with a 95% confidence interval and null hypothesis stating that the averaged *LLE* value equals 0. In all but one case, the null hypothesis was rejected, thus the statistical significance of the results was confirmed. Only results for $<0 \ldots 500>$ period in horizontal direction turned to be not significant, therefore this segment will be omitted in the further comparison. Studying charts in Figs. 4a and b it may be noticed that chaotic behaviour - represented by positive *LLE* values - was revealed in the first part of the fixation duration $<0 \ldots 200>$ in both movement directions. However, in the case of horizontal one chaos was slightly stronger. During the remaining periods, negative *LLE* values exhibit asymptotic stability of eye movements. The greatest stability was discovered in the fifth of analysed periods: $< 200 \ldots 700 >$ and slightly greater stability was presented in horizontal movements. On the other hand, in the remaining four segments, greater stability regarded vertical movements. Compounding results from all periods it may be reasoning that after changing a stimulus position eye movements is characterised by a chaotic behaviour for first 200 ms, however, after that, this characteristic changes to the convergent one. It is visible in both $<0 \ldots 700>$ and $<0 \ldots 1500>$ scopes. Introducing shorter periods - $< 500 \ldots 1500 >$, $< 700 \ldots 1500 >$ - made the deeper analysis possible, from which it may be inferred that during periods defined between 500 and 1500 ms, eyes strive toward a fixed point. However, it must be remembered that

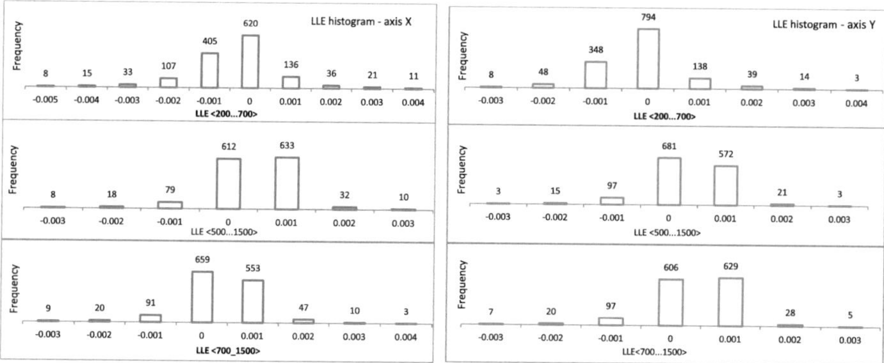

Fig. 5. Histograms of *LLE* values within various time scopes - X-axis [10] on the left and Y-axis on the right

values presented on charts are averaged over all time series, thus it may also mean the return to the chaotic behaviour for some of them. When analysing charts presented in Fig. 5, it is visible that in each time period there are some time series for which the chaotic behaviour is still observable. In the first scope $<200\ldots700>$, contribution of positive LLE values in both movement directions are similar and amounts approximately 15%. In two next periods this percentage is much higher and - averaged over directions - it is, in both cases, about 45%. It clearly indicates, that in the case of near the half of time series eyes, after achieving stability, yield again chaotic nature, which may be related to a stimuli location, yet a further studies are required to explore this.

5 Summary

The analysis of nonlinear system dynamic proves helpful in revealing system's characteristic, especially when its state equations are unknown. In the presented research this methods was used to investigate eye movement nature based on its velocity measured in the vertical direction. The results obtained during the studies were compared to the outcomes evaluated in [10] for the same measurements but taking horizontal velocity into account. This comparison showed that both variables provide the similar representation of the underlying dynamics, thus they may be treated as equivalent in a system dynamic reconstruction [2].

Both evaluated parameters - *the time lag* and *embedding dimension* - took comparable values in both analysed directiones. Although, their distributions presented in the form of a histogram differed slightly, finally the nature of eye movement signal, exposed on their basis, was almost the same. The only dissimilarities visible in the Largest Lyapunov Exponent values were related to the strength of chaotic behaviour. However, it is worth emphasising that its existence was detected in the same fixation's periods and determined differences were very small as well as overall chaos strength.

As it was mentioned above, for the experiment purpose participants with normal vison were involved, thus knowledge about eye movement characteristic, gained during described studies, may be utilized for a comparison with other time series as reference values, especially in medical diagnostic concerning people affected by various diseases. Furthermore, the participants were subjects aged between 22 and 24, therefore, the investigated signal characteristic was related to young people. It would be interesting to check whether this pattern changes during lifetime. Additionally, because the results were obtained with the usage of the particular eye tracker and experimental setup, the application of the proposed method in other environments would also be valuable to analyse. Both of issues are planned as the future work.

Acknowledgments. The research presented in this paper was partially supported by the Silesian University of Technology Rector's Pro-Quality Grant 02/020/RGJ17/0103.

References

1. Abarbanel, H.D.I.: Analysis of observed chaotic data. Institute for Nonlinear Science. Springer, New York (1996)
2. Aguirre, L.A., Letellier, C.: Observability of multivariate differential embeddings. J. Phys. A: Math. Gen. **38**(28), 6311 (2005)
3. Astefanoaei, C., Creanga, D., Pretegiani, E., Optican, L., Rufa, A.: Dynamical complexity analysis of saccadic eye movements in two different psychological conditions. Rom. Rep. Phys. **66**(4), 1038–1055 (2014)
4. Darrien, J.H., Herd, K., Starling, L.J., Rosenberg, J.R., Morrison, J.D.: An analysis of the dependence of saccadic latency on target position and target characteristics in human subjects. BMC Neurosci. **2**(1), 1–8 (2001)
5. Fiedler, S., Glockner, A.: The dynamics of decision making in risky choice: an eye-tracking analysis. Front. Psychol. **3**, 335 (2012)
6. Foster, T.E., Ardoin, S.P., Binder, K.S.: Underlying changes in repeated reading: an eye movement study. Sch. Psychol. Rev. **42**(2), 140 (2013)
7. Fraser, A.M., Swinney, H.L.: Independent coordinates for strange attractors from mutual information. Phys. Rev. A **33**, 1134–1140 (1986)
8. van Gog, T., Scheiter, K.: Eye tracking as a tool to study and enhance multimedia learning. Learn. Instr. **20**(2), 95–99 (2010)
9. Goldberg, J.H., Stimson, M.J., Lewenstein, M., Scott, N., Wichansky, A.M.: Eye tracking in web search tasks: design implications. In: Proceedings of the 2002 Symposium on Eye Tracking Research & Applications, ETRA 2002, pp. 51–58. ACM, New York (2002)
10. Harezlak, K.: Eye movement dynamics during imposed fixations. Inf. Sci. **384**, 249–262 (2017)
11. Harezlak, K., Kasprowski, P.: Evaluating quality of dispersion based fixation detection algorithm. In: Czachorski, T., Gelenbe, E., Lent, R. (eds.) Information Sciences and Systems 2014, pp. 97–104. Springer (2014)
12. Harezlak, K., Kasprowski, P., Stasch, M.: Towards accurate eye tracker calibration - methods and procedures. Procedia Comput. Sci. **35**, 1073–1081 (2014)

13. Harezlak, K., Rzeszutek, J., Kasprowski, P.: The eye tracking methods in user interfaces assessment. In: Intelligent Decision Technologies: Proceedings of the 7th KES International Conference on Intelligent Decision Technologies (KES-IDT 2015), pp. 325–335. Springer, Cham (2015)
14. Holmqvist, K., Nyström, M., Andersson, R., Dewhurst, R., Jarodzka, H., Van de Weijer, J.: Eye tracking: a comprehensive guide to methods and measures. OUP Oxford (2011)
15. Hyona, J., Lorch, R.F., Kaakinen, J.K.: Individual differences in reading to summarize expository text: evidence from eye fixation patterns. J. Educ. Psychol. **94**(1), 44–55 (2002)
16. Novo, J.: Ober Consulting (2015). http://www.ober-consulting.com/9/lang/1/
17. Kennel, M.B., Brown, R., Abarbanel, H.D.I.: Determining embedding dimension for phase-space reconstruction using a geometrical construction. Phys. Rev. A **45**, 3403–3411 (1992)
18. Komogortsev, O., Gobert, D., Jayarathna, S., Koh, D.H., Gowda, S.: Standardization of automated analyses of oculomotor fixation and saccadic behaviors. IEEE Trans. Biomed. Eng. **57**(11), 2635–2645 (2010)
19. Liang, J.R., Moshel, S., Zivotofsky, A.Z., Caspi, A., Engbert, R., Kliegl, R., Shlomo, H.: Scaling of horizontal and vertical fixational eye movements. Phys. Rev. E **71**(3), 031909 (2005)
20. Martinez-Conde, S., Otero-Millan, J., Macknik, S.L.: The impact of microsaccades on vision: towards a unified theory of saccadic function. Nat. Rev. Neurosci. **14**, 83–96 (2013)
21. Martinez-Conde, S., Macknik, S.L., Hubel, D.H.: The role of fixational eye movements in visual perception. Nat. Neurosci. **5**, 229–240 (2004)
22. Otero-Millan, J., Troncoso, X.G., Macknik, S.L., Serrano-Pedraza, I., Martinez-Conde, S.: Saccades and microsaccades during visual fixation, exploration, and search: foundations for a common saccadic generator. J. Vis. **8**(14), 21 (2008)
23. Porterfield, W.: An essay concerning the motions of our eyes. Part I. Of their external motions. Edinb. Med. Essays Obs. **3**, 160–263 (1737)
24. Rosenstein, M.T., Collins, J.J., De Luca, C.J.: A practical method for calculating largest Lyapunov exponents from small data sets. Physica D **65**(1–2), 117–134 (1993)
25. Salvucci, D.D., Goldberg, J.H.: Identifying fixations and saccades in eye-tracking protocols. In: Proceedings of the 2000 Symposium on Eye Tracking Research & Applications, ETRA 2000, pp. 71–78. ACM, New York (2000)
26. Sprott, J.C.: Chaos and Time-Series Analysis. Oxford University Press, Oxford (2003)
27. Takens, F.: Detecting strange attractors in turbulence. Lecture Notes in Mathematics, vol. 898, p. 366 (1981)
28. Vallières, B.R., Chamberland, C., Vachon, F., Tremblay, S.: Insights from eye movement into dynamic decision-making research and usability testing. In: International Conference on Human-Computer Interaction, pp. 169–174. Springer (2013)

Dispersed System with Dynamically Generated Non–disjoint Clusters – Application of Attribute Selection

Małgorzata Przybyła–Kasperek(✉)

Institute of Computer Science, University of Silesia,
Będzińska 39, 41–200 Sosnowiec, Poland
malgorzata.przybyla-kasperek@us.edu.pl

Abstract. The main aim of this article is to apply the selection of attributes method in a dispersed decision–making system, which was proposed by the author in a previous work. The selection of attributes method, that is used, is based on the rough sets theory. At first, reducts of sets of attributes for local knowledge bases are generated and then the attributes that do not occur in the reducts are removed from the local bases. In the study, the accuracy of classification for the system without the use of attributes selection was compared with the results obtained for the system with attributes selection. The experiments were performed using two data sets from the UCI Repository.

Keywords: Dispersed knowledge · Decision–making system · Attribute selection

1 Introduction

In this study, knowledge that is accumulated in a dispersed form, i.e. in the form of separate knowledge bases, is considered. In order to refer to the real situations and applications, it is assumed that the local bases are collected independently and limitations are not imposed on form of local bases.

The aim of the paper is to use the method of selection of attributes in a system that copes with such a dispersed knowledge. For this purpose the method of selection of attributes that is based on the rough sets theory was used. Rough set theory was proposed by Pawlak [6] as a mathematical approach to deal with the vague and incomplete knowledge. One of the most important applications of this theory is a selection of the most significant attributes. This method allows to select a subset of the original attributes that contains the same information as the original one.

The concept of a dispersed decision–making system has been researched by the author for several years. In recent papers a system with a dynamic structure has been proposed [8–10] and various fusion methods in the system have been analyzed [11]. In this paper an approach proposed in the article [8] is used. The main novelty of this paper is to examine the impact of the method of

I. Czarnowski et al. (eds.), *Intelligent Decision Technologies 2017*,
Smart Innovation, Systems and Technologies 72, DOI 10.1007/978-3-319-59421-7_12

attribute selection on the effectiveness of the inference in this system. The main contributions of this paper are listed below:

- performing tests using the method of attribute selection and a dispersed system with dynamically generated non–disjoint clusters,
- performing a comparison of the approaches with and without attribute selection.

The issue of combining classifiers is a very important aspect in the literature. Many models and methods have been proposed in the issue of multiple model approach [4,5,7]. Also, areas such as the distributed decision–making [12,14] and group decision making [1,2] deal with this issue. This study presents a completely different approach to solution of the problem. The main differences appear in the assumptions related to the form of dispersed data. In the models proposed in the literature usually sets of objects/attributes of local bases are equal or disjoint. In the approach that is considered in the paper, such assumptions are not fulfilled. In addition, identification whether in the local knowledge bases the same objects are included, is not possible.

2 Reduction of the Set of Attributes

Methods of feature selection are used to automatically select the attributes in data that are most useful or most relevant for the problem that is considered. There are various methods to determine the most relevant attributes. The method used in the paper is based on the rough set theory proposed by Pawlak in 1982 [6]. This reduction method is very popular and widely used [3,13,16]. The two main definitions of this reduction methods are given below.

Definition 1. *Let $D = (U, A, d)$ be a decision table, where U is the universe, A is a set of conditional attributes, d is a decision attribute. Let $x, y \in U$ be given objects and $B \subseteq A$ be a subset of conditional attributes. We say that objects x and y are discernible by B if and only if there exists $a \in B$ such that $a(x) \neq a(y)$.*

Definition 2. *Let $D = (U, A, d)$ be a decision table. A set of attributes $B \subseteq A$ is called a reduct of decision table D if and only if:*

1. *for any objects $x, y \in U$ if $d(x) \neq d(y)$ and x, y are discernible by A, then they are also discernible by B,*
2. *B is the minimal set that satisfies the condition 1.*

The reduct of the decision table is a minimal subset of conditional attributes that provides exactly the same possibility of the classification of objects to the decision classes as the original set of conditional attributes. Sometimes in the decision table, not all attributes are required to classify objects to the decision classes. In such cases, the reduction allows to remove unnecessary attributes by generating a subtable in which a set of attributes is limited to reduct.

In a situation that is considered in the study we are dealing with dispersed data sets – set of local knowledge bases. Selection of attributes is particularly important in this case. Large data sets require high computational power. Reduction of the set of attributes to the necessary minimum is often a solution to this problem. In the approach that is proposed in this paper set of reducts is generated for each local decision table. Then, one reduct, which contains the smallest number of attributes is selected from this set. The conditional attributes that do not occur in the reduct are removed from the local decision table. Based on modified local decision tables, decisions are taken with the use of a dispersed system that is discussed in the next section. The aim of this study is to analyze the results obtained using the attribute selection in a dispersed decision–making system.

3 Dispersed Decision–making System with Dynamically Generated Non–disjoint Clusters

A dispersed decision–making system, which is used in the paper, was proposed by the author in the article [8]. The classification process that is realized by the dispersed system is as follows. First, based on each local knowledge base a classifier is constructed, which generates a vector of probabilities for the test object. Then, based on these vectors certain relationships between classifiers are defined. Two types of relations are used: conflict and friendship relation. Next, clusters, which consist of classifiers which are in friendship relation, are generated. For each cluster an aggregated decision table is created based on local knowledge bases included in the cluster. Decisions within the cluster are determined based on such aggregated knowledge. In the last stage, the global decisions are determined using the density algorithm.

The basic concepts and definitions of the system are very briefly described below. A detailed discussion can be found in the paper [8].

We assume that the knowledge is available in a dispersed form, which means in a form of several decision tables. The set of local knowledge bases that contain data from one domain is pre–specified. The only condition which must be satisfied by the local knowledge bases is to have common decision attributes. We assume that each local decision table $D_{ag} = (U_{ag}, A_{ag}, d_{ag})$ is managed by one agent, which is called a resource agent ag. We want to designate homogeneous groups of resource agents. The agents who agree on the classification for a test object into the decision classes will be combined in the group. First, for each agent we determine a vector of probabilities using the m_1–nearest neighbors classifier. Based on the vector of probabilities a vector of ranks is generated. Then the function $\phi^x_{v_j}$ for the test object x and each value of the decision attribute $v_j \in V^d$ is defined; $\phi^x_{v_j} : Ag \times Ag \to \{0, 1\}$

$$\phi^x_{v_j}(ag_i, ag_k) = \begin{cases} 0 & \text{if} \;\; r_{i,j}(x) = r_{k,j}(x) \\ 1 & \text{if} \;\; r_{i,j}(x) \neq r_{k,j}(x) \end{cases}$$

where $ag_i, ag_k \in Ag$, Ag is a set of agents, $[r_{i,1}(x), \ldots, r_{i,c}(x)]$ is the vector of ranks for agent ag_i.

We also define the distance between agents ρ^x for the test object x: $\rho^x : Ag \times Ag \to [0,1]$

$$\rho^x(ag_i, ag_k) = \frac{\sum_{v_j \in V^d} \phi^x_{v_j}(ag_i, ag_k)}{card\{V^d\}},$$

where $ag_i, ag_k \in Ag$.

Definition 3. *We say that agents $ag_i, ag_k \in Ag$ are in a friendship relation due to the object x, which is written $R^+(ag_i, ag_k)$, if and only if $\rho^x(ag_i, ag_k) < 0.5$. Agents $ag_i, ag_k \in Ag$ are in a conflict relation due to the object x, which is written $R^-(ag_i, ag_k)$, if and only if $\rho^x(ag_i, ag_k) \geq 0.5$.*

The cluster due to the classification of object x is the maximum subset of resource agents $X \subseteq Ag$ such that

$$\forall_{ag_i, ag_k \in X} \quad R^+(ag_i, ag_k).$$

For each cluster that contains at least two resource agents, a superordinate agent is defined, which is called a synthesis agent, as_j, where j is the number of cluster. The synthesis agent, as_j, has access to knowledge that is the result of the process of inference carried out by the resource agents that belong to its subordinate group. As is a finite set of synthesis agents.

Then for each synthesis agent an aggregated decision table is generated. In previous papers the approximated method of the aggregation of decision tables have been used for this purpose [8,9]. In this paper, we also use this method. In the method, objects of the aggregated table are constructed by combining relevant object from decision tables of the resource agents that belong to one cluster. In order to define the relevant objects parameter m_2 is used.

Based on the aggregated table of synthesis agent a vector of probabilities is determined. Then, some transformations are performed on these vectors (described in the paper [9]) and the set of decisions is generated. In this set there are decisions which have the highest support among all agents. This set is generated using the DBSCAN algorithm, description of the method can be found in the papers [8–10]. The DBSCAN algorithm was used to search for decisions that are closest to the decision with the greatest support among agents.

4 Experiments

In the experimental part, at first the reduction of the sets of conditional attributes of local knowledge bases was applied, and then the reduced knowledge bases were used in a dispersed decision–making system with dynamically generated non–disjoint clusters. The results obtained in this way were compared with the results obtained for a dispersed system and full sets of conditional attributes.

The author does not have access to dispersed data that are stored in the form of a set of local knowledge bases and therefore some benchmark data that are stored in a single decision table were used. The division into a set of decision tables was made for the used data.

The data from the UCI repository were used in the experiments: Soybean data set and Landsat Satellite data set. Both data sets are available in the UCI repository in the form divided into a training set and a test set. Table 1 presents a numerical summary of the data sets.

Table 1. Data set summary

Data set	# The training set	# The test set	# Conditional attributes	# Decision classes
Soybean	307	376	35	19
Landsat Satellite	4435	1000	36	6

For both sets of data the training set, that was originally written in the form of a single decision table, was dispersed, which means divided into a set of decision tables. The dispersed decision–making system with five different versions (with 3, 5, 7, 9 and 11 decision tables) were considered. The following designations are used:

- WSD^{dyn}_{Ag1} – 3 decision tables;
- WSD^{dyn}_{Ag2} – 5 decision tables;
- WSD^{dyn}_{Ag3} – 7 decision tables;
- WSD^{dyn}_{Ag4} – 9 decision tables;
- WSD^{dyn}_{Ag5} – 11 decision tables.

The division into a set of decision tables was made in the following way. The author defined the number of conditional attributes in each of the local decision tables. Then the attributes from the original table were assigned to the local tables randomly. As a result of this division, some local tables have common conditional attributes. The universes of the local tables are the same as the universe of the original table, but in the local tables identifiers of object are not stored.

The measures for determining the quality of the classifications were:

- *estimator of classification error* e in which an object is considered to be properly classified if the decision class that is used for the object belonged to the set of global decisions that were generated by the system;
- *estimator of classification ambiguity error* e_{ONE} in which an object is considered to be properly classified if only one correct value of the decision was generated for this object;
- *the average size of the global decisions sets* $\overline{d}_{WSD^{dyn}_{Ag}}$ that was generated for a test set.

In the first stage of experiments for each local decision tables reducts of the set of conditional attributes were generated. For this purpose, a program Rough Set Exploration System (RSES [15]) was used. The program was developed at the University of Warsaw, in Faculty of Mathematics, Informatics and Mechanics under the direction of Professor Andrzej Skowron. For both analyzed data sets, for each version of the dispersion (3, 5, 7, 9 and 11 decision tables) and for each local decision table separately a set of reducts was generated. For this purpose, the following settings of the RSES program were used: Discernibility matrix settings – Full discernibility, Modulo decision; Method – Exhaustive algorithm. For certain decision tables many reducts were generated. For example, for the Landsat Satellite data set and the dispersed system with 3 local decision tables for one of the tables 1469 reducts were obtained, and for another table 710 reducts were obtained. If more than one reduct was generated for the table, one reduct was randomly selected, from the reducts that had the smallest number of attributes. Table 2 shows the number of conditional attributes that were deleted from the local decision tables by the reduction of knowledge. In the table the following designations were applied: $\#Ag$ – is the number of local decision tables (the number of agents) and A_{ag_i} – a set of conditional attributes of the ith local table (of the ith agent). As can be seen for the Landsat Satellite data set and the dispersed system with 7, 9 and 11 local decision tables, the reduction of the set of conditional attributes did not bring any changes. Therefore in the rest of the paper these systems will no longer be considered.

Table 2. The number of conditional attributes removed as a result of the reduction of knowledge

Data set, $\#Ag$	A_{ag_1}	A_{ag_2}	A_{ag_3}	A_{ag_4}	A_{ag_5}	A_{ag_6}	A_{ag_7}	A_{ag_8}	A_{ag_9}	$A_{ag_{10}}$	$A_{ag_{11}}$
Soybean, 3	4	5	2	–	–	–	–	–	–	–	–
Soybean, 5	4	0	0	1	0	–	–	–	–	–	–
Soybean, 7	0	0	0	1	1	0	0	–	–	–	–
Soybean, 9	0	0	0	0	0	1	1	0	0	–	–
Soybean, 11	0	0	0	0	0	0	0	0	1	0	0
Landsat Satellite, 3	1	9	8	–	–	–	–	–	–	–	–
Landsat Satellite, 5	0	0	1	0	0	–	–	–	–	–	–
Landsat Satellite, 7	0	0	0	0	0	0	0	–	–	–	–
Landsat Satellite, 9	0	0	0	0	0	0	0	0	0	–	–
Landsat Satellite, 11	0	0	0	0	0	0	0	0	0	0	0

In the second stage of experiments, for both sets of data and each version of a dispersed system, an optimization of the system's parameters was carried out according with the following test scenario.

In the first step, the optimal values of parameters:

- m_1 – parameter which determines the number of relevant objects that are selected from each decision class of the decision table and are then used in the process of cluster generation;
- m_2 – parameter of the approximated method of the aggregation of decision tables;

were determined. This is done by performing tests for different parameter values: $m_1, m_2 \in \{1, \ldots, 10\}$ for the Soybean data set and $m_1, m_2 \in \{1, \ldots, 5\}$ for the Landsat Satellite data set. In order to determine the optimal values of these parameters, a dispersed decision–making system with dynamically generated non–disjoint clusters and voting method instead of the DBSCAN algorithm were used. Then, the minimum value of the parameters $m_1$and m_2 are chosen, for which the lowest value of the estimator of the classification error is obtained. In the second step parameters ε of the DBSCAN algorithm are optimized. For this purpose, the optimal parameter values m_1 and m_2 that were previously set were used. Parameter ε is optimized by performing a series of experiments with different values of this parameter. Then, the values that indicate the greatest improvement in the quality of classification are selected. The best results that were obtained for the optimal values of the parameter m_1, m_2 and ε are presented below.

The results of the experiments with both data sets and with conditional attributes selection are presented in Table 3. In the table the following information is given: the name of dispersed decision–making system (System); the selected, optimal parameter values (Parameters); the estimator of classification error e; the estimator of classification ambiguity error e_{ONE} and the average size of the global decisions sets $\overline{d}_{WSD^{dyn}_{Ag}}$.

For comparison, Table 4 contains results for the same data sets and dispersed systems but without the use of conditional attributes selection. These results were taken from the paper [8].

Comparing the results given in Tables 3 and 4, it can be noted that in the case of the Soybean data set, after applying the selection of attributes poorer quality of the classification was usually obtained. Exceptions to this statement are only the systems with nine WSD^{dyn}_{Ag4} and eleven WSD^{dyn}_{Ag5} resource agents. For the Landsat Satellite data set, selection of attributes brought an improvement in the quality of the classification.

The biggest improvement in the quality of the classification (the biggest difference in the values of the measure e) was observed for the Soybean data set and the dispersed system with nine resource agents WSD^{dyn}_{Ag4}. The largest deterioration in the results was observed for the Soybean data set and the dispersed system with three resource agents WSD^{dyn}_{Ag1}. Which means that for the Soybean data set using the attribute selection very unpredictable results were obtained. For the Landsat Satellite data set the improvement in the quality of the classification was obtained. This situation may be related to the occurrence of missing values. In the Soybean data set there is a lot of missing values while in the

Table 3. Results with conditional attributes selection

Experiments results with the Soybean data set				
System	Parameters	e	e_{ONE}	$\overline{d}_{WSD_{Ag}^{dyn}}$
WSD_{Ag1}^{dyn}	$m_1 = 3$, $m_2 = 1$, $\varepsilon = 0.0087$	0.069	0.245	1.617
WSD_{Ag2}^{dyn}	$m_1 = 3$, $m_2 = 1$, $\varepsilon = 0.015$	0.016	0.314	1.731
	$m_1 = 3$, $m_2 = 1$, $\varepsilon = 0.012$	0.032	0.285	1.545
WSD_{Ag3}^{dyn}	$m_1 = 6$, $m_2 = 1$, $\varepsilon = 0.0175$	0.019	0.295	1.902
	$m_1 = 6$, $m_2 = 1$, $\varepsilon = 0.0135$	0.021	0.266	1.590
WSD_{Ag4}^{dyn}	$m_1 = 7$, $m_2 = 1$, $\varepsilon = 0.0205$	0.019	0.261	1.779
	$m_1 = 7$, $m_2 = 1$, $\varepsilon = 0.017$	0.032	0.231	1.551
WSD_{Ag5}^{dyn}	$m_1 = 2$, $m_2 = 2$, $\varepsilon = 0.0225$	0.045	0.309	1.614
	$m_1 = 2$, $m_2 = 2$, $\varepsilon = 0.0115$	0.080	0.258	1.287
Experiments results with the Landsat Satellite data set				
System	Parameters	e	e_{ONE}	$\overline{d}_{WSD_{Ag}^{dyn}}$
WSD_{Ag1}^{dyn}	$m_1 = 2$, $m_2 = 5$, $\varepsilon = 0.0034$	0.017	0.413	1.757
WSD_{Ag2}^{dyn}	$m_1 = 2$, $m_2 = 2$, $\varepsilon = 0.00529$	0.014	0.394	1.731
	$m_1 = 2$, $m_2 = 2$, $\varepsilon = 0.00235$	0.037	0.237	1.268

Table 4. Results without conditional attributes selection

Experiments results with the Soybean data set				
System	Parameters	e	e_{ONE}	$\overline{d}_{WSD_{Ag}^{dyn}}$
WSD_{Ag1}^{dyn}	$m_1 = 2$, $m_2 = 1$, $\varepsilon = 0.00885$	0.019	0.277	1.864
WSD_{Ag2}^{dyn}	$m_1 = 3$, $m_2 = 1$, $\varepsilon = 0.01365$	0.008	0.266	2.000
	$m_1 = 3$, $m_2 = 1$, $\varepsilon = 0.01215$	0.016	0.239	1.527
WSD_{Ag3}^{dyn}	$m_1 = 6$, $m_2 = 1$, $\varepsilon = 0.0188$	0.016	0.298	2.016
	$m_1 = 6$, $m_2 = 1$, $\varepsilon = 0.0134$	0.019	0.271	1.590
WSD_{Ag4}^{dyn}	$m_1 = 2$, $m_2 = 1$, $\varepsilon = 0.01775$	0.027	0.287	1.761
	$m_1 = 2$, $m_2 = 1$, $\varepsilon = 0.01625$	0.035	0.242	1.519
WSD_{Ag5}^{dyn}	$m_1 = 2$, $m_2 = 2$, $\varepsilon = 0.01825$	0.035	0.319	1.910
	$m_1 = 2$, $m_2 = 2$, $\varepsilon = 0.00825$	0.082	0.255	1.274
Experiments results with the Landsat Satellite data set				
System	Parameters	e	e_{ONE}	$\overline{d}_{WSD_{Ag}^{dyn}}$
WSD_{Ag1}^{dyn}	$m_1 = 1$, $m_2 = 4$, $\varepsilon = 0.0032$	0.021	0.382	1.751
WSD_{Ag2}^{dyn}	$m_1 = 2$, $m_2 = 1$, $\varepsilon = 0.0052$	0.019	0.402	1.739
	$m_1 = 2$, $m_2 = 1$, $\varepsilon = 0.0022$	0.042	0.219	1.231

Landsat Satellite data set missing values do not exist. This can be crucial for the results that were obtained.

Of course, according to the well known principle, it can not be expected that one method will be suitable for all classes of problems (all data). "No free lunch" theorem also refers to this study. However, it can be said that for some data sets, the method of selection of attributes that was applied in a dispersed system with dynamically generated non–disjoint clusters results in improved of the quality of classification. It should also be noted that the selection of attributes, in both data sets, leads to reduction of the number of conditional attributes. This in turn improves the readability of data and leads to reduce the effort that must be incurred on the processing of data.

5 Conclusions

In this study, a dispersed system with dynamically generated non–disjoint clusters, which has been proposed in the paper [8], is considered. In this system the method of selection of attributes, which is based on the rough sets theory, was applied. The experiments were performed on two data sets from the UCI Repository. The data are dispersed in five different ways. Based on the results, it was found that for one of the data sets the improvement in the quality of classification was obtained after applying the method of selection of attributes. For the second data set very unpredictable results were obtained. For one version of dispersion noticeable improvement was achieved, and for another version of dispersion significant deterioration in the quality of the classification was noted. Probably it is related to missing values that occur in this data set. For data without missing values the application of the method of attribute selection in a dispersed system with dynamically generated non–disjoint clusters improves the quality of the classification.

References

1. Bregar, A.: Towards a framework for the measurement and reduction of user-perceivable complexity of group decision-making methods. IJDSST **6**(2), 21–45 (2014)
2. Cabrerizo, F., Herrera-Viedma, E., Pedrycz, W.: A method based on PSO and granular computing of linguistic information to solve group decision making problems defined in heterogeneous contexts. Eur. J. Oper. Res. **230**(3), 624–633 (2013)
3. Demri, S., Orlowska, E.: Incomplete information: structure, inference, complexity, Monographs in Theoretical Computer Science. An EATCS Series. Springer (2002)
4. Gatnar, E.: Multiple-Model Approach to Classification and Regression. PWN, Warsaw (2008)
5. Kuncheva, L.I.: Combining Pattern Classifiers Methods and Algorithms. Wiley, Chichester (2004)
6. Pawlak, Z.: Rough sets. Int. J. Comput. Inf. Sci. **11**, 341–356 (1982)
7. Polikar, R.: Ensemble based systems in decision making. IEEE Circ. Syst. Mag. **6**(3), 21–45 (2006)

8. Przybyła-Kasperek, M., Wakulicz-Deja, A.: Global decision-making system with dynamically generated clusters. Inf. Sci. **270**, 172–191 (2014)
9. Przybyła-Kasperek, M., Wakulicz-Deja, A.: A dispersed decision-making system - The use of negotiations during the dynamic generation of a systems structure. Inf. Sci. **288**, 194–219 (2014)
10. Przybyła-Kasperek, M., Wakulicz-Deja, A.: Global decision-making in multi-agent decision-making system with dynamically generated disjoint clusters. Appl. Soft Comput. **40**, 603–615 (2016)
11. Przybyła-Kasperek, M.: Selected methods of combining classifiers, when predictions are stored in probability vectors, in a dispersed decision-making system. Fundam. Inform. **147**(2–3), 353–370 (2016)
12. Schneeweiss, C.: Distributed decision making–a unified approach. Eur. J. Oper. Res. **150**(2), 237–252 (2003)
13. Skowron, A.: Rough sets and vague concepts. Fundam. Inform. **64**(1–4), 417–431 (2005)
14. Skowron, A., Wang, H., Wojna, A., Bazan, J.G.: Multimodal classification: case studies, pp. 224–239 (2006)
15. Skowron, A.: Rough Set Exploration System. http://logic.mimuw.edu.pl/~rses/. Accessed 31 Jan 2017
16. Susmaga, R., Słowiński, R.: Generation of rough sets reducts and constructs based on inter-class and intra-class information. Fuzzy Sets Syst. **274**, 124–142 (2015)

Generational Feature Elimination to Find All Relevant Feature Subset

W. Paja(✉)

Department of Computer Science, Faculty of Mathematics and Natural Sciences, University of Rzeszow, Rzeszow, Poland
wpaja@ur.edu.pl

Abstract. The recent increase of dimensionality of data is a target for many existing feature selection methods with respect to efficiency and effectiveness. In this paper, the all relevant feature selection method based on information gathered using generational feature elimination was introduced. The successive generations of feature subset were defined using *DTLevelImp* algorithm and in each step the subset of most important features were eliminated from the primary investigated dataset. This process was executed until the most important feature reach importance value on the level similar to importance of the random shadow features. The proposed method was also initially tested on well-know artificial and real-world datasets and the results confirm its efficiency. Thus, it can be concluded that selected attributes are relevant.

Keywords: Feature selection · Feature ranking · Dimensionality reduction · All relevant feature selection

1 Introduction

The main task of the feature selection (FS) process is to determine which predictors should be included in a model to make the best prediction results. It is one of the most critical questions as data are becoming increasingly high-dimensional. Specifically in business research to find important relationships between customers and products, in pharmaceutical research to define relationship between structure of molecules and their activity, in the wide area of biology application to analysis of different genetic data aspects to find relationships with diseases [1–5].

Basically, the feature selection methods could be divided into two categories: unsupervised in which predictions are not considered during FS process, and supervised in which quality of predictions are the main aspect to estimate importance of features. Feature selection is primarily focused on removing non-informative or redundant predictors from the information systems. Thus, the development of efficient methods for significant feature selection is valid.

The feature selection methods are also typically categorized into three classes depends on how they combine the selection algorithm and the model building: filter, wrapper and embedded FS methods. Filter methods select features regardless of the model. They are based only on general features like the correlation with the variable to predict, they are unsupervised. These methods select only the most interesting

I. Czarnowski et al. (eds.), *Intelligent Decision Technologies 2017*,
Smart Innovation, Systems and Technologies 72, DOI 10.1007/978-3-319-59421-7_13

attributes that will be part of a classification model [6]. However, some redundant, but relevant features could not be recognized. Wrapper methods evaluate subsets of features which allows to detect the possible interactions between variables [4, 7, 8]. But, the increasing overfitting risk when the number of observations is insufficient is also possible. Additionally, the significant increase of computation time is observed when the number of variables is large. The third type called embedded methods try to combine the advantages of both previous methods. Thus, the learning algorithm takes advantage of its own FS algorithm. So, it needs to know initially what a good selection is, which limits their exploitation [9]. During the selection process two main goals are identified [2]:

- *Minimal Optimal Feature Selection* (*MOSF*), where the goal means identification of minimal subset of features with optimal classification quality;
- *All Relevant Feature Selection* (*ARFS*), where the main goal is to discover all informative features, even that with minimal importance value [8].

In this study proposed approach could be treated as the second type of FS. Some motivation for this methodology was *Recursive Feature Elimination* (*RFE*) algorithm in which by application of external estimator specific weights values are assigned to each features [10, 11]. This procedure is repeated recursively, and in each step, features whose weights are the smallest are removed from the current set of features. It works until the desired set of features to select is eventually reached. In *RFE* approach a number of feature to select should be initially defined. In turn, in this research, to distinguish between relevant and irrelevant features the *contrast variable* concept [8] have been applied. It is a variable that does not carry information on the decision variable by design that is added to the system in order to discern relevant and irrelevant variables. Here, it is obtained from the real variables by random permutation of values between objects of the analyzed dataset. The use of contrast variables was for the first time proposed by Stoppiglia [12] and then by Tuv [13]. Summarizing, the goal of proposed methodology is to simplify and improve feature selection process by relevant feature elimination during recursive generation of decision tree. The hypothesis is that by removing subsets of relevant features in each step gradually all-relevant feature subset could be discovered.

The paper is divided into three main sections besides Introduction. In the second section, methods and algorithms are described in details. Then investigated datasets are briefly characterized and the last section contains gathered results of experiments and also some conclusions about efficiency of proposed approach.

2 Methods and Algorithms

2.1 The *DTLevelImp* Algorithm

During research the *DTLevelImp* algorithm [14] is used to define ranking values for each investigated feature. This algorithm is based on the presence of different feature in the decision tree structure generated from dataset. Thus, three different ranking measures could be defined. The first one, called ***L****evel-****W****eight based* ***I****mportance*

(LWI) takes into consideration weight factor W_j assigned to a given level j of the tree in which feature A_i occurs (Eq. 1).

$$LWI_{A_i} = \sum_{j=1}^{l} \sum_{\text{node}=1}^{x} \mathrm{W_j} \cdot \{\mathrm{A_i}\} \tag{1}$$

where l is the number of levels inside the model, x is the number of nodes inside the given level j and $\{A_i\}$ describe the presence of the ith attribute, usually *1* (feature occurred) or *0* (feature didn't occur). The weight W of the level j is defined as follow:

$$W_j = \begin{cases} 1 & j = 1, j \in N; \\ \frac{W_{j-1}}{2} & 1 < j \leq l \end{cases} \tag{2}$$

In turn, the second measure, called ***L**evel-**P**ercentage based **I**mportance* (LPI) takes into account the percentage of cases *p(node)* occurring in a given tree node at each level, where there is examined attribute A_i (Eq. 3).

$$LPI_{A_i} = \sum_{j=1}^{l} \sum_{\text{node}=1}^{x} \mathrm{p(node)} \cdot \{\mathrm{A_i}\} \tag{3}$$

The third measure, called ***L**evel-**I**nstance based **I**mportance* (LII) takes into account the number of cases *Inst(node)* which are classified in a given tree node at each level, where there is examined attribute A_i (Eq. 4).

$$LII_{A_i} = \sum_{j=1}^{l} \sum_{\text{node}=1}^{x} \mathrm{Inst(node)} \cdot \{\mathrm{A_i}\} \tag{4}$$

Additionally, a sum of LPI + LWI measures for each examined attribute was also investigated. In this way, the next measure was introduced.

2.2 The Experimental Procedure

The experimental procedure, initially called *Generational Feature Elimination*, is presented in form of pseudo-code below as an *Algorithm 1*. Generally, this algorithm apply four importance measures (*impMeasure*) during the feature ranking process. The original data (*originalData*) are extended by adding contrast variables (*contrastData*) which are the result of permutation of original predictors. While $x = 0$ algorithm recursively builds classification tree model (*treeModel*) from current dataset (*currentSet*). Then, based on applied importance measure the set of important features (*impDataSet*) is defined from this features which have rank value (*featureRankValue*) greater than maximal rank value of contrast feature (*maxCFRankValue*). Next, extracted feature set is pruned from current data. These iterations execute until no one original feature have greater ranking value than contrast feature ranking value, the *impDataSet* is ∅. Finally, a table with selected relevant features for each importance measure $finalSet_{impMeasure}$ is created.

Algorithm 1. The experimental procedure used in the main body of the current study.

Input: *originalData* - an investigated dataset; *ncross* is a number of cross validation steps; *impMeasure∈{LWI, LPI, LII, LW+PI}* is a parameter for selection of importance measure used to rank variables.
Output: Table that contains all relevant feature for each *impMeasure*.

```
contrastData ← permute(originalData(predictors))
mergedPredictors ← cbind(originalData(predictors),contrastData)
mergedData ← cbind(mergedPredictors, originalData(decisions))

for each impMeasure∈{LWI, LPI, LII, LW+PI} do
   finalSet_impMeasure = ∅
   for each n=1,2,..,ncross do
        maxCFRankValue = 0
        featureRankValue = 0
        cvDataSet ← selectCVData(mergedData-n)
        currentSet ← cvDataSet
        x = 0
        while x = 0 do
             treeModel ← treeGeneration(currentSet)
             featureRank ← DTLevelImp(treeModel, impMeasure)
             maxCFRankValue ← maxValue(featureRank(contrastData))
             impDataSet = ∅
                for each featureRankValue ∈ featureRank(originalData)
                   if featureRankValue > maxCFRankValue then
                        impDataSet ← impDataSet ∪ feature(featureRankValue)
                end
             currentSet ← currentSet - impDataSet
             if impDataSet = ∅ then x++
        end
        resultImpDataSet[n] ← cvDataSet - currentSet
   end
   for each feature ∈ resultImpDataSet
        z ← countIfImp(feature)
        if z ≥ ncross/2 then
             finalSet_impMeasure ← finalSet_impMeasure ∪ feature
   end
   return finalSet_impMeasure
end
```

3 Investigated Dataset

During experiments well-known Madelon dataset in the domain of feature selection was investigated. Madelon is an artificial data set, which was one of the Neural Information Processing Systems challenge problems in 2003 (called NIPS2003) [15]. The data set contains 2600 objects (2000 of training cases + 600 of validation cases) corresponding to points located in 32 vertices of a 5-dimensional hypercube. Each vertex is randomly assigned to the one of two classes: −1 or +1, and the decision of each object is a class of its vertex. Each object is described by 500 features which were constructed in the following way: 5 of them are randomly jittered coordinates of points, the other 15 attributes are random linear combinations of the first 5. The rest of the data is a uniform random noise. The goal is to select 20 important attributes from the system without false attributes selection.

Additionally, dataset that predict climate model simulation crashes and determine the parameter value combinations that cause the failures was also investigated as an example of real-world problem. This dataset contains records of simulation crashes encountered during climate model uncertainty quantification (UQ) ensembles [16]. Ensemble members were constructed using a Latin hypercube method in LLNL's UQ Pipeline software system to sample the uncertainties of 18 model parameters within the Parallel Ocean Program (POP2) component of the Community Climate System Model (CCSM4). Data contain 540 of simulation and 46 of them failed for numerical reasons at combinations of parameter values. The goal is to predict simulation outcomes: fail or succeed. It was initially investigated using SVM models in [16] by Lucas et al., and additional analysis were performed by the author in [4] using random forest approach to estimate importance of variables. Both datasets are available for download from UCI Machine Learning Repository.

4 Results and Conclusions

To illustrate the proposed methodology only experimental results for 2nd sample fold of Madelon dataset using LII measure are presented (Table 1). As it was shown, four iterations of the algorithm were utilized. During the first iteration, the classification tree (the first generation of tree) has been built based on all input data. Using the LII importance measure values, the subset of 10 features could be treated as important one (grey cells marked), according to decreased value of LII parameter calculated from developed decision tree structure. The eleventh feature, *f335_1*, which is contrast feature, defines threshold for selection between relevant and irrelevant subset. Next, the subset of selected features is removed from dataset. Then, in the 2nd iteration of algorithm the next subset of six probably important features is selected using LII parameter calculated from the tree built on the reduced dataset. The seventh feature, *f115_1*, which is contrast feature defines threshold for selection of the next important subset. This subset is also removed from dataset. Later, in the 3rd iteration of algorithm the next subset of six probably important features is selected using LII measure calculated from the tree built on the subsequently reduced dataset. The seventh feature, *f291_1*, defines threshold for selection of the next important subset. This subset is therefore removed from dataset. Finally, in the 4th iteration the subset of important features is empty, because the highest value of LII measure is reached by contrast feature *f187_1*. Thus, the algorithm quits, and the subset of 22 features is defined as an important one. The truly important features in Madelon dataset are written in bold. It could be observed that three attributes: *f85*, *f256* and *f112*, were also included in discovered subset. However their importance is very random and unique what is clearly presented in Table 2, where these attributes are only single-selected. They reach ≥ 0.8 probability of attribute removing threshold during 10-fold cross-validation of investigated Madelon dataset.

Table 1. The example of results gathered in 2nd fold using LII measure of importance. Bold names denote truly relevant features, other denotes irrelevant features. Additionally, names with *_1* index denote contrast features. The grey colored cells contain feature set found as important and removed from the data for the next iteration.

1st iteration		2nd iteration		3rd iteration		4th iteration	
feature name	LII value	feature name	LII value	feature name	LII value	feature name	LII value
f476	2 368.05	**f242**	2 353.78	**f337**	2 359.33	f187_1	2 340.66
f339	625.03	**f129**	764.19	**f454**	1 104.63	f5	1 148.00
f154	603.13	**f319**	552.59	**f452**	752.50	f231_1	456.00
f379	455.50	**f473**	206.94	f256	92.50	f223_1	150.88
f29	171.78	**f282**	150.00	**f65**	54.06	f291_1	118.00
f443	83.54	**f434**	128.80	f112	23.13	f414	77.13
f106	82.94	f115_1	124.25	f291_1	23.13	f86_1	51.19
f85	39.38	f65	83.53	f258_1	23.00	f267_1	49.25
f494	23.40	f409_1	60.63	f365_1	14.50	f11	32.75
f49	22.75	f452	39.22	f119	12.71	f178_1	28.44
f335_1	19.50	f200_1	23.69	f385_1	12.59	f264_1	23.88
f454	13.94	f423	18.81	f164	12.06	f411_1	22.00
f337	13.03	f163	16.19	f468_1	12.00	f17	20.38
f319	11.74	f454	14.98	f157_1	10.69	f459	14.25
f74	11.69	f442	11.50	f368_1	10.14	f135	12.50
f322_1	10.38	f473_1	7.25	f307_1	9.72	f51_1	11.47
f176_1	6.06	f293	6.63	f406	8.69	f56	9.86
f33_1	5.47	f491	6.53	f20	8.13	f307	9.35
f432	4.97	f214	4.69	f375_1	5.97	f109	8.00
...	...	...	...	...	...	...	...

Table 2. The number of selections of each feature during 10 folds CV using LII and LWI + LPI importance measures. Bold names denote 20 known relevant features, other denotes irrelevant features. The grey colored cells contain feature set found as important.

LII measure						LWI+LPI measure					
relevant feature name	# of selec-tions	removing probability	**irrelevant feature name**	# of selec-tions	removing probability	**relevant feature name**	# of selec-tions	removing probability	**irrelevant feature name**	# of selec-tions	removing probability
f29	10	0.0	f5	4	0.6	**f29**	10	0.0	f53	1	0.9
f49	10	0.0	f177	1	0.9	**f49**	10	0.0	f128	1	0.9
f65	7	0.3	f359	1	0.9	**f65**	8	0.2	f325	1	0.9
f106	10	0.0	f414	1	0.9	**f106**	10	0.0	f344	1	0.9
f129	10	0.0	f85	1	0.9	**f129**	10	0.0	f264	1	0.9
f154	10	0.0	f256	1	0.9	**f154**	9	0.1	f20	1	0.9
f242	10	0.0	f112	1	0.9	**f242**	10	0.0	f142	1	0.9
f282	10	0.0	f286	2	0.8	**f282**	9	0.1	f278	1	0.9
f319	10	0.0	f216	1	0.9	**f319**	10	0.0	f291	1	0.9
f337	10	0.0	f292	2	0.8	**f337**	9	0.1	f314	1	0.9
f339	10	0.0	f343	1	0.9	**f339**	10	0.0	f148	1	0.9
f379	10	0.0	f74	1	0.9	**f379**	10	0.0	f19	1	0.9
f434	8	0.2	f148	1	0.9	**f434**	8	0.2	f425	1	0.9
f443	10	0.0	f472	1	0.9	**f443**	10	0.0	f222	1	0.9
f452	10	0.0	f203	1	0.9	**f452**	10	0.0	f406	1	0.9
f454	6	0.4	f211	1	0.9	**f454**	7	0.3	f286	1	0.9
f456	5	0.5	f304	1	0.9	**f456**	6	0.4	f18	1	0.9
f473	10	0.0	f7	1	0.9	**f473**	10	0.0	f441	1	0.9
f476	10	0.0	f440	1	0.9	**f476**	10	0.0	f145	1	0.9
f494	6	0.4	f323	1	0.9	**f494**	7	0.3	f332	1	0.9
			f245	1	0.9				f497	1	0.9
									f200	1	0.9
									f335	1	0.9

For some comparison, experiments with real-world dataset about Climate Model Simulation Crashes [16] were performed. The results (see Table 3) are very similar to earlier investigations by Lucas et al. [16] and Paja et al. [4]. Here, the same subset of variables are confirmed as relevant *{V1, V2, V9, V13, V14, V16}* using 10-fold CV. Lucas et al. found that eight parameters *{V1, V2, V4, V5, V13, V14, V16, V17}* were important, however, Paja et al. during deep analysis found that only six of them

Table 3. The results of generational feature selection using dataset about Climate Model Simulation Crashes. The grey colored cells contain feature set found as important.

Feature name	LII measure		LWI+LPI measure	
	# of selections	removing probabilty	# of selections	removing probabilty
V1	10	0.0	9	0.1
V2	10	0.0	8	0.2
V3	0	1.0	0	1.0
V4	2	0.8	1	0.9
V5	0	1.0	0	1.0
V6	3	0.7	1	0.9
V7	0	1.0	0	1.0
V8	0	1.0	0	1.0
V9	7	0.3	5	0.5
V10	1	0.9	0	1.0
V11	0	1.0	1	0.9
V12	0	1.0	0	1.0
V13	10	0.0	5	0.5
V14	10	0.0	5	0.5
V15	1	0.9	0	1.0
V16	9	0.1	5	0.5
V17	2	0.8	3	0.7
V18	0	1.0	1	0.9

{V1, V2, V9, V13, V14, V16} are truly relevant and *V4* could be treated as weekly relevant feature.

Presented initial results are promising. However, it could be identified problem with unequivocal definition of the threshold used to separate truly important feature from other non-informative. For example, in case of Madelon dataset, the *f5* feature which is random noise was indicated *4* times during 10-fold cross-validation (see Table 2), thus their probability estimator for pruning is *0.6*. If we define non-informative features with estimator ≥ 0.8 then *f5* would be treated as important in case of using *LII* importance measure. However, in the case of using *LWI + LPI* all relevant feature were identified properly.

The proposed procedure of generationally feature elimination seems to be effective and let to find weakly relevant important attributes due to sequential elimination of strongly relevant attributes. There is strong need to clearly define the not only statistical way of separation between relevant and irrelevant features.

Acknowledgment. This work was supported by the Center for Innovation and Transfer of Natural Sciences and Engineering Knowledge at the University of Rzeszów.

References

1. Kuhn, M., Johnson, K.: Applied Predictive Modeling, pp. 487–500. Springer, New York (2013)
2. Nilsson, R., Peña, J.M., Björkegren, J., Tegnér, J.: Detecting multivariate differentially expressed genes. BMC Bioinform. **8**, 150 (2007)
3. Phuong, T.M., Lin, Z., Altman, R.B.: Choosing SNPs using feature selection. In: Proceedings of the IEEE Computational Systems Bioinformatics Conference, CSB 2005, pp. 301–309 (2005)
4. Paja, W., Wrzesień, M., Niemiec, R., Rudnicki, W.R.: Application of all-relevant feature selection for the failure analysis of parameter-induced simulation crashes in climate models. Geoscientific Model Dev. **9**, 1065–1072 (2016)
5. Pancerz, K., Paja, W., Gomuła, J.: Random forest feature selection for data coming from evaluation sheets of subjects with ASDs. In: Ganzha, M., Maciaszek, L., Paprzycki, M. (eds.) Proceedings of the 2016 Federated Conference on Computer Science and Information Systems (FedCSIS), pp. 299–302 (2016)
6. Bermingham, M.L., Pong-Wong, R., Spiliopoulou, A., Hayward, C., Rudan, I., Campbell, H., Wright, A.F., Wilson, J.F., Agakov, F., Navarro, P., Haley, C.S.: Application of high-dimensional feature selection: evaluation for genomic prediction in man. Sci. Rep. **5**, 10312 (2015)
7. Kohavi, R., John, G.H.: Wrappers for feature subset selection. Artif. Intell. **97**, 273–324 (1997)
8. Rudnicki, W.R., Wrzesień, M., Paja, W.: All relevant feature selection methods and applications. In: Stańczyk, U., Jain, L. (eds.) Feature Selection for Data and Pattern Recognition. Studies in Computational Intelligence, vol. 584, pp. 11–28. Springer-Verlag, Germany (2015)
9. Zhu, Z., Ong, Y.S., Dash, M.: Wrapper-filter feature selection algorithm using a memetic framework. IEEE Trans. Syst. Man Cybern. Part B Cybern. **37**, 70–76 (2007)
10. Guyon, I., Weston, J., Barnhill, S., Vapnik, V.: Mach. Learn. **46**, 389–422 (2002)
11. Johannes, M., Brase, J.C., Frohlich, H., Gade, S., Gehrmann, M., Falth, M., Sultmann, H., Beiflbarth, T.: Integration of pathway knowledge into a reweighted recursive feature elimination approach for risk stratification of cancer patients. Bioinformatics **26**(17), 2136–2144 (2010)
12. Stoppiglia, H., Dreyfus, G., Dubois, R., Oussar, Y.: Ranking a random feature for variable and feature selection. J. Mach. Learn. Res. **3**, 1399–1414 (2003)
13. Tuv, E., Borisov, A., Torkkola, K.: Feature selection using ensemble based ranking against artificial contrasts. In: International Symposium on Neural Networks, pp. 2181–2186 (2006)
14. Paja, W.: Feature selection methods based on decision rule and tree models. In: Czarnowski, I., Caballero, A.M., Howlett, R.J., Jain, L.C. (eds.) Intelligent Decision Technologies 2016: Proceedings of the 8th KES International Conference on Intelligent Decision Technologies (KES-IDT 2016) - Part II, pp. 63–70. Springer, Cham (2016)
15. Guyon, I., Gunn, S., Ben-Hur, A., Dror, G.: Result analysis of the NIPS 2003 feature selection challenge. Adv. Neural Inf. Process. Syst. **17**, 545–552 (2013)
16. Lucas, D.D., Klein, R., Tannahill, J., Ivanova, D., Brandon, S., Domyancic, D., Zhang, Y.: Failure analysis of parameter-induced simulation crashes in climate models. Geosci. Model Dev. Discuss. **6**, 585–623 (2013)

Optimization of Exact Decision Rules Relative to Length

Beata Zielosko(✉)

Institute of Computer Science, University of Silesia in Katowice, 39, Będzińska Street, 41-200 Sosnowiec, Poland
beata.zielosko@us.edu.pl

Abstract. In the paper, an idea of modified dynamic programming algorithm is used for optimization of exact decision rules relative to length. The aims of the paper are: (i) study a length of decision rules, and (ii) study a size of a directed acyclic graph (the number of nodes and edges). The paper contains experimental results with decision tables from UCI Machine Learning Repository and comparison with results for dynamic programming algorithm.

Keywords: Decision rules · Optimization · Length · Dynamic programming

1 Introduction

Decision rules are one of the most popular form of knowledge representation. The rule length is important measure from the point of view of understanding and interpreting of knowledge encoded by rules.

In the paper, an idea of modified dynamic programming algorithm [13,14] is used for optimization of exact decision rules relative to length. The choice of this measure is connected with the Minimum Description Length principle introduced by Rissanen [9]: the best hypothesis for a given set of data is the one that leads to the largest compression of the data. So, construction and optimization of decision rules relative to length can be considered as important task for knowledge representation [8,10–12].

Unfortunately, the problem of construction of rules with minimum length is NP-hard [4,6]. The most part of approaches, with the exception of brute-force, Boolean reasoning [5,7], and dynamic programming [2,4], cannot guarantee the construction of rules with the minimum length.

This paper can be considered as possibility of finding a heuristic, modification of a dynamic programming approach that allows one to obtain the values of length of exact decision rules close to optimal ones.

There are two aims for a proposed algorithm: (i) study the length of exact rules and comparison with the length of decision rules constructed by the dynamic programming algorithm [2], (ii) study the size of a directed acyclic

I. Czarnowski et al. (eds.), *Intelligent Decision Technologies 2017*,
Smart Innovation, Systems and Technologies 72, DOI 10.1007/978-3-319-59421-7_14

graph (the number of nodes and edges) and comparison with the size of a directed acyclic graph constructed by the dynamic programming algorithm.

For a given decision table T a directed acyclic graph $\Delta^*(T)$ is constructed. Nodes of this graph are subtables of a decision table T described by descriptors (pairs *attribute* = *value*). In comparison with classical algorithm presented in [2], subtables of the graph $\Delta^*(T)$ are constructed for one attribute with the minimum number of values, and for the rest of attributes from T – the most frequent value of each attribute (value of an attribute attached to the maximum number of rows) is chosen. So, the size of the graph $\Delta^*(T)$ is smaller than the size of the graph constructed by the dynamic programming algorithm. This fact is important from the point of view of scalability. Based on the graph $\Delta^*(T)$, sets of decision rules for rows of table T, are described. Then, using procedure of optimization of the graph $\Delta^*(T)$ relative to length, for each row r of T, a decision rule with the minimum length is obtained.

The paper consists of six sections. Section 2 contains main notions. In Sect. 3, modified dynamic programming algorithm for construction of a directed acyclic graph is presented. Section 4 contains a description of a procedure of optimization relative to length. Section 5 contains experimental results with decision tables from UCI Machine Learning Repository, and Sect. 6 – conclusions.

2 Main Notions

In this section, notions corresponding to decision table and decision rules are presented.

A *decision table* is a table which elements belong to the set $\{0, 1\}$. Columns of this table are labeled with attributes $f_1, \ldots, f_n$. Rows of the table are pairwise different, and each row is labeled with a natural number (a decision), interpreted as a value of the decision attribute. Formally, decision table is defined [7] as $T = (U, A \cup \{d\})$, where $U = \{r_1, \ldots, r_x\}$ is nonempty, finite set of objects (rows), $A = \{f_1, \ldots, f_n\}$ is nonempty, finite set of attributes, $f : U \rightarrow V_f$ is a function, for any $f \in A$, V_f is the set of values of an attribute f. Elements of the set A are called conditional attributes, $d \notin A$ is a distinguished attribute, called a decision attribute.

The table T is called *degenerate* if T is empty or all rows of T are labeled with the same decision.

A table obtained from T by the removal of some rows is called a *subtable* of the table T. Let T be nonempty, $f_{i_1}, \ldots, f_{i_m} \in \{f_1, \ldots, f_n\}$ and $a_1, \ldots, a_m$ be nonnegative integers. The subtable of the table T that contains only rows that have numbers $a_1, \ldots, a_m$ at the intersection with columns $f_{i_1}, \ldots, f_{i_m}$ is denoted by $T(f_{i_1}, a_1) \ldots (f_{i_m}, a_m)$. Such nonempty subtables (including the table T) are called *separable subtables* of T.

An attribute $f_i \in \{f_1, \ldots, f_n\}$ is *not constant* on T if it has at least two different values. An attribute's value attached to the maximum number of rows in T is called *the most frequent value* of f_i, $i = 1, \ldots, n$.

By $E(T)$ is denoted a set of attributes from $\{f_1, \ldots, f_n\}$ which are not constant on T, by $E(T, r)$ is denoted a set of attributes from $E(T)$ attached to the row r.

For each attribute $f_i \in E(T)$, let us define a set $E^*(T, f_i)$ of its values. If f_i is an attribute with minimum number of values, and it has the minimum index i among such attributes (this attribute will be called the *minimum attribute for* T), then $E^*(T, f_i)$ is the set of all values of f_i on T. Otherwise, $E^*(T, f_i)$ contains only the most frequent value of f_i on T.

The expression

$$f_{i_1} = a_1 \wedge \ldots \wedge f_{i_m} = a_m \rightarrow d \tag{1}$$

is called a *decision rule over* T if $f_{i_1}, \ldots, f_{i_m} \in \{f_1, \ldots, f_n\}$, and $a_1, \ldots, a_m, d$ are nonnegative integers. It is possible that $m = 0$. In this case (1) is equal to the rule

$$\rightarrow d. \tag{2}$$

Let $r = (b_1, \ldots, b_n)$ be a row of T. The rule (1) is called *realizable for* r, if $a_1 = b_{i_1}, \ldots, a_m = b_{i_m}$. If $m = 0$ then the rule (2) is realizable for any row from T. The rule (1) is *true for* T if each row of T for which the rule (1) is realizable has the decision d attached to it. The considered rule is a *rule for* T *and* r if this rule is true for T and realizable for r.

Let τ be a decision rule over T and τ be equal to (1). The *length* of τ it is the number of descriptors (pairs *attribute* = *value*) from the left-hand side of rule, it is denoted by $l(\tau)$.

3 Modified Algorithm for Directed Acyclic Graph Construction

In this section, a modified dynamic programming algorithm which construct, for a given decision table T, a *directed acycylic graph* $\Delta^*(T)$, is presented. Based on this graph it is possible to describe the set of decision rules for T and each row r of T. Nodes of the graph are separable subtables of the table T. During each step, the algorithm processes one node and marks it with the symbol * (see Algorithm 1). A node of this graph will be called *terminal* if there are no edges leaving this node. Note that a node Θ of $\Delta^*(T)$ is terminal if and only if Θ is degenerate.

Algorithm 1, in comparison with algorithm presented in [2] does not construct complete directed acyclic graph but its part: instead of all nonconstant attributes and all their values, an attribute with the minimum number of values and all its values are used, and for the rest of attributes – the most frequent value of each attribute is chosen. The size of this graph is smaller than the size of the graph constructed by the classical dynamic programming algorithm. However, the length of the rules obtained for this graph after the procedure of optimization,

Algorithm 1. Algorithm for construction of a graph $\Delta^*(T)$

Input : Decision table T with attributes $f_1,\ldots,f_n$
Output: Graph $\Delta^*(T)$.
A graph contains a single node T which is not marked as processed;
while *all nodes of the graph are not marked as processed* **do**
  Select a node (table) Θ, which is not marked as processed;
  if Θ *is degenerate* **then**
    The node is marked as processed;
  end
  if Θ *is not degenerate* **then**
    For each $f_i \in E(\Theta)$, draw a bundle of edges from the node Θ.
    Let $E^*(\Theta, f_i) = \{b_1, \ldots, b_t\}$ (note that $t \geq 2$ only for the minimum attribute for Θ). Then draw t edges from Θ and label these edges with pairs $(f_i, b_1), \ldots, (f_i, b_t)$, respectively. These edges enter to nodes $\Theta(f_i, b_1), \ldots, \Theta(f_i, b_t)$.
    If some of nodes $\Theta(f_i, b_1), \ldots, \Theta(f_i, b_t)$ are absent in the graph then add these nodes to the graph. Each row r of Θ is labeled with the set of attributes $E_{\Delta^*(T)}(\Theta, r) \subseteq E(\Theta)$.
    Mark the node Θ as processed;
  end
end
return Graph $\Delta^*(T)$;

usually is not far, from the length of the rules obtained for classical algorithm, after the procedure of optimization.

In the next section, a procedure of optimization of the graph $\Delta^*(T)$ relative to the length, will be presented. As a result, a graph G is obtained, with the same sets of nodes and edges as in $\Delta^*(T)$. The only difference is that any row r of each nondegenerate table Θ from G is labeled with a set of attributes $E_G(\Theta, r) \subseteq E_{\Delta^*(T)}(\Theta, r)$. It is possible also that $G = \Delta^*(T)$.

Now, for each node Θ of G and for each row r of Θ a set of decision rules $Rul_G(\Theta, r)$ will be described. Let us move from terminal nodes of G to the node T.

Let Θ be a terminal node of G: Θ is a degenerate table, so each row is labeled with the same decision d. Then

$$Rul_G(\Theta, r) = \{\rightarrow d\}.$$

Let now Θ be a nonterminal node of G such that for each child Θ' of Θ and for each row r' of Θ' the set of rules $Rul_G(\Theta', r')$ is already defined. Let $r = (b_1, \ldots, b_n)$ be a row of Θ. For any $f_i \in E_G(\Theta, r)$, let us define the set of rules $Rul_G(\Theta, r, f_i)$ as follows:

$$Rul_G(\Theta, r, f_i) = \{f_i = b_i \wedge \alpha \rightarrow d : \alpha \rightarrow d \in Rul_G(\Theta(f_i, b_i), r)\}.$$

Then

$$Rul_G(\Theta, r) = \bigcup_{f_i \in E_G(\Theta, r)} Rul_G(\Theta, r, f_i).$$

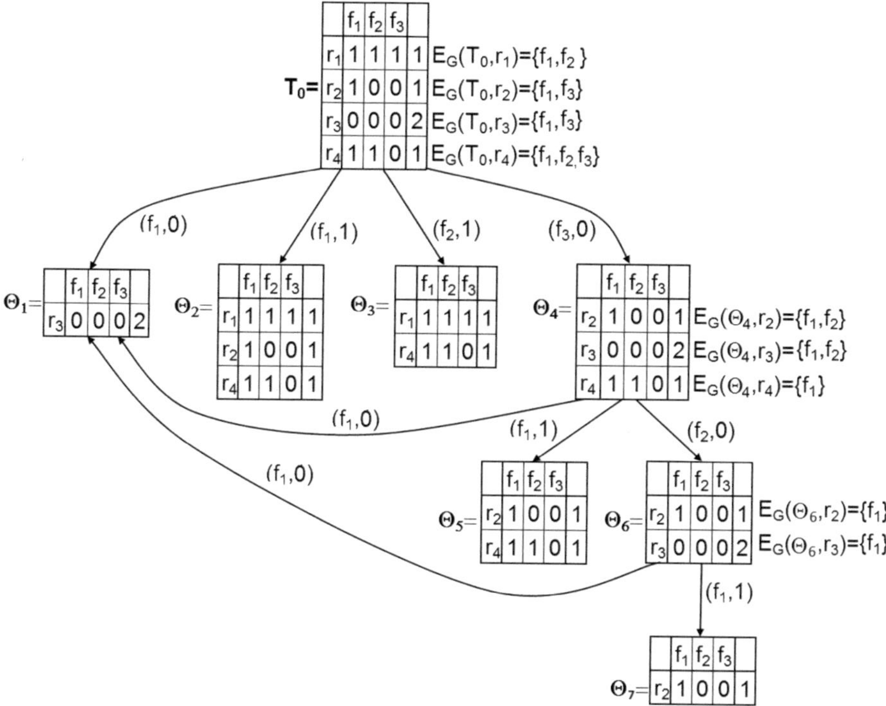

Fig. 1. A graph $\Delta^*(T_0)$

Figure 1 presents a directed acyclic graph $\Delta^*(T)$ constructed by the Algorithm 1 for simple decision table T_0. Based on the graph $\Delta^*(T_0)$ it is possible to describe, for each row r_i, $i = 1, \ldots, 4$, of the table T_0, the set $Rul_G(T_0, r_i)$ of decision rules for T_0 and r_i. Sets of rules attached to rows of decision table T_0 are the following:

$Rul_G(T_0, r_1) = \{f_1 = 1 \to 1, f_2 = 1 \to 1\}$,
$Rul_G(T_0, r_2) = \{f_1 = 1 \to 1, f_3 = 0 \wedge f_1 = 1 \to 1, f_3 = 0 \wedge f_2 = 0 \wedge f_1 = 1 \to 1\}$,
$Rul_G(T_0, r_3) = \{f_1 = 0 \to 2, f_3 = 0 \wedge f_1 = 0 \to 2, f_3 = 0 \wedge f_2 = 0 \wedge f_1 = 0 \to 2\}$,
$Rul_G(T_0, r_4) = \{f_1 = 1 \to 1, f_2 = 1 \to 1, f_3 = 0 \wedge f_1 = 1 \to 1\}$.

4 Procedure of Optimization Relative to Length

In this section, a procedure of optimization of the graph G relative to the length l is presented. For each node Θ in the graph G, this procedure assigns to each row r of Θ the set $Rul^l_G(\Theta, r)$ of decision rules with the minimum length from $Rul_G(\Theta, r)$ and the number $Opt^l_G(\Theta, r)$ – the minimum length of a decision rule from $Rul_G(\Theta, r)$.

Let us move from the terminal nodes of the graph G to the node T. Each row r of each table Θ has assigned the number $Opt^l_G(\Theta, r)$ and the set $E_G(\Theta, r)$

attached to the row r in the nonterminal node Θ of G is changed. The obtained graph is denoted by G^l.

Let Θ be a terminal node of G. Then each row r of Θ has assigned the number $Opt^l_G(\Theta, r) = 0$.

Let Θ be a nonterminal node of G and all children of Θ have already been treated. Let $r = (b_1, \ldots, b_n)$ be a row of Θ. The number

$$Opt^l_G(\Theta, r) = \min\{Opt^l_G(\Theta(f_i, b_i), r) + 1 : f_i \in E_G(\Theta, r)\}$$

is assigned to the row r in the table Θ and we set

$$E_{G^l}(\Theta, r) = \{f_i : f_i \in E_G(\Theta, r), Opt^l_G(\Theta(f_i, b_i), r) + 1 = Opt^l_G(\Theta, r)\}.$$

Figure 2 presents the directed acyclic graph G^l obtained from the graph G (see Fig. 1) by the procedure of optimization relative to the length.

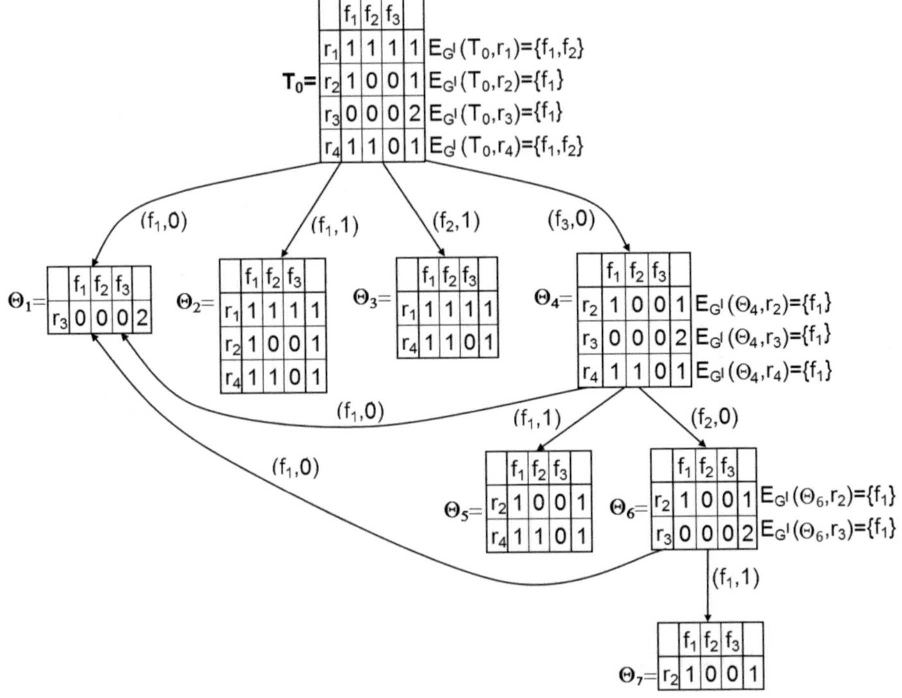

Fig. 2. Graph G^l

Based on the graph G^l it is possible to describe, for each row r_i, $i = 1, \ldots, 4$, of the table T_0, the set $Rul^l_G(T_0, r_i)$ of decision rules for T_0 and r_i with the minimum length. The value $Opt^l_G(T_0, r_i)$ is equal to the minimum length of decision rule for T_0 and r_i. This value was obtained during the procedure of

optimization of the graph G relative to the length. Sets of the shortest rules attached to rows of decision table T_0 are the following:
$Rul_G(T_0, r_1) = \{f_1 = 1 \rightarrow 1, f_2 = 1 \rightarrow 1\}$, $Opt^l_G(T_0, r_1) = 1$;
$Rul_G(T_0, r_2) = \{f_1 = 1 \rightarrow 1\}$, $Opt^l_G(T_0, r_2) = 1$;
$Rul_G(T_0, r_3) = \{f_1 = 0 \rightarrow 2\}$, $Opt^l_G(T_0, r_3) = 1$;
$Rul_G(T_0, r_4) = \{f_1 = 1 \rightarrow 1, f_2 = 1 \rightarrow 1\}$, $Opt^l_G(T_0, r_4) = 1$.

5 Experimental Results

Experiments were made on decision tables from UCI Machine Learning Repository [3] using system Dagger [1] created in King Abdullah University of Science and Technology. Some decision tables contain conditional attributes that take unique value for each row. Such attributes were removed. In some tables there were equal rows with, possibly, different decisions. In this case each group of identical rows was replaced with a single row from the group with the most common decision for this group. In some tables there were missing values. Each such value was replaced with the most common value of the corresponding attribute.

Let T be one of these decision tables. For each of the considered decision table T and for each row r of the table T, the minimum length of a decision rule for T and r was obtained. After that, for rows of T, the minimum (column *min*), average (column *avg*), and maximum (column *max*) values of rules with the minimum length were obtained. Column *Attr* contains the number of conditional attributes, column *Rows* - the number of rows. To make comparison with the optimal vales obtained by the dynamic programming algorithm, some experiments were repeated and columns *min_dp*, *avg_dp*, and *max_dp* present,

Table 1. Length of decision rules

Decision table	Rows	Attr	Modified algorithm			Dynamic programming		
			min	avg	max	min_dp	avg_dp	max_dp
Adult-stretch	16	4	1	1.75	4	1	1.25	2
Agaricus-lepiota	8124	22	1	2.04	5	1	1.18	2
Flags	193	26	2	3.91	6	1	1.93	3
Hause-votes	279	16	2	3.13	8	2	2.54	5
Lenses	10	4	1	2.13	4	1	1.40	3
Lymphography	148	18	2	3.14	7	1	1.99	4
Monks-1-train	124	6	2	2.42	3	1	2.27	3
Monks-2-test	432	6	3	4.70	6	3	4.52	6
Nursery	12960	8	1	4.39	8	1	3.12	8
Shuttle-landing	15	6	1	4.40	6	1	1.40	4
Spect-test	169	22	2	2.51	8	1	1.48	8
Teeth	23	8	2	3.35	4	1	2.26	4

Table 2. Comparison of length of decision rules

Decision table	Rows	Attr	min-diff	avg-diff	max-diff
Adult-stretch	16	4	**1.00**	1.40	2.00
Agaricus-lepiota	8124	22	**1.00**	1.73	2.50
Flags	193	26	2.00	2.02	2.00
Hause-votes	279	16	**1.00**	1.23	1.60
Lenses	10	4	**1.00**	1.52	1.33
Lymphography	148	18	2.00	1.58	1.75
Monks-1-train	124	6	2.00	1.07	**1.00**
Monks-2-test	432	6	**1.00**	1.04	**1.00**
Nursery	12960	8	**1.00**	1.41	**1.00**
Shuttle-landing	15	6	**1.00**	3.14	1.50
Spect-test	169	22	2.00	1.70	**1.00**
Teeth	23	8	2.00	1.48	**1.00**

respectively, minimum, average, and maximum values of length of rules with the minimum length.

Table 2 presents a difference regarding to minimum (column *min-diff*), average (column *avg-diff*) and maximum (column *max-diff*) length of rules. The cells of the table contain values which are equal to, respectively, minimum, average, and maximum length of rules constructed by modified dynamic programming algorithm divided by corresponding values obtained by dynamic programming algorithm. Values in bold presented in Table 2 show that the length of rules constructed by modified and classical dynamic programming algorithms, are equal. Should be also noticed, that usually, the average difference of length is small, especially for “Monks-1-train”, “Monks-2-test”, “Hause-votes” and “Nursery”.

Table 3 presents a size of the directed acyclic graph, i.e., number of nodes and number of edges in the graph constructed by the modified dynamic programming algorithm (column *Modified algorithm*) and dynamic programming algorithm (column *Dynamic programming*). The last column *Difference* presents the number of nodes/edges in the directed acyclic graph constructed by the modified algorithm divided by the number of nodes/edges in the directed acyclic graph constructed by the dynamic programming algorithm.

Presented results show that the size of the directed acyclic graph constructed by the modified dynamic programming algorithm is smaller than the size of the directed acyclic graph constructed by the dynamic programming algorithm. The biggest values of difference are in bold. In particular, for dataset “Nursery” the difference regarding to nodes is about six times, regarding to edges - about fifteen times, however the results of length are comparable (see Table 1). In the case of edges, for each dataset, the difference is at least one time.

Table 3. Size of a directed acyclic graph

Decision table	Rows	Attr	Modified algorithm		Dynamic programming		Difference	
			Nodes	Edges	Nodes	Edges	Nodes	Edges
Adult-stretch	16	4	36	37	72	108	2.00	2.92
Agaricus-lepiota	8124	22	75125	524986	149979	2145617	2.00	4.09
Flags	193	26	538999	9183204	631084	31662429	1.17	3.45
House-votes			123372	744034	176651	1981608	1.43	2.66
Lenses	10	4	59	66	105	204	1.78	3.09
Lymphography	148	18	26844	209196	40928	814815	1.52	3.89
Monks-1-train	124	6	442	807	1168	4592	2.64	5.69
Monks-2-test	432	6	638	1072	2719	10816	**4.26**	**10.09**
Nursery	12960	8	18620	27826	115200	434338	**6.19**	**15.61**
Shuttle-landing	15	6	78	257	85	513	1.09	2.00
Spect-test	169	22	443588	4271279	677002	10619692	1.53	2.49
Teeth	23	8	118	446	135	1075	1.14	2.41

6 Conclusions

In the paper, a modified dynamic programming algorithm for optimization of exact decision rules relative to the length was presented. Experimental results show that the size of the directed acyclic graph constructed by the modified algorithm is smaller than the size of the directed acyclic graph constructed by the dynamic programming algorithm, and in the case of edges, the difference is at least two times. For some datasets the results regarding to length are comparable, however the difference regrading to nodes is about six times, regarding to edges - about fifteen times.

In the future works, approximate decision rules, and comparison of length with greedy heuristics, will be studied.

References

1. Alkhalid, A., Amin, T., Chikalov, I., Hussain, S., Moshkov, M., Zielosko, B.: Dagger: a tool for analysis and optimization of decision trees and rules. In: Computational Informatics, Social Factors and New Information Technologies: Hypermedia Perspectives and Avant-Garde Experiences in the Era of Communicability Expansion, pp. 29–39. Blue Herons (2011)
2. Amin, T., Chikalov, I., Moshkov, M., Zielosko, B.: Dynamic programming approach for exact decision rule optimization. In: Skowron, A., Suraj, Z. (eds.) Rough Sets and Intelligent Systems - Professor Zdzisław Pawlak in Memoriam. Intelligent Systems Reference Library, vol. 42, pp. 211–228. Springer (2013)
3. Asuncion, A., Newman, D.J.: UCI Machine Learning Repository (2007). http://www.ics.uci.edu/~mlearn/
4. Moshkov, M., Zielosko, B.: Combinatorial Machine Learning - A Rough Set Approach. Studies in Computational Intelligence, vol. 360. Springer, Heidelberg (2011)

5. Nguyen, H.S.: Approximate boolean reasoning: foundations and applications in data mining. In: Peters, J.F., Skowron, A. (eds.) T. Rough Sets. LNCS, vol. 4100, pp. 334–506. Springer (2006)
6. Nguyen, H.S., Ślęzak, D.: Approximate reducts and association rules - correspondence and complexity results. In: Zhong, N., Skowron, A., Ohsuga, S. (eds.) RSFDGrC. LNCS, vol. 1711, pp. 137–145. Springer (1999)
7. Pawlak, Z., Skowron, A.: Rough sets and boolean reasoning. Inf. Sci. **177**(1), 41–73 (2007)
8. Przybyla-Kasperek, M., Wakulicz-Deja, A.: Application of reduction of the set of conditional attributes in the process of global decision-making. Fundam. Inform. **122**(4), 327–355 (2013)
9. Rissanen, J.: Modeling by shortest data description. Automatica **14**(5), 465–471 (1978)
10. Skowron, A.: Rough sets in KDD - plenary talk. In: Shi, Z., Faltings, B., Musen, M. (eds.) Proceedings of the 16th IFIP. World Computer Congress, pp. 1–14. Publishing House of Electronic Industry (2000)
11. Stańczyk, U.: Decision rule length as a basis for evaluation of attribute relevance. J. Intell. Fuzzy Syst. **24**(3), 429–445 (2013)
12. Stefanowski, J., Vanderpooten, D.: Induction of decision rules in classification and discovery-oriented perspectives. Int. J. Intell. Syst. **16**(1), 13–27 (2001)
13. Zielosko, B.: Coverage of decision rules. In: Decision Support Systems, pp. 183–192. University of Silesia (2013)
14. Zielosko, B.: Optimization of approximate decision rules relative to coverage. In: Kozielski, S., Mrózek, D., Kasprowski, P., Małysiak-Mrózek, B., Kostrzewa, D. (eds.) BDAS 2014. CCIS, vol. 424, pp. 170–179. Springer (2014)

Evaluating Importance for Numbers of Bins in Discretised Learning and Test Sets

Urszula Stańczyk(✉)

Institute of Informatics, Silesian University of Technology,
Akademicka 16, 44-100 Gliwice, Poland
urszula.stanczyk@polsl.pl

Abstract. The paper presents research on the influence of the numbers of bins, found for attributes in supervised discretisation for input sets, on classifiers performance. Firstly, the variables were divided into categories defined by numbers of bins, and for these categories several decision systems were tested. Secondly, for features with single bins, unsupervised discretisation was executed and the resulting performance studied. The experiments show usefulness of characterisation of variables by numbers of bins, and cases of improvement of solutions by combining supervised with unsupervised discretisation.

Keywords: Supervised discretisation · Unsupervised discretisation · Attribute · Bin · Classification

1 Introduction

In supervised discretisation continuous values of features are transformed into nominal by grouping them into ranges called bins, while observing how such loss of information influences considered classes. It can be done by evaluation of entropy as happens in Fayyad and Irani's method [9], where numbers of bins are found by following Minimal Description Length (MDL) principle, with static treatment of features, that is for each independently on others [11]. Numbers of established bins can vary, what is more, for some variables only a single bin is found, suggesting that these features are unnecessary for observing recognised classes. On the other hand, for other variables higher numbers of bins are required, indicating that their values have to be studied closer.

Information about importance of particular features for a task is useful for any kind of processing, and it can be established by variety of approaches, measures, procedures [15,16]. In the research presented in this paper this importance was evaluated in relation to the numbers of bins found for them in supervised discretisation, and variables were grouped into corresponding categories, for which performance of several decision systems was tested.

When discretisation is executed independently for sets, characteristics of attributes are limited to the considered set. Thus for the same variables in different sets varying numbers of bins can be found. This fact is particularly

I. Czarnowski et al. (eds.), *Intelligent Decision Technologies 2017*,
Smart Innovation, Systems and Technologies 72, DOI 10.1007/978-3-319-59421-7_15

problematic in case of evaluating systems performance by test sets [3]. In the research special attention was given to 1-bin variables found for learning and test sets, firstly by processing while preserving all characteristics of sets as established by supervised procedures, and then by combining supervised with unsupervised discretisation, applied to all attributes with single bins in both types of sets.

The tests were executed for two tasks of binary authorship attribution [13] with balanced classes and the results indicate that associating the importance of features with numbers of bins found in supervised discretisation shows some merit, furthermore, in some cases combining supervised with unsupervised discretisation approaches can improve classification accuracy.

The paper is organised as follows. Section 2 describes background information on discretisation approaches, Sect. 3 provides some details of the experimental setup, and tests results are commented in Sect. 4. Section 5 concludes the paper.

2 Background

In discretisation, by observing characteristics of features values, the continuous input space is transformed into granular, with granules called bins, within which specific values are indiscernible. Changing attributes form the continuous into the discrete space always means loss of information, yet it can bring advantages of more general definitions of described concepts, ability to employ processing techniques working only for nominal variables, or even improved performance.

Dynamic discretisation methods study interdependencies among attributes, while static treat each variable independently. Numbers of bins are established either with disregarding class information in unsupervised approaches, or with taking this information into account in supervised discretisation [11].

Equal width and equal frequency binning are two popular unsupervised discretisation approaches. The former divides the whole range of values for each attribute into a specific, given as an input parameter, number of bins with equal width [8]. In the latter approach occurring values are grouped into some number of bins with the previously specified numbers of occurrences in them.

Fayyad and Irani's method [9] belongs with supervised procedures, as it studies class entropy to evaluate candidate partitions of the input ranges of values. $Ent(a,T;S)$ denotes the entropy of the set S of N instances for a attribute, with T a partition boundary for feature a,

$$Ent(a,T;S) = \frac{|S_1|}{|S|} Ent(S_1) + \frac{|S_2|}{|S|} Ent(S_2). \tag{1}$$

In the recursive processing as the stopping criterion for establishing final numbers of bins for attributes, Minimum Description Length (MDL) principle is used, where k is the number of classes in the set S_i,

$$\begin{gathered} Gain(a,T,S) = Ent(S) - Ent(a,T;S) \leq \tfrac{log_2(N-1)}{N} + \tfrac{\Delta(a,T;S)}{N}, \\ \Delta(a,T;S) = log_2(3^k - 2) - [k \cdot Ent(S) - k_1 \cdot Ent(S_1) - k_2 \cdot Ent(S_2)]. \end{gathered} \tag{2}$$

With such attitude, the numbers of bins found for variables can vary, in particular it is possible that for some features all occurring values are transformed into a single bin. Such situation can be interpreted as attribute irrelevancy from class information point of view, and a strong suggestion that these variables can be in fact disregarded without degrading class entropy.

Since a single bin reflects the information that no distinction of attribute values is required to preserve class entropy, it can be reasoned that the higher the number of bins established for some variable, the more detailed study of its values is needed for classification, thus such feature should be considered as more important than these with fewer bins found. As a result, considered attributes can be grouped into categories corresponding to numbers of bins they require.

In data mining, in case of evaluating systems performance by popular cross-validation there are no additional problems caused by using discretised input datasets. For evaluation by testing sets the question arises how these sets should be discretised [3]. If they are discretised independently, to find characteristics for attributes, discretised procedures can depend only on information from the considered sets, which means that established numbers of bins for the same variables in separate sets can vary, possibly causing worsened performance.

3 Experimental Setup

The first step within experiments was the choice of classification problem and construction of input datasets, then supervised discretisation for these sets was performed, with special attention given to numbers of bins found for all attributes. Next, several tests were executed to observe the influence of numbers of bins on classifiers performance, as described in detail in the following sections.

3.1 Input Datasets

Authorship attribution is considered as the most important of stylometric tasks [1]. Its fundamental notion of describing writing styles by employed textual markers [5] allows to treat such problems as recognition of classes corresponding to considered authors, as long as a set of characteristic features is decided upon. Typically, these features reflect lexical or syntactic elements of style, such as frequencies of occurrences for words, or punctuation marks. The particular choice depends to some extent on text genre, as for each different rules apply [6].

In the presented research the base texts were literary works. As female authors tend to exhibit different linguistic characteristics than male [14], for comparison there was selected one pair of female and one pair of male writers. Their longer works were divided into several smaller parts with comparable sizes. For these parts there were calculated frequencies of usage for 17 function words (selected from the list of the most popular words in English language), and 8 punctuation marks, resulting in the set of 25 stylistic characteristic features.

With construction of learning samples as described, evaluation of a decision system performance by cross-validation is unreliable, as it tends to return significantly higher recognition when compared to testing by independent sets [2].

That is why this latter approach was chosen in the presented research. For testing sets separate literary works were selected and processed.

3.2 Characterisation of Attributes by Supervised Discretisation

For both female and male writer datasets, including training and testing samples in separate sets, supervised discretisation with Fayyad and Irani's method was applied, returning characteristics for all attributes as listed in Table 1. For both datasests the same minimal and maximal numbers of bins needed were found, but the particular cardinalities of variable categories vary, not only between datasets but between training and testing sets as well.

Table 1. Categories of attributes established by supervised discretisation for separate learning (labelled L) and test (labelled T) sets

Nr of bins	Female writer dataset			
	Label	Attributes	Label	Attributes
1	L1	and in with of what from if . !	T1	but in with at of that what from if (
2	L2	but not at this as that by for to , ? (–	T2	and on as by for to . , ? ! ; –
3	L3	on ; :	T3	not this :
Nr of bins	Male writer dataset			
	Label	Attributes	Label	Attributes
1	L1	on of this . , : (	T1	in with at this as for , ? ! :
2	L2	but not in with as to if ? ! ; –	T2	but and not on of that what from by to . ; (–
3	L3	and at that what from by for	T3	if

Independent discretisation for training and testing samples has the advantage of being unbiased by characteristics of other sets, but at the cost of possible worsening of performance caused by varying bin numbers. When considerations involve 2- or 3-bin variables, such cost can be found acceptable, as such attributes are still present in the system and take active part in classification. When 1-bin features are not identical, as is the case shown in Table 1, the situation becomes more complex, and it is addressed in the next section.

3.3 1-Bin Variables in Learning and Test Sets

The characteristics of attributes returned from supervised discretisation were analysed between input sets, with overlapping parts for all categories shown in

Table 2. It was immediately apparent that for female writer dataset training and testing samples shared similar characteristics, because categories had significant intersection of attributes. On the other hand, for male writer dataset these characteristics were not so close, as there were fewer variables appearing in the same categories. Which was even more problematic, there were some attributes with 3 bins for the learning set (at, for) that for the test set had only 1 bin. With such characteristics of input sets higher prediction for female writers should be expected than for male authors.

Table 2. Intersection of attributes categories for learning and test sets

Number of bins found	Attributes			
	Female writer dataset		Male writer dataset	
1	in with of what from if	(6)	this , :	(3)
2	as by for to , ? –	(7)	but not to ; –	(5)
3	:	(1)		(0)

To study the influence of these 1-bin variables for both types of sets with respect to systems performance, in experiments executed the input sets were used in 3 versions, LA and TA respectively meant fully automated supervised discretisation for learning and test sets. By 1S2U there were denoted sets with additionally employed unsupervised discretisation of 1-bin variables, using equal width binning, with defining for them 2 bins, and 1S3U with binning into 3 ranges. Such processing was implemented for both training and testing samples.

3.4 Employed Classification Systems

To observe the influence of numbers of bins for variables on systems performance, tests involving several predictors were executed and their results analysed.

Firstly, a group of classifiers capable of operating on both continuous and discrete data was tested (all implemented in WEKA [10]), and they included: Naive Bayes (NB), Multi-layer Perceptron (MLP), Radial Basis Function Network (RBF), k-Nearest Neighbour (kNN), PART (a variant of C4.5), and Random Forest (RF). They were chosen because of their popularity in applications, in particular in comparisons of obtained results. All predictors were used with default settings, without any fine-tuning specific for each type. The performance for these classifiers was noted for both continuous and discretised data.

Secondly, for Classical Rough Set Approach (CRSA) [12], that requires nominal input attributes, RSES system [4] was used. It offers induction of rules with four algorithms: Lem2, covering, genetic and exhaustive, from which the first two were rejected as giving rather poor results due to too few rules generated.

4 Test Results

The first stage of experiments was dedicated to observing performance of classifiers capable to work in continuous space, with following categories of attributes obtained from discretisation. In the second step the same set of classifiers was employed in discrete domain. And in the third step CRSA algorithms, requiring discrete attributes, were induced and their effectiveness evaluated.

4.1 Tests for Continuous Datasets

While still operating on continuous values of attributes, for all categories of input characteristic features, selected decision systems were employed, and their performance, shown in Table 3, provided a reference point for further comparisons. Category L denotes the complete set of variables, then respectively L1, L2, and L3 variables with 1, 2, and 3 bins. L23 is a composition of L2 and L3 categories, or a category L with excluding from it elements of L1.

Table 3. Classifiers performance for categories of attributes in continuous domain [in %]

Classifier	Categories of attributes									
	Female writer dataset					Male writer dataset				
	L	L2	L3	L1	L23	L	L2	L3	L1	L23
NB	92.22	92.22	88.89	70.00	92.22	81.67	76.67	86.67	55.00	90.00
MLP	90.00	88.89	90.00	68.89	91.11	85.00	80.00	66.67	58.33	88.33
RBF	93.33	91.11	91.11	65.56	92.22	76.67	78.33	91.67	53.33	91.67
kNN	92.22	91.11	84.44	53.33	94.44	85.00	76.67	78.33	58.33	80.00
PART	87.78	86.67	82.22	57.78	87.78	73.33	51.67	80.00	46.67	71.33
RF	88.89	88.89	86.67	65.56	91.11	80.00	71.67	80.00	60.00	80.00

Some of observations confirm previous expectations, such as the generally higher difficulty of recognition for male writers. Also the classification for L1 category was always the worst. On the other hand, differences between results for L and L23 categories were inconsistent, for some systems the performance remained the same, for others there was increased recognition, while still for others decreased. Comparison of categories L2 and L3 did not always show enhanced prediction ratio for increased number of bins, in particular for female writers. But in this case L3 category included only three attributes and for such small number of variables accuracy should be considered as reasonably high. For male authors, for all classifiers but one, prediction for L3 was higher than for L2.

4.2 Tests in Discrete Space for Selected Classifiers

The next step of tests involved using discrete input datasets for selected predictors, and the results are displayed in Table 4. Classifiers performance is given with relation to category of inputs considered, but also for versions of test sets used in evaluation. Categories L1S2U and L1S3U denote the complete set of input variables, in which for these attributes that belonged to 1-bin category there was executed unsupervised discretisation with equal width binning to respectively 2 or 3 bins. The same procedure was adapted to test sets, with TA fully automatically obtained from supervised discretisation, and T1S2U and T1S3U with equal width binning transforming 1-bin attributes into 2- and 3-bin features. For female authors for L3 category only one testing set was always used as none of attributes in this category for the test set belonged to T1 category, therefore no additional unsupervised discretisation was executed. As L1 category in discrete form would mean a single value for all occurrences and all attributes included, this category was excluded from tests concerning discrete sets.

Analysis of effectiveness for decision systems studied for the same sets in their continuous and the discrete forms brings conclusions that changes in performance were not uniform. Taking into account all categories of variables and testing with base TA sets, for female authors as the best predictor for discrete data Multi-layer Perceptron should be nominated, as even for the very small L3 category it gave comparable recognition as for other categories, unlike other systems. The worst performance was indicated by Radial Basis Function network, for which degraded but close to acceptable prediction accuracy was only for L3 category. Putting aside sets created in combined supervised-unsupervised discretisation, the overall highest recognition was 96.67% of either Naive Bayes or MLP, achieved for L2, L3, or L23 categories respectively.

For male authors for categories other than L3, the best performing classifier was Naive Bayes, the lowest position belonged to PART algorithm, and one before last was RBF, for both of which recognition fell even below 50%. For this dataset results for L3 category were significantly higher than for L2, in some cases (RBF, kNN, PART) outperforming all other categories.

Study of results for test sets modified by additional unsupervised discretisation for learning sets without such modifications for female writers brings conclusion that they never offered enhanced recognition, only the same or worse. This observation did not hold true for male authors, where some improvements could be found several times in L3 category, but also in other columns, for kNN for LA and L23 categories, or for Naive Bayes in L2.

Modifications of learning data by unsupervised binning for 1-bin variables either resulted in the same or degraded recognition with only few exceptions, but then they were quite significant for Random Forest working for female writers, where there was detected the highest recognition of all executed tests for these decision systems, both in continuous and discrete domain, at the level of 98.89%.

Table 4. Classifiers performance for different categories of attributes in discrete domain [in %]

Female writer dataset							
Classifier	Test set	Categories of attributes					
		LA	L2	L3	L23	L1S2U	L1S3U
NB	TA	95.56	96.67	77.78	95.56	95.56	95.56
	T1S2U	95.56	95.56	—	95.56	95.56	95.56
	T1S3U	88.89	87.78	—	88.89	86.67	88.89
MLP	TA	95.56	94.44	96.67	96.67	96.67	95.67
	T1S2U	95.56	94.44	—	95.56	95.56	95.56
	T1S3U	95.56	93.33	—	95.56	95.56	97.78
RBF	TA	65.56	65.56	77.78	65.56	66.67	66.67
	T1S2U	60.00	60.00	—	60.00	60.00	60.00
	T1S3U	53.33	53.33	—	53.33	53.33	53.33
kNN	TA	94.44	92.22	77.78	94.44	95.56	94.44
	T1S2U	93.33	86.67	—	93.33	90.00	93.33
	T1S3U	93.33	91.11	—	93.33	91.11	94.44
PART	TA	85.56	93.33	67.78	85.56	85.56	85.56
	T1S2U	85.56	93.33	—	85.56	85.56	85.56
	T1S3U	85.56	93.33	—	85.56	85.56	85.56
RF	TA	93.33	66.67	77.78	82.22	98.89	90.00
	T1S2U	92.22	66.67	—	75.56	93.33	91.11
	T1S3U	91.11	66.67	—	71.11	88.89	92.22
Male writer dataset							
Classifier	Test set	Categories of attributes					
		LA	L2	L3	L23	L1S2U	L1S3U
NB	TA	86.67	63.33	81.67	86.67	86.67	85.00
	T1S2U	83.33	66.67	90.00	83.33	85.00	81.67
	T1S3U	83.33	78.33	91.67	83.33	85.00	85.00
MLP	TA	86.67	60.00	65.00	86.67	81.67	80.00
	T1S2U	78.33	61.67	73.33	78.33	78.33	80.00
	T1S3U	80.00	65.00	71.67	80.00	75.00	80.00
RBF	TA	43.33	38.33	95.00	43.33	56.67	58.33
	T1S2U	45.00	41.67	83.33	45.00	51.67	51.67
	T1S3U	56.67	51.67	70.00	56.67	53.33	51.67
kNN	TA	75.00	56.67	80.00	75.00	70.00	56.67
	T1S2U	80.00	56.67	85.00	80.00	68.33	73.33
	T1S3U	80.00	58.33	90.00	80.00	81.67	75.00
PART	TA	33.33	38.33	90.00	33.33	63.33	33.33
	T1S2U	33.33	36.67	90.00	33.33	63.33	33.33
	T1S3U	33.33	53.33	70.00	33.33	48.33	33.33
RF	TA	81.67	46.67	73.33	83.33	45.00	60.00
	T1S2U	73.33	48.33	83.33	83.33	55.00	68.33
	T1S3U	51.67	46.67	66.67	73.33	65.00	75.00

4.3 Tests for CRSA Classifiers

Classical rough set approach (CRSA) relies on indiscernibility relation, thus to be able to employ it for data that is continuous, either some modification of this relation is needed [7] or discretisation procedures are considered as part of pre-processing for input sets. Decision rules can be induced in variety of approaches, resulting in rule sets with significantly different cardinalities and performance.

For all studied categories of discretised input variables two algorithms were induced, genetic and exhaustive, with results shown in Table 5. For rule classifiers many parameters can be listed to describe performance, and this performance can be given without any additional processing of generated rules or with standard filtering with respect to support required of rules to achieve the highest classification accuracy for fewest rules. This latter approach to presentation was employed in the paper, that is only the highest obtained percentage of correctly classified samples was given.

Table 5. Performance of CRSA classifier for different categories of attributes [in %]

	Induction algorithm	Test set	Categories of attributes					
			LA	L2	L3	L23	L1S2U	L1S3U
Female writers	Genetic	TA	96.67	81.11	38.89	98.89	96.67	97.78
		T1S2U	95.56	74.46	—	95.56	95.56	95.56
		T1S3U	94.44	74.44	—	94.44	94.44	95.56
	Exhaustive	TA	98.89	87.78	77.78	98.89	98.89	98.89
		T1S2U	94.44	76.64	—	94.44	94.44	95.56
		T1S3U	94.44	75.56	—	94.44	93.33	94.44
Male writers	Genetic	TA	85.00	76.67	80.00	83.33	83.33	86.67
		T1S2U	81.67	65.00	88.33	81.67	85.00	81.67
		T1S3U	75.00	60.00	93.33	73.33	76.67	71.67
	Exhaustive	TA	80.00	76.67	80.00	80.00	81.67	80.00
		T1S2U	78.33	63.33	88.33	78.33	78.33	75.00
		T1S3U	71.67	60.00	93.33	71.67	71.67	66.67

For female writer dataset better results were obtained for exhaustive algorithm than for genetic, while for male authors the situation was reversed, but the observed differences were minor. For female L3 category the recognition was rather low due to the fact that only 3 attributes worked for it. On the other hand, male L3 category shows significant advantage over L2, and for test sets modified with unsupervised discretisation, results for this category surpassed all others. Also modifications of the learning sets caused some cases of increased classification accuracy.

5 Conclusions

The paper described research works focused on numbers of bins found for characteristic features within procedures of supervised discretisation. The attributes were grouped into categories reflecting bin numbers and for such sets and subsets of variables performance of several decision systems was studied, firstly for continuous domain, and then for discretised training and test sets, which were used for evaluation of systems. Results from experiments show that when both training and test sets share significant similarities in characterisation of features by discretisation, the observed performance is higher. Also, variables with more bins provide better base for classifiers than those with fewer bins. Experiments on combining supervised with unsupervised discretisation for variables with single bins in some cases brought improvement, especially when there were attributes with higher numbers of bins in learning sets that were considered as single bin variables for test sets. Future research will involve analysis of the influence on the number of bins found in supervised discretisation on construction of feature rankings, and weighting of decision rules by quality measures based on numbers of bins for attributes used as conditions.

Acknowledgments. In the research there was used RSES system, developed at the Institute of Mathematics, Warsaw University (http://logic.mimuw.edu.pl/~rses/), and WEKA workbench [10]. The research was performed at the Silesian University of Technology, Gliwice, within the project BK/RAu2/2017.

References

1. Argamon, S., Burns, K., Dubnov, S. (eds.): The Structure of Style: Algorithmic Approaches to Understanding Manner and Meaning. Springer, Berlin (2010)
2. Baron, G.: Comparison of cross-validation and test sets approaches to evaluation of classifiers in authorship attribution domain. In: Czachórski, T., Gelenbe, E., Grochla, K., Lent, R. (eds.) Proceedings of the 31st International Symposium on Computer and Information Sciences, Communications in Computer and Information Science, vol. 659, pp. 81–89. Springer, Cracow (2016)
3. Baron, G.: On approaches to discretization of datasets used for evaluation of decision systems. In: Czarnowski, I., Caballero, A., Howlett, R., Jain, L. (eds.) Intelligent Decision Technologies 2016, Smart Innovation, Systems and Technologies, vol. 56, pp. 149–159. Springer (2016)
4. Bazan, J., Szczuka, M.: The rough set exploration system. In: Peters, J.F., Skowron, A. (eds.) Transactions on Rough Sets III. LNCS, vol. 3400, pp. 37–56. Springer, Heidelberg (2005)
5. Burrows, J.: Textual analysis. In: Schreibman, S., Siemens, R., Unsworth, J. (eds.) A Companion to Digital Humanities. Blackwell, Oxford (2004)
6. Craig, H.: Stylistic analysis and authorship studies. In: Schreibman, S., Siemens, R., Unsworth, J. (eds.) A Companion to Digital Humanities. Blackwell, Oxford (2004)
7. Cyran, K., Stańczyk, U.: Indiscernibility relation for continuous attributes: application in image recognition. In: Kryszkiewicz, M., Peters, J., Rybiński, H., Skowron,

A. (eds.) Rough Sets and Emerging Intelligent Systems Pardigms. LNAI, vol. 4585, pp. 726–735. Springer, Berlin (2007)
8. Dougherty, J., Kohavi, R., Sahami, M.: Supervised and unsupervised discretization of continuous features. In: Machine Learning Proceedings 1995: Proceedings of the 12th International Conference on Machine Learning, pp. 194–202. Elsevier (1995)
9. Fayyad, U., Irani, K.: Multi-interval discretization of continuous valued attributes for classification learning. In: Proceedings of the 13th International Joint Conference on Artificial Intelligence, vol. 2, pp. 1022–1027. Morgan Kaufmann Publishers (1993)
10. Hall, M., Frank, E., Holmes, G., Pfahringer, B., Reutemann, P., Witten, I.: The WEKA data mining software: an update. SIGKDD Explor. **11**(1), 10–18 (2009)
11. Kotsiantis, S., Kanellopoulos, D.: Discretization techniques: a recent survey. GESTS Int. Trans. Comput. Sci. Eng. **32**(1), 47–58 (2006)
12. Pawlak, Z.: Rough sets and intelligent data analysis. Inf. Sci. **147**, 1–12 (2002)
13. Peng, R., Hengartner, H.: Quantitative analysis of literary styles. Am. Stat. **56**(3), 15–38 (2002)
14. Stańczyk, U.: Recognition of author gender for literary texts. In: Czachórski, T., Kozielski, S., Stańczyk, U. (eds.) Man-Machine Interactions 2, Advances in Intelligent and Soft Computing, vol. 103, pp. 229–238. Springer, Berlin (2011)
15. Stańczyk, U.: Weighting of attributes in an embedded rough approach. In: Gruca, A., Czachórski, T., Kozielski, S. (eds.) Man-Machine Interactions 3, Advances in Intelligent and Soft Computing, vol. 242, pp. 475–483. Springer, Berlin (2013)
16. Stańczyk, U.: Weighting of features by sequential selection. In: Stańczyk, U., Jain, L. (eds.) Feature Selection for Data and Pattern Recognition. SCI, vol. 584, pp. 71–90. Springer, Berlin (2015)

Decision Support Systems

Decision Making Beyond Pattern Recognition: Classification or Rejection

Wladyslaw Homenda[1,2(✉)], Agnieszka Jastrzebska[1], and Piotr Waszkiewicz[1]

[1] Faculty of Mathematics and Information Science, Warsaw University of Technology, ul. Koszykowa 75, 00-662 Warsaw, Poland
{homenda,jastrzebskaa}@mini.pw.edu.pl, waszkiewiczp@student.mini.pw.edu.pl

[2] Faculty of Economics and Informatics in Vilnius, University of Bialystok, Kalvariju G. 135, LT-08221 Vilnius, Lithuania

Abstract. In the paper we discuss the issue of contaminated data sets that contain improper patterns apart from proper ones. To distinguish between those two kinds of patterns we use terms: native (proper) patterns and foreign (garbage) patterns. To deal with contaminated datasets we propose to build decision mechanism based on a collection of classification and regression models that together classify native patterns and reject foreign. The developed approach is empirically evaluated based on a set of handwritten digits as native patterns and handwritten letters playing role of foreign patterns.

Keywords: Pattern recognition · Pattern rejection · Foreign pattern · Native pattern

1 Introduction

Pattern recognition is a classical machine learning problem. Its aim is to form a model (called classifier) that, after initial training, can assign correct class label to a new pattern. It is important to have in mind that patterns in their original form are usually some sort of a signal, for instance images or voice recordings. Due to the fact, that original patterns are often collected automatically with some signal-acquiring device, patterns that should not be accounted to any of proper classes may be recorded. Such situation may happen, when a device that is used to acquire data was automatically reset due to a power outage and a poor default calibration hinders signal segmentation process. The main problem with such erroneous patterns, is that their characteristics cannot be acquired and therefore cannot be included in the model training process. Rejecting foreign patterns is a very natural variant of the decision-making problem beyond standard pattern recognition.

The motivation for our study is to provide algorithmic approaches used for distinguishing proper patterns (called *native patterns*) from garbage and unwanted ones (called *foreign patterns*) by using only classifiers for both rejection and classification. The design assumption is to provide decision methods

I. Czarnowski et al. (eds.), *Intelligent Decision Technologies 2017*,
Smart Innovation, Systems and Technologies 72, DOI 10.1007/978-3-319-59421-7_16

based on native patterns only so that the approach could be truly versatile and could be easily adapted to any pattern recognition problem in an uncertain environment, where foreign patterns may appear. The study focuses on a multi-class pattern recognition problem, for which we employ specifically constructed classification models that provide rejecting/classifying decision mechanism as addition to standard classification process. It should be emphasised that novelty of the contribution presented in this paper is in the application of well-known methods to a new area of study.

The remainder of this paper is organized as follows. Section 2 presents the background knowledge on foreign patterns detection present in the literature. Section 3.1 presents the backbone algorithms, known in the literature, that were used to construct our models. Section 3.6 presents the proposed approach. Section 4 discusses a series of experiments. Section 5 concludes the paper and highlights future research directions.

2 Literature Review

The problem of contaminated data sets has been reappearing in research on pattern recognition in various contexts. First studies, which we may account as tightly related to foreign patterns rejection, were dealing with outliers detection. Outliers are proper native patterns, whose characteristics significantly differ from the majority of data. Outliers, due to their atypical character, may cause problems at a model training stage. Therefore, we are often advised to apply a variety of methods to detect them and, if needed, to modify or remove from the set of native patterns before further processing. Literature provides us with a selection of statistical tests, for instance discussed in [5], which act as decision rules for outlier detection. Apart from statistical tests, we find more convoluted approaches. In particular, there is a wide variety of methods related to the notion of distance. Those methods evaluate either proximity of data points or their density in order to detect outliers. Popular example of such approach was presented in [2]. We shall stress that foreign patterns are not outliers. When it comes to outliers, we can actually determine their proper class belongingness, even though it is difficult. In contrast, foreign patterns do not belong to any class and should be removed from the data.

An area of studies, which is even more similar to foreign patterns rejection, is the so called novelty detection. Novelty elements are patterns that belong to extremely infrequent native classes. A typical domain, in which we recognize the importance of novelty elements, is text processing. Even large corpora of texts often contain meaningful, but rare key phrases. Studies on novelty detection typically utilize either a probabilistic approach or are based on the notion of distance, [8]. Among successful probabilistic approaches we can find kernel density estimators, as for example reported in [6], and mixture models (for instance discussed in [10]). The goal of these methods is to perform probability density estimation of the data. In contrast, distance-based methods for novelty detection rely on the assumption that novel elements are located far from majority of data. In this line of study, we find methods resolving to various sophisticated distance measures, for instance [4] or even methods based on well-known clustering algorithms, as for example in [11].

Foreign elements detection, alike novelty detection task, has to assume that native patterns, alike majority class, is the only available information. Therefore, model construction has to rely on native patterns only. In this paper we present a method for foreign patterns rejection based on classifying models. The approach that we work on is fit to solve multi-class classification problems (when more than two native classes are present). The proposed technique uses a collection of specifically trained binary classifiers, which set together to form a specific structure, provide capability not only to classify native patterns, but also to reject foreign patterns.

3 Preliminaries

The issue of contaminated data sets frequently emerges from a substantial automation of data acquisition and processing, that results in poor data quality. The research that we present in this paper has been motivated by the need for algorithmically-aided methods for data sets purging: accepting proper patterns and rejecting foreign ones. In the study addressed in this paper we concentrated on enhancing well-known classifiers in order to perform a rejection task. We focused on regression models, but we also applied other popular classification algorithms, such as Support Vector Machines (SVM), random forest and k-Nearest Neighbours (kNN). In what follows we discuss applied algorithms.

3.1 The Task of Classification with Rejection

The task of classification aims at categorising unknown patterns to their appropriate groups. The procedure is based on quantifiable characteristics obtained from the source signal. Those characteristics, i.e. features, are gathered in a feature vector (of independent variables) and each pattern is described with one vector of features' values. We expect that patterns accounted to the same category are in a relationship with one another. In other words, patterns accounted to the same category are expected to be in some sense similar. There are many mathematical models that can be used as classifiers, such as SVM, random forest, kNN, regression models, or Neural Networks. Their main disadvantage lies in their need to be trained prior to usage, which makes them unable to recognize elements from a new class, not present during the training process.

3.2 Support Vector Machines

Support Vector Machines (SVM) are a collection of supervised learning methods used for classification, regression and outliers detection. The SVM algorithm relies on a construction of hyperplane with a maximal margin that separates patterns of two classes, [3]. SVMs are effective in high-dimensional spaces, memory efficient, and quite versatile with many kernel functions that can be specified for the decision function. Although in some cases, where the number of features is much greater than the number of samples, this method can give poor results, and is not cost-efficient when calculating probability estimates.

3.3 Random Forest

Random forest is a popular ensemble method. The main principle behind ensemble methods, in general, is that a group of "weak learners" can come together to form a "strong learner". In the random forest algorithm [1] the weak learners are decision trees, which are used to predict class labels. For a feature vector representing one pattern a decision tree calculates its class label by dividing value space into two or more subspaces. After a relatively large number of trees is generated, they vote for the most popular class. Random forests join few important benefits: (a) they are relatively prone to the influence of outliers, (b) they have an embedded ability of feature selection, (c) they are prone to missing values, and (d) they are prone to overfitting.

3.4 K-Nearest Neighbors

The k-Nearest Neighbours algorithm, denoted as kNN, is an example of a "lazy classifier", where the entire training dataset is the model. There is no typical model building phase, hence the name. Class membership is determined based on class labels encountered in k closest observations in the training dataset. In a typical application, the only choice that the model designer has to make is selection of k and distance metrics. Both are often determined experimentally with a help of supervised learning procedures.

3.5 Regression

Regression analysis is a statistical method for describing relationships between variables. It is applied in order to predict future values of a given dependent variable based on known values of other variables (independent variables). Regression model construction resolves to formation of a function that describes how expected value of the dependent variable is related with one or more independent variables. The function can literally be a mathematical formula, but it can also be an algorithm taking us from space of independent variables values to space of depended variables values. Examples of algorithms building a regression model are: regression tree, neural network, etc.

Typically, we deal with parametric regression models, where the shape of the regression function is known beforehand and our goal is to determine its components (parameters). When we apply the same set of parameters for all input variables, no matter their particular values, we have the case of global parametric regression. A general formula for parametric regression is as follows:

$$Y = f(X, \beta) + \epsilon \tag{1}$$

where Y is the dependent variable, X is the vector of independent variables, β is the vector of regression coefficients, f is regression function with values in real numbers, ϵ is random error. The estimation target is a function of the independent variables f. The goal is to find such a function that minimizes loss.

There are many different regression models: linear regression, polynomial regression, logistic regression, Bayesian Ridge regression, etc. In this paper we will focus mainly on two regression variants: logistic and polynomial regression (please note that linear regression is in fact polynomial regression using polynomial of degree 1).

Linear regression is the simplest regression model. It describes the relationship between scalar dependent variable and one or more explanatory variables. For example, if we have three independent variables, multiple regression model looks as follows:

$$Y = \beta_0 + \beta_1 x_1 + \beta_2 x_2 + \beta_3 x_3 \tag{2}$$

Polynomial regression can be viewed as a model that uses linear regression extended by constructing polynomial features from the coefficients. This approach maintains the generally fast performance of linear methods, while allowing them to fit to a much wider range of data. If we have three independent variables, polynomial regression quadratic model looks as follows:

$$Y = \beta_0 + \beta_1 x_1 + \beta_2 x_1^2 + \beta_3 x_2 + \beta_4 x_2^2 + \beta_5 x_3 + \beta_6 x_3^2 \tag{3}$$

In general, regression is used to predict continuous values, what does not match the objective of classification. In classification we aim at assigning discrete class labels. A very naive trick applied in this paper to employ regression in a binary classification problem is to define a threshold for the output of regression function that serves as a binary decision making tool. If a given pattern scores below the threshold when processed with a regression function, we assign it one class label, otherwise we assign it the other class label.

Logistic regression is a more sophisticated regression variant applicable to a case, when dependent variable is dichotomous (assumes only two values). However, independent variables do not need to be dichotomous, they can be numerical, categorical, interval-based, etc. Logistic regression, instead of using "classical" definition of probability, uses odds. Say that p is "classical" probability of a success, then $odds = p/(1-p)$. Logarithm of odds assumes values from the $(-\infty, +\infty)$, and thanks to this we can use any regression method that we like (not restricted to $[0, 1]$) in order to estimate the logarithm of odds. In other words, technically the logistic regression is a Generalized Linear Model with logit link function.

3.6 Rejection Mechanism for Classifiers

The proposed approach is based on an assumption that at the stage of model construction we do not have information about foreign patterns. We aim at constructing a model relying only on native patterns only. The model shall be able to:

- reject foreign patterns,
- classify native patterns.

We are investigating a multi-class classification scheme, what means that we have more than two distinct classes, say class_1, class_2, ..., class_c, c is the number of all classes.

We propose two methods based on an ensemble of specifically trained binary classifiers:

- "one-versus-all",
- "one-versus-one".

The "one-versus-all" method requires creating an array of c binary classifiers constructed using specially selected training data. Together, they provide a way to classify and reject. Training data set for each classifier in this method consists of two sets: the first set (denoted as "class_i") holds all training data for certain i-th native class, and the second one (denoted as "rest") being union of remaining classes. The actual classification with rejection is performed by presenting the unknown pattern to each of the classifiers from the array. When any classifier recognizes this element as a native one (belonging to class_i), then the pattern is treated as a recognized one, and it is assumed to be native. In a case when all classifiers reject a pattern (all binary classifiers say that it belongs to set "rest"), it is treated as a foreign pattern and it is rejected. It is worth to notice that there is a possibility that more than one classifier recognizes the pattern as a native element. Such case requires further processing and, since our study is not aimed on such the case, we assign an actual class label randomly.

The "one-versus-one" method requires preparing an array of classifiers, but this time it consists of $\binom{c}{2}$ classifiers, where c is the number of native classes. Each classifier is trained on data consisting of two sets: the first one (denoted as class_i) holding all training data entries for i-th native class, and the second one (denoted as class_o) holding all training data entries for some other class (not the same as class_i). In the end, there is one classifier for each pair of classes: 1 vs. 2, 1 vs. 3, ..., 1 vs. c, ..., $(c-1)$ vs. c. Classification with rejection mechanism is based on presenting unknown pattern to each classifier in the vector and remembering their classification decisions (e.g. classifier constructed for 1 vs. c classes can classify the pattern as belonging to class 1 or class c). In the end, those answers can be summarized and for each pattern we can form a c-elements array with numbers saying how many times this pattern was classified as belonging to class $1, 2, 3, \ldots, c$. The pattern is rejected when the difference between two biggest values in the result array is less than two. In such case, it is assumed that classifiers were highly uncertain as to which class should this unknown element belong to. Otherwise, the pattern is classified as an element belonging to the class, which had the biggest value in the result array.

The modified "one-versus-one" method is based on the"one-versus-one" method discussed in the previous paragraph. The difference between those two methods lies in a rejection mechanism. In this method an unknown pattern is treated as a foreign element if the biggest value in the result array is smaller than $(c-1)$. What actually means, that there must be a certain class that has always been chosen by a classifier from the vector providing a more strict decision rule.

3.7 Quality Evaluation

In order to evaluate the quality of the proposed methods we use a set of measures. Below we list basic notions applied in the formulas for those measures, while measures themselves are placed in Table 1.

- *Correctly Classified* – number of native patterns classified as native with a correct class label.
- *True Positives* – number of native patterns classified as native (no matter, into which native class).
- *False Negatives* – number of native patterns incorrectly classified as foreign.
- *False Positives* – number of foreign patterns incorrectly classified as native.
- *True Negatives* – number of foreign patterns correctly classified as foreign.

Table 1. Quality measures for classification with rejection.

$$\text{Native Precision} = \frac{\text{TP}}{\text{TP+FP}} \qquad \text{Accuracy} = \frac{\text{TP+TN}}{\text{TP+FN+FP+TN}}$$

$$\text{Foreign Precision} = \frac{\text{TN}}{\text{TN+FN}} \qquad \text{Strict Accuracy} = \frac{\text{CC+TN}}{\text{TP+FN+FP+TN}}$$

$$\text{Native Sensitivity} = \frac{\text{TP}}{\text{TP+FN}} \qquad \text{Fine Accuracy} = \frac{\text{CC}}{\text{TP}}$$

$$\text{Foreign Sensitivity} = \frac{\text{TN}}{\text{TN+FP}} \qquad \text{Strict Native Sensitivity} = \frac{\text{CC}}{\text{TP+FN}}$$

$$\text{F–measure} = 2 \cdot \frac{\text{Precision} \cdot \text{Sensitivity}}{\text{Precision} + \text{Sensitivity}}$$

4 Experiments

4.1 Presentation of Datasets

Let us now move to the empirical evaluation of the proposed models. In what follows we present a study on handwritten digits recognition.

The native dataset consisted of 10,000 handwritten digits with approximately 1,000 observations in each class ($c = 10$) taken from publicly available MNIST database [7]. We split each class in proportion ca. 7:3 and as a result we got two

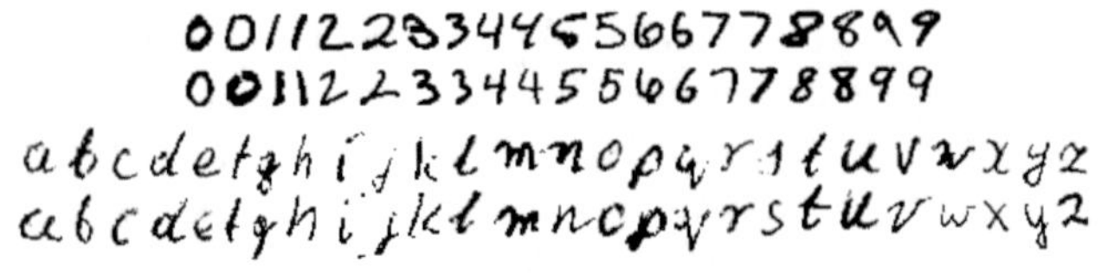

Fig. 1. Samples of native patterns – handwritten digits and foreign patterns – handwritten Latin alphabet letters.

sets. The first one included 6,996 patterns and was used for training. The test set, contained 3,004 patterns.

In order to provide foreign patterns we used a set of handwritten Latin alphabet letters (26 symbols from a to z). The dataset of foreign patterns contained 26,383 handwritten Latin letters, ca. 1,000 letters in each class. This dataset was created by 16 students, writing about 80 copies of each letter.

The justification to assume such foreign dataset for testing purposes is that appearance of other real-world symbols, not belonging to any proper class, is a common issue in a character recognition problem. Let us stress again, that foreign patterns do not participate in the model building phase. The entire scheme is trained on native patterns set only and the foreign set is involved after the model has been built. We use it for quality evaluation only. Samples of processed patterns are displayed in Fig. 1.

All patterns were normalized and feature vectors comprising of 106 numerical features were created. Numerical features were normalized to the range $[0, 1]$. Best first search for the optimal feature subset was performed using the FSelector R package, [9]. Next, we performed analysis of variance what led to further reduction of the feature set. Finally, the ready-to-use feature vector contained 24 elements.

4.2 Experimental Settings

We have formed the described three rejection mechanisms based on popular classification methods: SVM, random forest, kNN and we compare results obtained with those methods with results obtained with regression models: logistic regression and polynomial regression. The experiment presented in this paper has been conducted in Python programming language, using NumPy and scikit libraries. Final results were obtained for SVM model with the Radial Basis Function kernel, cost parameter was tuned to be eight and $\gamma = 0.5$. When using Random Forests a total of 100 trees were used. kNN method used five neighbours when classifying.

4.3 Model Quality

Experimental results are summarized in Figs. 2, 3 and 4. Comparison between proposed three methods for composing rejection mechanism clearly shows that

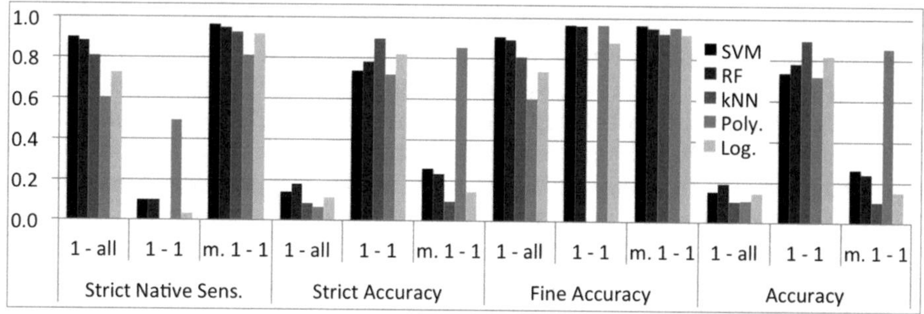

Fig. 2. Classification quality obtained with ensembles of binary classifiers based on SVM, random forests (RF) and kNN and regression models (polynomial and logistic) for the test set of native patterns.

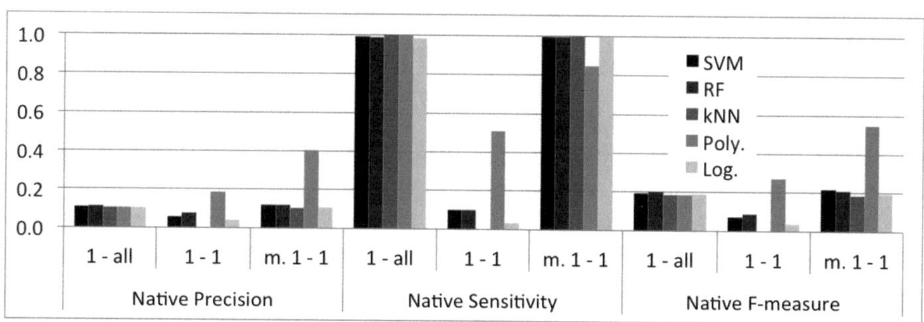

Fig. 3. Identification of native patterns measured with Native Precision, Native Sensitivity, and Native F–measure for models constructed based on three different schemes: "one-versus-all", "one-versus-one", and modified "one-versus-one".

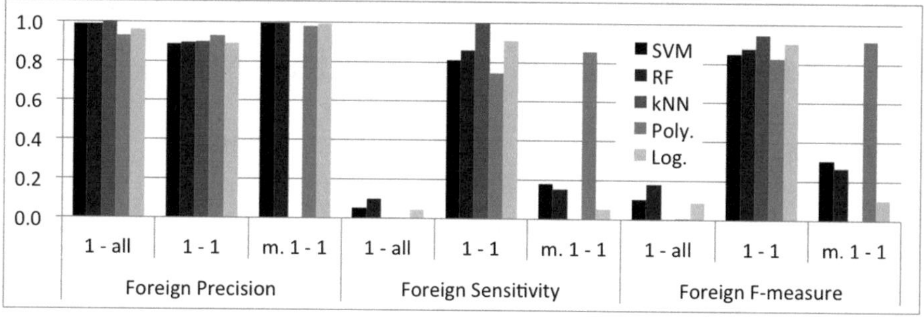

Fig. 4. Quality of foreign patterns rejection obtained using different models.

the "one-versus-all" and modified "one-versus-one" schemes work best when applied to distinguished between native and foreign elements. The basic variant of the "one-versus-one" method performs poorly as it tends to reject the majority of native patterns. This can be easily observed with very low bars in the plots corresponding to the "one-versus-one" scheme. On the other hand, the "one-versus-all" approach tends to accept a lot of foreign elements.

The described behaviour derives from the way of how we construct classifying/rejecting models. The first approach ("one-versus-all") rejected input only after c failed tests which makes rejection less probable when the number of tests gets larger (more native classes means more tests). Let us remind that in this scheme a new pattern is treated as native and labelled as an element from the first class that assumed it as native. Because tests were done in a predefined order (from class 0 to class 9) we observed increased number of classified elements being treated as members from the first few classes.

The second method ("one-versus-one") counted votes produced by all binary classifiers. In order to account a new element as native it was required that a certain native class gained significant advantage (its label was appearing more frequently than other classes).

The third method (modified "one-versus-one") provided some sort of balance between classification and rejection by modifying the second scheme. Elements were rejected only when there was no strong evidence that they belong to certain class. This evidence was in fact the number of times pattern was assigned certain class label. If there was no class with required number of labels the element was identified as a foreign one.

Comparing models constructed using various machine learning algorithms (SVM, random forest, kNN, logistic and polynomial regression) shows that the quality of results is comparable. Nonetheless, we are able to draw a few some general comments. The first is that polynomial regression performed better than logistic regression in the task of discriminating between native and foreign elements. The second conclusion is that even though kNNs perform relatively well in the task of differentiating between native and foreign elements (as it is shown in Figs. 3 and 4), it, performs poorly in the task of classification (what could be seen in Fig. 2). Among superior models we find the one developed with the use of polynomial regression – it demonstrates its high quality with relatively high Native and Foreign F–measure (as showed on Figs. 3 and 4).

5 Conclusion

In the paper we showed that employing regression models in the task of classification joint with foreign elements rejection yields satisfying results. Among five different base models applied to process handwritten digits contaminated with handwritten letters, polynomial regression provided the best results.

All in all, proposed method provides moderately satisfying results. We are aware that to truly confirm obtained results, test should be repeated on different data sets. The set described in this paper consisting of letters and digits, although being very large, might not match wide spectrum of problems.

Acknowledgments. The research is partially supported by the National Science Center, grant No 2012/07/B/ST6/01501, decision no DEC-2012/07/B/ST6/01501.

References

1. Breiman, L.: Random forests. Mach. Learn. **45**(1), 5–32 (2001)
2. Breunig, M.M., Kriegel, H.P., Ng, R.T., Sander, J.: OPTICS-OF: identifying local outliers. In: European Conference on Principles of Data Mining and Knowledge Discovery (1999)
3. Cortes, C., Vapnik, V.: Support-vector networks. Mach. Learn. **20**(3), 273–297 (1995)
4. Ghoting, A., Parthasarathy, S., Otey, M.: Fast mining of distance-based outliers in high-dimensional datasets. Data Min. Knowl. Discov. **16**(3), 349–364 (2008)
5. Grubbs, F.E.: Sample criteria for testing outlying observations. Ann. Math. Stat. **21**(1), 27–58 (1950)
6. Kapoor, A., Grauman, K., Urtasun, R., Darrell, T.: Gaussian processes for object categorization. Int. J. Comput. Vis. **88**(2), 169–188 (2010)
7. LeCun, Y., Cortes, C., Burges, C.: The MNIST database of handwritten digits. http://yann.lecun.com/exdb/mnist
8. Pimentel, M.A.F., Clifton, D.A., Clifton, L., Tarassenko, L.: A review of novelty detection. Sig. Process. **99**, 215–249 (2014)
9. Romanski, P., Kotthoff, L.: Package FSelector. http://cran.r-project.org/web/packages/FSelector/FSelector.pdf
10. Song, X., Wu, M., Jermaine, C., Ranka, S.: Conditional anomaly detection. IEEE Trans. Knowl. Data Eng. **19**(5), 631–645 (2007)
11. Sun, H., Bao, Y., Zhao, F., Yu, G., Wang, D.: CD-trees: an efficient index structure for outlier detection. In: Advances in Web-Age Information Management, vol. 3129, pp. 600–609 (2004)

Assessing the Similarity of Situations and Developments by Using Metrics

Peeter Lorents, Erika Matsak, Ahto Kuuseok(✉), and Daniil Harik

Department of Information Technology,
Estonian Business School, Tallinn, Estonia
peeter.lorents@ebs.ee, erikamatsak@gmail.com,
ahto.kuuseok@ttu.ee, daniil.harik@eesti.ee

Abstract. In this article we observe numerical evaluations of situations and developments. Similarities will be elaborated from structural and descriptive aspects. The basis of the structural similarity is homomorphism of algebraic systems. The bases of descriptive similarities are claims in descriptions of situations, which are expressible with formulas of calculations of the predictions. Developments will be addressed by binary precedency and sequential links. Mentioned connections will be observed as sets, elements of which are ordered pairs of situations. The procedure for numeric evaluation of similarities is based on metrics of relative difference of final sets. The numeric rating of similarity will be associated with plausibility: the more similar or close the objects are, the more plausible it is, that the acquired knowledge about one object will be valid for other similar object. The method of numeric evaluation of similarities may have interesting and beneficial implementation options for many areas.

Keywords: Descriptive similarities · Situations, developments · The procedure of evaluation of similarity or difference · Similarity and plausibility · Unsuccessful IT-projects · Situations and developments in security world · The comparison of content of documents · Values in business management

1 Introduction

In many areas and situations with the lack of reliable statistics, time series etc., we see the endeavour to rely on similarities, trying to manage situations and developments. More specifically, there is a reliance on these examples, which are sufficiently studied and also similar to managed situations or developments. This particular approach often helps, before (plausible, justified) decision making process (A) to analyse available data in a plausible way (B), to explain possible ways in which situations may have happened previously, and (C) analyse and predict situations or developments which may conceivably occur in sequences. Detection of similarities in observed situations and already well known situations allows us to plausibly suppose the existence of objects and circumstances, which are in particular observed in situation yet unnoticed, but which are observed in some other known and examined situations or developments. Specifically – in such a situation or development which is similar to one under observation.

I. Czarnowski et al. (eds.), *Intelligent Decision Technologies 2017*,
Smart Innovation, Systems and Technologies 72, DOI 10.1007/978-3-319-59421-7_17

These, initially overlooked (and often emerged thanks to looking for similarities) facts may significantly change situations and developments ensued in our decisions.

In case of developments often deemed to be plausible, similar situations could have had a similar (recent) history and may have a similar (near) future. Some examples:

Let's assume that Robert and William have similar houses (built with the same Project, using same building materials etc.) and both are upset about excessive heating costs. According to the expert, hired by William, the loss of heat is due to the large size of the first floor windows. After one extraordinarily freezing winter night Robert discovered that a frozen pipe had burst inside a wall on the second floor. Informed of this accident, and determined that the same may happen in he's house, William took some action.

Another example: after bankruptcy of the Lehman Brothers Holdings Inc. in September 2008, and when the economic crisis began to gather momentum, many leading financial and economic analysts started to warn that the same will take place in Europe, including Estonia. Answering the question- why they think so, their responses revealed the following arguments and discussions: in Europe, we have had similar developments as in the USA. Therefore, it is plausible that in Europe (and in our country) similar situations and developments are likely to take place.

We can argue if it is correct to make decisions based on beliefs and plausibility's. At the same time the plausibility is based on similarities. The purpose of this work is not to provide a devastating assessment to the processes which decision makers have done time after time. At the antiquity and also today, starting with personal life and following to state level or International high levels of control.

The purpose of this article is to explain, how it applies the plausibility to similarities and how it is possible to make numerical evaluations of similarities and developments. At this point, it does not provide a "direct" interest in the elements forming the systems, the characteristics of the elements and relationships between the elements. The more particular of interest is "slightly more indirect" material. Namely-objections about the elements forming the systems, the characteristics of the elements and relationships between them. In this article we call the mentioned two approaches respectively; structural and descriptive. In the context of structural approaches, we consider two systems to be similar. If between them particular conformities are found, where respective elements have corresponding characteristics, then these elements are in corresponding relationships. The example of this similarity is homomorphism and isomorphism. Unfortunately, the availability of homomorphism and even less isomorphism is not an "everyday occurrence", and verification of these particular types of similarities is not easy task either. One of the goals of this work is to find possibilities of treatments of similarities in the frames of descriptive approaches. Such treated similarity we call, hereinafter a descriptive similarity.

As mentioned above, in this work we treat developments throughout systems. Specifically, we observe developments as Algebraic systems formed of situations. In systems, situations are elements and relations are (usually normally reflect depending of time) relations between situations. In current work we confine ourselves with just one, binary relation between situations (reflecting sequence, for example temporal sequence).

Intrinsically in such case it is limited with relatively easy and widespread application type systems: systems with one binary type of relations. They include for example; all types of equivalence classes, but also several sequences. Including rankings, meaning sequences based on "importance", "priorities", etc. [1, 2].

With this particular approach we may replace the question "is perception similarity (sufficiently) plausible", with "is the similarity of adequately defined and verified." If we could decide to limit ourselves with structural approach which remains in frames of algebraic theory of systems [3–6] then, unfortunately, we should build theory on homomorphism of systems (in special cases on isomorphism).

Approving homomorphism (in special cases isomorphism) of systems is frequently complicated and laborious. Thereby determining for example, if systems expressed with particular formulas are homomorphic or downright isomorphic (in the latter case the mentioned set of formulas is examined in logics (See [7] §5.6), with categorical theories.

Hereby, this signals the question: in which case results from descriptive similarity to structural similarity? And the other way around: in which case results structural similarity to descriptive similarity? Before examining the questions mentioned it would be purposeful to find particular approaches to treat similarities, using the frames it would be possible to evaluate, in addition to the presence of similarity, also evaluate numerically how similar are objects Under observation. In the frame of traditional structural approaches, we have only two evaluations we can call as follows: absolutely similar (it means isomorphic) and quite similar (homomorphic). Unfortunately, this is not enough. Specifically, we should, evaluate and manage situations based on quite an inadequate "Picture" and lack of details and come to terms- is the current situation similar enough to already known situation to decide: what will we do in the current situation that is exactly as it was in the already known situation.

This paper observes methods for assessing similarities of systems which bind certain distances, in special cases metrics. Mentioned distances [2, 8–12, 31] are tied in one way or another to the Works of Swiss botanist Paul Jaccard, more specifically with numeric value used by Jaccard- coefficient de communauté [13, 32]. In our work, this particular approach is applied to analyse special situations (and developments) in field of economy, security and text analysis.

2 Assessment of Similarity and Difference of Finite Sets

The basis of differentiating sets relies on observing elements that are found in one set, but not in the other. Connecting all the elements from one side found in set X, but not in set Y and from the other side elements that are found in set Y, but not set X, we get a set that is denoted as

$$X \Delta Y = (X - Y) \cup (Y - X) = (X \cup Y) - (X \cap Y). \tag{1}$$

Denoting any number of elements E(A) in the finite set A, we can take figure

$$E(X \Delta Y) = E(X) + E(Y) - 2E(X \cap Y) \tag{2}$$

as final collection of these elements which differentiate two finite set X and Y. It is obvious that the higher is the figure $E(X \Delta Y)$, the greater is the difference between sets

X and Y. And also vice versa – the smaller is the figure E(XΔY), that more similar are sets X and Y. In doing so, it can be proved that figure E(XΔY) satisfies both axioms, scale and distance. This is the reason why this figure is named a measure of the difference between the sets [14] also named as distance e.g. – for example distance of Frechet-Nicodym-Aronszajn [15]. Unfortunately, this kind of figure does not present proportions of differences. For example- 20 distinctive elements in set with thousands elements are not so dramatic as 2 distinctive elements in set with 10 elements together. In order to avoid such a "dramas" is expedient to use for assessment either similarities or differences, ratio expedient. As many authors have independently (also obviously unknowingly each other's work and repeatedly) approved, satisfies this rate distance (metrics) axiom [10, 16, 17, 29]. Mentioned rate have found use in many other fields like biology [10] implementation of computers [9], assessment of differences of concepts and definitions [18] assessing different perceptions in "problematic" organisations [11]. At the same time, we can use for assessment of differences and also similarities the rate

$$1 - E(X\Delta Y) : E(X \cup Y). \tag{3}$$

Indeed, the closer are the sets, meaning the closer to figure "0" is the rate E(XΔY): E(X∪Y), the closer to Fig. 1 is

$$1 - E(X\Delta Y) : E(X \cup Y) = E(X \cap Y) : E(X \cup Y). \tag{4}$$

Incidentally, probably was the swiss botanist and plant physiologist Paul Jaccard [13] the first one who used rate

$$E(X \cap Y) : E(X \cup Y) \tag{5}$$

for assessment of similarities of certain plant communities. Obviously, due to this fact the rate E(X∩Y): E(X∪Y) is referred in the literature, not only as Marczewski-Steinhausi, or Tanimoto Distance, or Lorents Metrics, but also as Jaccard distance [15].

In this current work we rely on the implementation of ideas and experience of the rate

$$E(X\Delta Y) : E(X \cup Y) \tag{6}$$

or E(X∩Y): E(X∪Y) in order to assess similarities, or differences of situations. For this we introduce *the procedure of assessment of descriptive similarities* created by P. Lorents, which phases of execution is relied on expertise or Expert System.

Observation of two situations S_1 and S_2.

Stage 1. Using the appropriate language to present descriptions (a language similar to first-order predicate calculus) [19, 26]. All required statements describing the situation S_1 are written down. For example: "Officer C belongs to a set". "Officer C is a male". "Officer C is injured". "The situation also includes a weapon - grenade G". "Grenade G is not owned by Officer C", etc. We denote all of the aforementioned statements with the symbol Des(S_1).

Stage 2. Analogous to the first stage, the relevant statements are presented to describe situation S_2. The denotation for this group of statements would be Des(S_2).

Stage 3. Consulting an expert, it is decided which of the statement written down in the first two stages should be likened. We denote the resulting set as

$$\varepsilon(\mathrm{Des}(S_1), \mathrm{Des}(S_2). \tag{7}$$

A good general idea here is to liken statements that can be presented with logical equivalent formulas. We could equate statement from the description of the first system like "C is male" and "C is not female". Likewise, it would be possible to equate the statement from the first system that "it is certain that C1 commands a medium-sized car" and the assentation from the second system that "it is not possible that C2 has no access to a regular sedan-type car".

Note: The set ε(Des(S_1), Des(S_2)) is not always the same as set Des(S_1)∩Des(S_2)! This explains what was already mentioned – "we **base** our approach on applying magnitudes E(XΔY): E(X∪Y) or E(X∩Y): E(X∪Y) on ideas and experiences". This means we do not (and cannot) limit ourselves to only applying these magnitudes.

Stage 4. Magnitudes

$$\mathrm{E}(\mathrm{Des}(S_1)), \mathrm{E}(\mathrm{Des}(S_2)), \tag{8}$$

$$\mathrm{E}(\varepsilon(\mathrm{Des}(S_1)),\ (\mathrm{Des}(S_2))) \tag{9}$$

are found, as well as magnitudes that assess the descriptive difference of situations.

$$[\mathrm{E}(\mathrm{Des}(S_1)) + \mathrm{E}(\mathrm{Des}(S_2)) - 2\mathrm{E}(\varepsilon(\mathrm{Des}(S_1),\ \mathrm{Des}(S_2)))] \\ : [\mathrm{E}(\mathrm{Des}(S_1)) + \mathrm{E}(\mathrm{Des}(S_2) - \mathrm{E}(\varepsilon(\mathrm{Des}(S_1), \mathrm{Des}(S_2)))], \tag{10}$$

or the descriptive similarity of situations

$$1 - [\mathrm{E}(\mathrm{Des}(S_1)) + \mathrm{E}(\mathrm{Des}(S_2)) - 2\mathrm{E}(\varepsilon(\mathrm{Des}(S_1), \mathrm{Des}(S_2)))] : [\ \mathrm{E}(\mathrm{Des}(S_1))\ +\ \mathrm{E}(\mathrm{Des}(S_2)) - \mathrm{E}(\varepsilon(\mathrm{Des}(S_1), \mathrm{Des}(S_2)))] \\ = \mathrm{E}(\varepsilon(\mathrm{Des}(S_1), \mathrm{Des}(S_2))) : [\ \mathrm{E}(\mathrm{Des}(S_1))\ +\ \mathrm{E}(\mathrm{Des}(S_2)) - \mathrm{E}(\varepsilon(\mathrm{Des}(S_1), \mathrm{Des}(S_2)))]. \tag{11}$$

With that, if all other approaches fail, it is possible to use the following assumption or principle **P1:** ***The smaller the difference assessment (or bigger the similarity assessment) in situations, the more believable it is that the situations at hand share a similar past and a similar future.***

3 Assessment of Similarities and Differences in Finite Orders

Non-recurring situational development in this paper is a system in which a set is comprised of situations and the signature of which includes exactly one binary equation with a subsequent relationship is called a development. During the course of this paper we limit our approach to developments that include a finite amount of situations. In a case like this, the binary equation is set up in a way that the first, as the second pair is one of the aforementioned situations.

Assessment of differences or similarities in two developments D1 and D2 works by a method created by P. Lorents by going through the following stages:

Stage 1. An expert (or expert system) has to decide on which situational developments of D1 should be likened (equated) on situational developments of D2. As a result of this, situations that have been equated are then represented in a single copy.

Note. Equating situations doesn't mean that for example situation S, that was once preceded by a situation S' couldn't also be followed by the same situation S' later.

Stage 2. In the frame of two developments, all ordered pairs are formed if they answer to the requirement that the first pair, situation S_a precedes (in observational development) to pair two, situation $S_{b.}$ The opposite is also true - situation S_b must follow situation $S_{a.}$ In accordance with this, if one development presents a pair $\langle S_a, S_b \rangle$ along with a second development that presents a pair $\left\langle S_p, S_q \right\rangle$ and it is found that S_a is equivalent to S_p and S_b is equivalent to S_q, then both aforementioned pairs are equated. As a result of this equated pairs are therefore presented in a single copy.

Stage 3. Assessing the descriptive difference or similarity of sets with the aforementioned technology, we can calculate and rate the difference and similarity of both development pair sets.

Note: This is essentially how the values of various human communities are analyzed (e.g. Brazil, China and China business folk) [1, 2].

With that, if all other approaches fail, it is possible to use the following assumption or principle **P2:** ***The smaller the difference assessment (or bigger the similarity assessment) of developments, the more believable it is that the mentioned developments started from similar situations and that their conclusion will come to similar situations.***

Note: Ideas which relate to situational similarity and the believability of developments were established by private conversations between P. Lorents and G. Jakobson. In these conversations, G. Jakobson introduced his approach to the essence of believability and methods of application [28]. P. Lorents then recommended to advance with algebraic systems and logic models to relate believability with similarity.

And so began the search, by P. Lorents and G. Jakobson, of pertinent mathematical methods that allow the equating or likening of similarity and believability, mainly in the branch of logics and algebra, but also in regards to info technology solutions.

4 Some Mathematical Aspects

Like mentioned in the introduction, in this paper we approach the description of systems, situations and developments as a relevant collection of statements. Statements however are viewed as formulas that are written down on the basis of a previously fixed logical language (for formulas pertaining to second-order predicate calculus or even intuitionistic calculus).

This approach is explained by the following meaning attribution diagram.

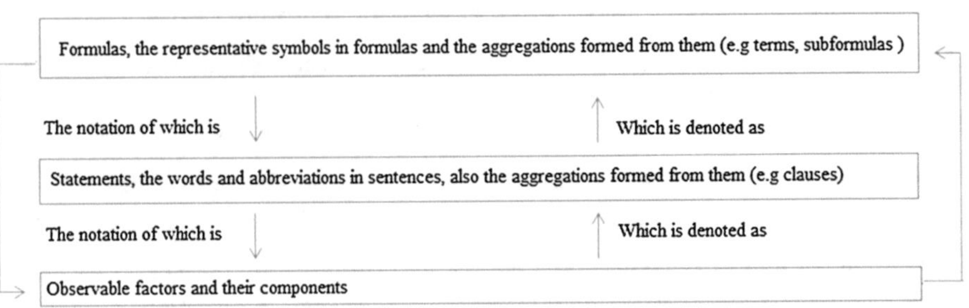

Note. The shortcuts in the above diagram (from formulas to descriptive factors and vice versa) are based on the transitive relation of fundamental relationships of denotations and definitions [25]. Unfortunately, realizing of these shortcuts for theoretical and also software solutions still needs some work. Then again, a lot of work has already been done on the roads connecting the upper columns of the diagram [20, 21].

Transforming statements into formulas allows for the necessary explanation of decisions made when processing situations to transform into proof constructed according to rules of (mathematical) logic. In such a case it is guaranteed that deriving from that which is true, we make our way in a legitimate result which is also true. Making sure of being correct is left to an interpretative procedure in accordance to previously fixed logic. An important step in interpretation is the previously mentioned attribution of definitions. If thereat we have agreed that definitions must come from a system, for example system S, then system S is part of one possible model for the observable (true) collection of formulas. This is not hard to prove.

Theorem. If there is found an interpretation, in which formula W is true in system M and M is homomorphic to system M', then an interpretation' is found, in which formula W is true in system M' (see and compare Maltsev 1970 work [5]).

Proof Idea. Based on induction by length of formulas. Emanating from atomic formulas, the interpretation of which in system M' is handled with a homomorphic function f and composition of interpretation.

Relying on the theorem presented it is possible to answer one of the questions above. In what case does structural similarity become descriptive similarity? Here we can acknowledge the fact that the homomorphism of systems M and M' allows us to say that statements which are true for system M are also true for system M'.

The best-case scenario in comparing systems would be without a doubt their isomorphism. This however isn't really a likely scenario or daily happenstance, so we must deal with homomorphism instead. A question arises here: How similar to each other are homomorphic systems? The distance of isomorphic systems should be 0, but what about homomorphism? There is room for more enquiries.

5 Examples from Studied Situations and Developments

The Failure of Ordered IT-Projects

Many failed IT-projects by various Estonian government agencies, but also by private enterprises (including the Ministry of Social Affairs, one of the largest clothes producers "Baltika", "Töötukassa" and so on) have been observed. To study the situation and put it into practice, an approach was made by Daniil Harik, where a list of statements was made to characterize the situations of different projects [23–25]. The list was compartmentalized into different parts (e.g. team, leadership and process and so on). To describe the team, eight statements were used – for example the team has Scrum experience [30]; the team has a competent Scrum Master; the team has worked together before, and so on. For leadership and process, twelve statements were used – for example tasks have been added in the middle of sprint; the backlog has been constantly updated; time estimates were mostly correct, and so on. The different parts of these lists can be approached as situations (e.g. Team situation, Leadership and process situation).

Using the assessment procedure of descriptive similarities, it was evaluated how similar or how different the situations pertaining failed projects by different organizations were. For example, in the cases of Ministry of Social Affairs and "Baltika" the leadership and process situation difference was quite large: 0.71. Difference between statements describing the team was also quite substantial: 0.6. Due to this, it is not believable that in comparing these organizations and the failure of their projects is due to similarity in descriptions of team or leadership and process.

Security Situations and Developments

Situations and developments (pertaining to security) that can lead to armed conflict are observed. In cases such as this, the so-called key situation system compiled by Ahto Kuuseok is applied. The system can be used to observe/study developments that may lead to armed conflict (see also [27]). The two dimensions of the system are (I) list of influence (that could contain statements such as *creating new norms to influence international opinion; influencing society through economics)* and (II) list of distinctive situations or in other words a list of developmental stages (for example *destabilization phase, military invasion* and so on). Influences and ordered phase types are related to statements that can be true or false, depending on the observed (comparable) development. For example, in the destabilization phase there is a statement to influence politics and public opinion*: military exercises near a border are being orchestrated*; in a different case of military invasion there might be a statement: *there is damage being done to the flow of economic resources (energy carriers, raw materials, various goods.*

Using the assessment procedure of descriptive similarities, different development phases that had already taken place were rated and compared with each other on how similar or different they were. Among other things developments that occurred in 2007 in relation to Estonia, in 2008 relating to Georgia and presently in Ukraine were compared in this way. For instance, during the military invasion phase the difference between Estonia and Georgia and also Estonia and Ukraine was maximal –1. Indeed, military action in Estonia, unlike Georgia and Ukraine, did not take place. The fact that both Georgia and Ukraine present a phase related to military action does not mean that both countries had phases that were automatically characteristic of each other, as confirmed by the score of 0.44 in assessing similarity.

Assessing the Similarity of Documents
Text-based parts of a document that don't contain graphs, diagrams, pictures and so on are observed.

For comparative analysis of text documents, not only methods that are orientated towards word density and placement (e.g. various statistical methods, Markov chains, n-grams, various morphological and syntax structure analysis). In addition to these methods, it is prudent to explain how large or small of a part words that have the same meaning have in comparable texts. If it so happens that the texts being compared do not have a lot of words or concepts with a common meaning, then it is not likely the texts share a similar structure or rhythm. (This might not apply to deep metaphorical poems, Dadaistic or other such writings). In the end, ambiguous assessments such as "a lot" or "a tiny bit" are not of much use.

One possibility to replace this ambiguousness with numerical values is to use the aforementioned metrics, where equating elements presented in different sets plays an important role. Such an approach allows for assessing the initial similarity in content of documents. To go into details, we present a relevant list of actions:

- The index and other parts of the document not in relation to the body of the text are removed
- Two word grams and the frequency of their occurrence are found.
- The list of word grams is "cleaned up" and unnecessary words are removed. These could be certain words, letter combinations, numbers et cetera that do not play a significant role from a semantic aspect. For example, articles, auxiliary words, numbers that represent paragraphs. If a word turns out to be insignificant, all occurrences of the word are removed from the list of word grams.

Explanation: The statistical aspects of text can be analysed with the use of Markov chains. The least complicated links of these chains can be approached as ordered triplets such as s_1, s_2, p, where p is probability of the word s_2 following word s_1 in the observed text. Analysing the text, we can use a word gram based on n-grams (for example kfNgram, KwiCFinder.com) to find such successions and their probability. The frequency of two succeeding words is in a direct one-to-one relationship with probabilities mentioned in the Markov chain. In this paper, to measure similarities from now on, we use the priory mentioned order of triplets, but replace the last position of probability with frequency.

- Subsequently, words are grouped into groups of synonyms (e.g. synset aka. Synonym ring). The words presented in these groups are henceforth equated. To automate this process to a greater extent, we can use an known and used software – a lexical database for English WordNet (http://wordnet.princeton.edu/). Each group is designated a word (denotation), that is used to denote equivalent words in the word gram.

When all required steps are completed, data gathered from the two documents is put into accordance.

Example. Compare two documents: "National Security Concept of Estonia" (http://www.kmin.ee/files/kmin/nodes/9470_National_Security_Concept_of_Estonia.pdf) and "Germany: Defence Policy Guidelines 2011" (http://www.isn.ethz.ch/Digital-Library/Publications/Detail/?ots591=0c54e3b3-1e9c-be1e-2c24-a6a8c7060233 HYPERLINK "http://www.isn.ethz.ch/Digital-Library/Publications/Detail/?ots591=0c54e3b3-1e9c-be1e-2c24-a6a8c7060233&lng=en&id=157024"& HYPERLINK "http://www.isn.ethz.ch/Digital-Library/Publications/Detail/?ots591=0c54e3b3-1e9c-be1e-2c24-a6a8c7060233&lng=en&id=157024"lng=en HYPERLINK "http://www.isn.ethz.ch/Digital-Library/Publications/Detail/?ots591=0c54e3b3-1e9c-be1e-2c24-a6a8c7060233&lng=en&id=157024"& HYPERLINK "http://www.isn.ethz.ch/Digital-Library/Publications/Detail/?ots591=0c54e3b3-1e9c-be1e-2c24-a6a8c7060233&lng=en&id=157024"id=157024).

After discarding insignificant words, it seems that the first (EST) document contains 2913 different word grams and the second (GER) document contains 1705 different word grams.

Here we bring out only words with a frequency value above 5 and assess similarity on the account of this factor (Table 1).

Grouping with the help of similar words (synonyms): homeland security, our security: we denote homeland **security and the frequency of it is** $7 + 7 = 14$ (Table 2)

$$\mathbf{E(Common) = 54}$$
$$\mathbf{E(Differ1 + Differ2) = 91 + 293 = 384}$$
$$\mathbf{d\,(GER, EST) = 1 - 54/\,(384 + 54) = 0,876712329}$$

Note. To achieve a more accurate analysis, single representations of successions should also be observed

Completing the appropriate steps and calculations for our observed documents, we come to a difference value of **0,876712329.** This shows that there is a huge discrepancy between the compared documents. The same kind of assessment has also been given by experts that have studied the body of these documents in-depth.

Table 1. Significant words from the first (EST) and the second (GER) document with a frequency value above 5

Germany		Estonia					
armed forces	21	European union	51	psychological defence	9	security threats	7
security policy	11	security policy	30	security concept	9	bilateral relations	6
north Atlantic	10	national security	17	civil society	8	Estonia contributes	6
Atlantic alliance	8	military defence	16	co-operation between	8	euro-Atlantic area	6
European union	8	national defence	14	Estonia supports	8	foreign policy	6
national security	8	security environment	14	internal security	8	international organisations	6
defence policy	7	collective defence	13	international co-operation	8	relations between	6
homeland security	7	organised crime	13	member states	8		
our security	7	critical services	12	common foreign	7		
crisis management	6	Baltic sea	11	defence policy	7		
military capabilities	6	crisis management	10	democratic values	7		
united nations	6	International security	10	Estonia deems	7		
		united states	10	other countries	7		
		defence Estonia	9	Schengen area	7		

Table 2. Common relationships

Germany	Estonia	Common	Differ1
Security policy 11	Security policy 30	11	19
North atlantic 10	Euro-atlantic area 6	6	4
European union 8	European union 51	8	43
National security 8	National security 17	8	9
Defence policy 7	Defence policy 7	7	0
Homeland security 14	Internal security 8	8	6
Crisis management 6	Crisis management 10	6	4

6 Conclusions

Relying on similarity has most likely been one of the instruments of human decision for tens of thousands of years. In principle, it doesn't replace perfect logical deduction, but is still useful when we must come to terms with imperfect descriptions of situations that require decisions to be made when managing them. Approaching similarity in the frame of algebraic systems (in a lucky case as homomorphism, or in a very lucky case as isomorphism) leaves some flexibility to be desired and may be too austere. Algebraic systems dose not answer the question on how similar or different the situation being observed and a situation being compared to it. Still, to produce necessary numerical assessments on similarity, useful procedures have been developed. These procedures can be used to assess the descriptive similarity of various factors (descriptions formed from statements). A high assessment of similarity gives us believability required to make decisions. It also came to light that the procedure presented in this paper can be used to give an initial assessment of similarity when researching failed info technology projects, analysis of national security situations and also correlation in contents of text documents.

References

1. Alas, R., Gao, J., Lorents, P., Übius, Ü., Matsak, E., Carneiro, J.: Associations between ethics and cultural dimensions: similarities and differences concerning ethics in China, Brazil and Estonia. Megadigma J. **4**(2), 153–162 (2011)
2. Tuulik, K., Alas, R., Lorents, P., Matsak, E.: Values in institutional context. Probl. Perspect. Manag. **9**(2), 8–20 (2011). (SECTION 1. Macroeconomic processes and regional economies management)
3. Cohn, P.M.: Universal Algebra, Harper & Row. Evanstone, New York (1965)
4. Cohn, P.M.: Universal Algebra. Springer, London (1981)
5. Maltsev, A.I., Мальцев А.И.: Алгебраические системы. "Наука". Москва (1970)
6. Mac Lane, S., Birkhoff, G.: Algebra. AMS Chelsea Publishing (1999)
7. Shoenfield, J.R.: Mathematical Logic. Addiso-Wesley Publishing Company, Boston (1967)
8. Tanimoto, T.T.: An Elementary Mathematical theory of Classification and Prediction. Internal IBM Technical Report (1957)
9. Tanimoto, T.T.: An elementary mathematical theory of classification and prediction, IBM Report (November, 1958). Cited in: Salton, G.: Automatic Information Organization and Retrieval, p. 238. McGraw-Hill (1968)
10. Marczewski, E., Steinhaus, H.: On a certain distance of sets and the corresponding distance of functions. Colloq. Math. **6**(1), 319–327 (1958)
11. Lorents, P., Lorents, D.: Applying difference metrics of finite sets in diagnostics of systems with human participation. In: Proceedings of the International Conference on Artificial Intelligence, IC – AI' 2002, vol. 3, pp. 1297–1301 (2002)
12. Lorents, P.: Denotations, knowledge and lies. In: Proceedings of the International Conference on Artificial Intelligence, 2007, Las Vegas, USA, vol. 2, pp. 324–329. CSREA Press (2007)
13. Jaccard, P.: Étude comparative de la distribution florale dans une portion des Alpes et des Jura. Bulletin de la Soc. Vaud. des Sci. Nat. **37**, 547–579 (1901)

14. Lorents, P., Lorents D.: Applying difference metrics of finite sets in diagnostics of systems with human participation. In: Proceedings of the International Conference on Artificial Intelligence, IC – AI' 2002, pp. 704–708. CSREA Press (2002)
15. Deza, E., Deza, M.M.: Encyclopedia of Distances. Springer, Heidelberg (2009)
16. Lipkus, A.H.: A proof of the triangle inequality for the Tanimoto distance. J. Math. Chem. **26**, 263–265 (1999)
17. Lorents, P.: Mathematical aspects of lies. Theory of Lie. Schola Biotheoretica. 31th Spring School on Theoretical Biology. Eesti Loodusuurijate Selts 2005, pp. 16–40 (2005). ISBN 9985-9591-1-6
18. Jents, M.: Evaluation the difference of knowledge and concepts using the metrics of Lorents. In: Proceedings of the International Conference on Artificial Intelligence. IC – AI' 2004, vol. 1, pp. 338–341 (2004)
19. Matsak, E., Lorents, P.: Decision-support systems for situation management and communication through the language of algebraic systems. In: 2012 IEEE International Multi-Disciplinary Conference on Cognitive Methods in Situation Awareness and Decision Support (CogSIMA 2012), CogSima 2012, pp. 301–307. IEEE Press (2012)
20. Matsak, E.: Dialogue system for extracting logic constructions in natural language texts. In: The 2005 International Conference on Artificial Intelligence, Las Vegas, USA (2005)
21. Matsak, E.: Discovering Logical Constructs from Estonian Children Language. Lambert Academic Publishing, Germany (2010)
22. Arnuphaptrairong, T.: Top ten lists of software project risks: evidence from the literature survey. In: International Multi Conference of Engineers and Computer Scientists, Hong Kong, pp. 1–6 (2011)
23. Kniberg, H.: Get Agile With Crisp. Scrum Checklist 04 October 2010. http://www.crisp.se/gratis-material-och-guider/scrum-checklist
24. Cohn, M.: Serena. Toward a Catalog of Scrum Smells (2003). http://www.serena.com/docs/agile/papers/Scrum-Smells.pdf
25. Lorents, P.: Knowledge and information. In: Proceedings of the 2010 International Conference on Artificial Intelligence. pp. 209–215. CSREA Press (2010)
26. Lorents, P., Matsak, E.: Applying time-dependent algebraic systems for describing situations. In: 2011 IEEE Conference on Cognitive Methods in Situation Awareness and Decision Support (CogSIMA 2011), Miami Beach, FL, 2011, (IEEE Catalog Number: CFP11COH-CDR), pp. 25–31. IEEE Press (2011)
27. Balzacq, T.: Securitization Theory. How Security Problems Emerge and Dissolve. Edited by Balzacq, T. Rotledge. London (2011)
28. Jakobson, G.: Extending situation modeling with inference of plausible future cyber situations. In: The 1st IEEE Conference on Cognitive Methods in Situation Awareness and Decision Support (CogSIMA 2011), Miami Beach, FL, USA (2011)
29. Levandowsky, M., Winter, D.: Distance between sets. Nature **234**(5), 34–35 (1971)
30. Levison, M.: Scrum Community Wiki. 11, July 2008. http://scrumcommunity.pbworks.com/w/page/10149004/Scrum%20Smells
31. Rogers, David J., Tanimoto, Taffee T.: A computer program for classifying plants. Science **132**(3434), 1115–1118 (1960). doi:10.1126/science.132.3434.1115
32. Jaccard, P.: Distribution de la flore alpine dans le bassin des Dranses et dans quelques régions voisines. Bulletin de la Soc. Vaud. des Sci. Nat. **37**, 241–272 (1901)

ISS-EWATUS Decision Support System - Overview of Achievements

Wojciech Froelich(✉) and Ewa Magiera

Institute of Computer Science, University of Silesia, Sosnowiec, Poland
{wojciech.froelich,ewa.magiera}@us.edu.pl

Abstract. Integrated Support System for Efficient Water Usage and Resources Management (ISS-EWATUS) is a project founded by the European Union's Seventh Framework Programme. Its main objective was to recognize and exploit an untapped potential to save water by using information and communication technology (ICT). After three years of the project life-time several tools for this purpose have been developed. In this paper we describe all these tools working together in synergy to conserve water. During the implementation of the presented tools, specific, scientific problems were faced. We explain how these problems were solved. In addition to the technical details, we make an overview of the research papers published as a result of the undertaken works.

Keywords: Decision support system · Water conservation

1 Introduction

As the water resource is becoming scarce, conservation of water has a high priority around the globe. Study on water management and conservation becomes an important research problem. In particular, the problem is important for countries with relatively low water resources. To meet the growing demand with conservation of water resources, novel and interdisciplinary solutions have to be implemented.

In this paper we describe the results obtained by the ISS-EWATUS project, an interdisciplinary effort of specialists from water management and ICT research. The project developed several innovative ICT tools enabling to exploit the untapped water-saving potential. This main goal has been achieved by developing an innovative, multi-factor system supporting water conservation [15,16,22].

Before going into the details of the developed system, let us recognize and overview the limitations of the existing decision support systems related to water management. We relate the recognized gaps to the contributions of the ISS-EWATUS. This review is given in Table 1. The research undertaken by the ISS-EWATUS goes beyond state of the art in every of the listed issues. A literature review related to the considered issues has been made in the papers referenced in the following sections devoted to specific tools and modules of the ISS-EWATUS.

I. Czarnowski et al. (eds.), *Intelligent Decision Technologies 2017*,
Smart Innovation, Systems and Technologies 72, DOI 10.1007/978-3-319-59421-7_18

Table 1. Overview of the ISS-EWATUS contributions

No	Issue	Contribution of the ISS-EWATUS
1	Not sufficiently efficient wireless water measurements.	Improved efficiency, more detailed information on water usage.
2	Lack of a universal spatio-temporal database for storing water related data.	Interpretable, ready to exploit information on water usage. Dedicated spatio-temporal, centralised database.
3	Inefficient control over water pressure and thus excessive leakages within water delivery system.	Novel methods enabling the control of water pressure within the water delivery system.
4	Lack of flexible, adaptive pricing schemes. Flat pricing schemes.	Decision support system for the evaluation of different pricing schemes.
5	Insufficient awareness by water consumers of water usage; lack of data analysis at household level (e.g. information on leaks).	Detection of excessive water usage by households. Low-cost, information and data analysis system run on mobile devices with the access to the water sensors installed in the household.
6	Lack of a decision support system for water-saving at household level.	Easy to use, low-cost, decision support system for water conservation for every type of household.
7	No possibility to observe positive examples of water-saving activities, no link between water producers and consumers, no possibility for consumers to get direct advice from water-saving experts.	Reinforcement of water-saving behaviour by social interactions between water stakeholders. Low-cost, social-media platform run on mobile devices for induction of water saving behaviour.

Also in the referenced papers, detailed scientific contributions of the project are reported.

From the general perspective, the ISS-EWATUS consists of several modules [15,16,29]. The structure and interdependence among them is shown in Fig. 1. The first considered module is the urban decision support system (DSS). It optimizes water pressure within water distribution system (WDS), dependent on the actual demand. It also monitors water flow within the WDS and presents it to the user in the convenient, graphical form, being thus a valuable tool for a water distribution company. The urban DSS is supported by the adaptive pricing system which evaluates and selects pricing schemes. In this way, it plays the role of an economic instrument helping to decrease water consumption. Another essential module of the ISS-EWATUS is the household DSS aiming to support water consumers in their attempts to conserve water at home. This system monitors and analyzes household water consumption not only for informative purposes but also and primarily to generate advices to the users. The generated advices together with the additionally implemented water diary help to optimize water consumption at households. An important module of the ISS-EWATUS is the

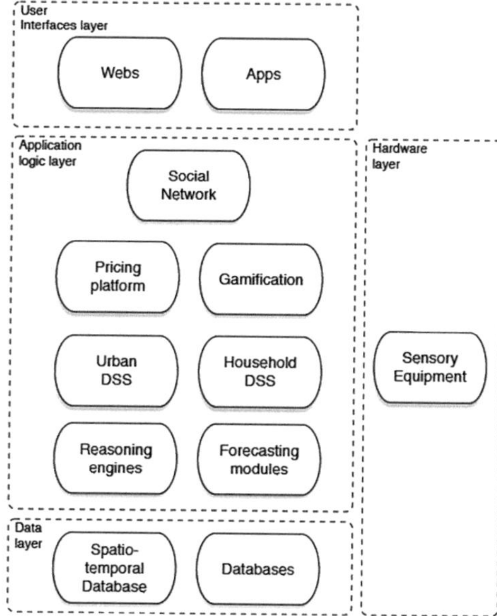

Fig. 1. Architecture of the ISS-EWATUS

social media platform by which water consumers exchange information and experiences related to water conservation. By participating in the specialized competition (game), the users encourage each other to save water.

In the following sections of the paper we present main characteristics of the above mentioned modules of the ISS-EWATUS.

2 Urban DSS

The urban DSS monitors and synthesizes heterogeneous information and displays them to the decision maker in order to assist her in controlling water resources. The input to the urban DSS are data acquired from measurement devices installed in the water distribution system. Real time monitoring and management of water pressure and flow is performed. The DSS integrates all water network devices that capture water pressure and flow in near-real time [19]. Research regarding the processing of the collected data has been made [10]. The DSS enables the water company expert to access and visualize all historical data. Consolidation of these data with meteorological, socioeconomic and touristic (arrivals to the city) data is performed [27,28]. Using the DSS, the residents are able to initiate alarms related to the network functionality and failures. The collected data are used to perform a consumption pattern analysis to provide evidence for leaks and trigger alerts. The developed urban DSS system is able to create heat maps of water demand for various seasonality-oriented parameters [13].

The urban DSS developed as a part of the ISS-EWATUS reduces water leaks within the water distribution system [14]. The leaks are reduced by controlling water pressure within the pipeline system. In a nutshell, during night periods or during weekends, when water demand is lower, its pressure is decreased using pressure reduction valves. When water demand increases during the working days, the pressure in the pipelines swells up. Such type of control has been accomplished by using appropriate hardware and software developed by the ISS-EWATUS.

To accomplish this, water demand modeling and forecasting models have been implemented. A set of various artificial intelligence techniques are integrated in the DSS to obtain accurate water demand (time series) forecasting. Accurate forecasting is crucial for the efficient control over pressure within the WDS. Researchers were confronted with the task to select the best forecasting model for the given water demand time series gathered at a particular place of the water distribution system. To select the best approach, a systematic comparison of numerous state-of-the-art predictive models was performed [5]. A multivariate analysis of daily water demand and an investigation of the Artificial Neuro-Fuzzy Inference System forecasting method were performed [18].

On-line availability of water demand time series has been assumed. This enabled day-by-day retraining of the predictive model. Under such assumption, the influence of missing data, outliers, and external variables on the accuracy of forecasting were investigated [5]. Among less known forecasting models, standard Bayesian networks [5,17], dynamic Gaussian Bayesian networks [7] and fuzzy cognitive maps [1,25] were considered. The other attempt towards improving the forecasting water demand was related to dealing with the seasonality encountered in water demand time series [6]. As a result, a new dynamic optimization approach for learning FCMs has been proposed [24]. The performed experiments shown that the proposed approach led to excellent results outperforming most of the competitive forecasting models. As an alternative, a new approach to the forecasting of the approximated (granular) time series has been developed [9]. After extensive research, the best forecasting model has been selected and applied [12]. The forecasted time series were inputed to the hydraulic model of the WDS.

According the requirements of the urban DSS, spatial and temporal disaggregation of water demand was performed. Daily water supply time series and consumer quarterly billing data for each water meter were used for this purpose. The applied disaggregation methodology was presented in [11].

The urban DSS also selects the most appropriate source of water exploitation according to demand predictions. Genetic algorithm is applied for this purpose.

3 Adaptive Pricing Module

One of the factors influencing water consumption is pricing. Economic instruments can be used to promote efficient water usage.

The adaptive pricing module of the ISS-EWATUS is an application developed for policy makers to assess the impact of pricing schemes on water consumption.

The module can be used to design pricing schemes that create incentives to reduce water consumption while maintaining the revenues, or to increase revenues while maintaining water consumption levels. The module identifies how difficult these objectives are to achieve and is able to generate optimal pricing schemes that satisfy predefined criteria. The adaptive pricing module is able to evaluate pricing strategies. It incorporates modeling techniques that offer a new perspective on water demand analysis and its dependency on pricing schemes. The module is thus an important tool for long-term, strategic water management applied by stakeholders who define water policies.

The adaptive pricing module also evaluates the effectiveness and cost efficiency of a water abstraction tax fitted to the scarcity of surface water resources. The modeling of the hypothetical consequences of the proposed taxation scheme were conducted using several datasets [3].

4 Household DSS

A conscious approach to water consumption in European households is unsatisfactory. In spite of many existing real-time oriented data gathering systems, the average water consumers are still unable to access real-time information on how their behavior influences their water consumption. This leads to a low level of motivation for water savings. In addition, those users who are motivated and would like to save water cannot rely on any unsupervised, widely available and cheap system capable of supporting and advising them regarding water conservation. Information on water-saving devices or advice regarding consumer habits are available solely through web-pages, books, newspapers or other traditional media. The influence of such media is abstracted from actual households and their daily water-usage profiles, and thus not having a direct influence on users. There is a lack of decision support system (DSS) recommending particular water saving devices or stimulating changes in people's behavior towards efficient water usage based on data gathered from households.

At household level, ISS-EWATUS proposes a low cost, mobile-device oriented set of tools supporting all household members in water conservation. The ISS-EWATUS makes users aware of their water consumption by providing near real-time access to their household water meters. On the basis of data gathered individually for every household, the ISS-EWATUS exploits water-saving potential and develop a mobile decision support system giving advice regarding water-saving behavior.

To measure water flow rate and water temperature, numerous sensors have been installed in volunteer homes [32]. A wireless transmission system sends data gathered by these sensors to the remote server. Each home is equipped with at least one tablet. The tablet provides access to the data stored on the server. The tablet allows users to view and interpret household's water consumption, broken down by appliance, across the past 24 h or at a daily, weekly or monthly level. The users of the household DSS are able to set a goal for reducing their overall water consumption. The developed software gives them feedback on their progress towards this goal.

The first achievement of the ISS-EWATUS related to the household DSS is the adaptive sampling strategy for wireless sensing systems aimed at reducing the number of data samples by sensing data only when a significant change of the signal is detected [2]. In addition, a benchmarking model for household water consumption based on Adaptive Logic Networks (ALNs) has been proposed. This new model takes account of the socio-demographical information as input and outputs a prediction on the average household water usage. Real world data collected by water consumption monitoring systems installed in Sosnowiec, Poland and Skiathos, Greece were respectively used to build a model for each considered city [4].

On the basis of the acquired water consumption data, a new water user classification function of residential water consumers has been proposed. This function was designed to harness personal value systems and wider social norms in order to promote water conservation [20].

An important research element considered during the implementation of the household DSS was of socio- and psychological nature. Three major elements pertinent to the behavior of domestic water consumers were investigated: end use behaviors; sociodemographic and property characteristics; and psychosocial constructs such as attitudes and beliefs [26]. Psychosocial and behavioural factors influencing consumers' intention to engage in everyday water saving actions around the home were investigated. An extended theory of planned behaviour perspective was used to model intention to engage in water saving actions around the home. Research hypotheses were constructed regarding the influence of attitudes, subjective norms, perceived behavioural control, information exposure and current engagement in water saving actions [21].

The household DSS generates messages to users. After comparing actual and forecasted water consumption, the DSS sends messages (tips) informing the user about the progress achieved in water conservation. In this case, among standard, statistical forecasting models, the Bayesian forecasting model was applied [8].

Water diary is the other part of the urban DSS. It prompts household members to identify their personal water usage in a fixed range of time, e.g., daily or weekly. The need to identify usage is signaled by an alarm. The diary user specifies the purpose for which the water was used for and who used the water.

5 Social-Media Platform

Besides technical, administrative and economic perspectives, there is still a relatively neglected, sociological aspect of water consumption. Positive examples of efficient water consumption should be propagated through interpersonal relationships among people. For this reason: recognition of positive water-saving actions (buying new water-saving devices, water-saving behaviour) should be carried out and then, as feedback, stimulate the entire communities to mimic or even improve the observed behavioral patterns.

A social-media platform (SMP) is a part of the ISS-EWATUS. It is able to support the promotion of water efficiency in a holistic approach [23].

This includes its impact on local, national and international levels across Europe and its target audiences of water stakeholders: individuals, households, water managers, researchers and policy makers. It aims to ease the communication and the creation of relationships between stakeholders and to produce a sustainable impact for the communities involved. The SMP allows users to share water tips and photos under different environmental scenes.

The proposed social-media platform enables water stakeholders to share experiences. The social-media platform can be used for interaction among different categories of water stakeholders in order to transmit feedback from those who were successful in reducing water consumption. This way the users in a certain category increase their water consumption awareness and help users in the same or other categories to manage water consumption better. By pushing the hi-tech envelope in a user-friendly way, even consumers who are tech-resistant and don't follow the trends understand the impact of their actions on consumption, and face the social challenge of supporting large scale behavioral change regarding water use across households. The software architecture of the developed social-media platform was presented in [31].

The ISS-EWATUS also proposes a social-centered gamification approach to improve household water usage efficiency. The approach uses a set of indicators to explicitly detect and monitor both online social network activities and offline water use activities. With this approach the gamification effectiveness can be better traced and evaluated [30]. Gamification enables the whole SMP to be used as a platform with gaming elements, which involve game task, competition and rewarding. The game tasks can be any user tasks on the social networks or any water use related offline activities such as recording down water use activities. Each of the user tasks can be rewarded upon its accomplishment.

6 Conclusions

In this paper we have presented main achievements of the ISS-EWATUS project. More information about the project is available on the project website: http://issewatus.eu. We encourage all readers to check the developed social media platform which is available on the website: http://watersocial.org.

Acknowledgments. The work was supported by ISS-EWATUS project which has received funding from the European Union's Seventh Framework Programme for research, technological development and demonstration under grant agreement no. 619228.

References

1. Ahmadi, S., Alizadeh, S., Forouzideh, N., Yeh, C., Martin, R., Papageorgiou, E.: ICLA imperialist competitive learning algorithm for fuzzy cognitive map: Application to water demand forecasting. In: IEEE International Conference on Fuzzy Systems, FUZZ-IEEE 2014, Beijing, China, 6–11 July 2014, pp. 1041–1048 (2014)

2. Al-Hoqani, N., Yang, S.H.: Adaptive sampling for wireless household water consumption monitoring. Procedia Eng. **119**, 1356–1365 (2015)
3. Berbeka, K., Palys, M.: An evaluation of the instruments aimed at poland's water savings. In: Intelligent Decision Technologies 2016: Proceedings of the 8th KES International Conference on Intelligent Decision Technologies (KES-IDT 2016), pp. 347–356. Springer International Publishing (2016)
4. Chen, X., Yang, S.H., Yang, L., Chen, X.: A benchmarking model for household water consumption based on adaptive logic networks. Procedia Eng. **119**, 1391–1398 (2015)
5. Froelich, W.: Daily urban water demand forecasting-comparative study. In: International Conference: Beyond Databases, Architectures and Structures, pp. 633–647. Springer International Publishing (2015)
6. Froelich, W.: Dealing with seasonality while forecasting urban water demand. In: Intelligent Decision Technologies 2015: Proceedings of the 7th KES International Conference on Intelligent Decision Technologies (KES-IDT 2015), pp. 171–180. Springer International Publishing (2015)
7. Froelich, W.: Forecasting daily urban water demand using dynamic Gaussian Bayesian network. In: International Conference: Beyond Databases, Architectures and Structures, pp. 333–342. Springer International Publishing (2015)
8. Froelich, W., Magiera, E.: Forecasting domestic water consumption using Bayesian model. In: Intelligent Decision Technologies 2016: Proceedings of the 8th KES International Conference on Intelligent Decision Technologies (KES-IDT 2016), pp. 337–346. Springer International Publishing (2016)
9. Froelich, W., Pedrycz, W.: Fuzzy cognitive maps in the modeling of granular time series. Knowl.-Based Syst. **115**, 110–122 (2017)
10. Jach, T., Magiera, E., Froelich, W.: Application of hadoop to store and process big data gathered from an urban water distribution system. Procedia Eng. **119**, 1375–1380 (2015)
11. Kofinas, D., Mellios, N., Laspidou, C.: Spatial and temporal disaggregation of water demand and leakage of the water distribution network in Skiathos, Greece. In: Proceedings of the 2nd International Electronic Conference on Sensors and Applications, pp. 1–6 (2015)
12. Kofinas, D., Mellios, N., Papageorgiou, E., Laspidou, C.: Urban water demand forecasting for the island of skiathos. Procedia Eng. **89**, 1023–1030 (2014)
13. Kokkinos, K., Papageorgiou, E.I., Poczeta, K., Papadopoulos, L., Laspidou, C.: Soft computing approaches for Urban water demand forecasting, pp. 357–367. Springer International Publishing (2016)
14. Laspidou, C.: Ict and stakeholder participation for improved urban water management in the cities of the future. Water Util. J. **8**, 79–85 (2014)
15. Magiera, E., Froelich, W.: Integrated support system for efficient water usage and resources management (iss-ewatus). Procedia Eng. **89**, 1066–1072 (2014)
16. Magiera, E., Froelich, W., Jach, T., Kurcius, Berbeka, K., Bhulai, S., Kokkinos, K., Papageorgiou, E., Laspidou, C., Yang, L., Perren, K., Yang, S.H., Capiluppi, A., El-Jamal, S., Wang, Z.: ISS-ewatus an example of integrated system for efficient water management. In: Proceedings of the Conference Computing and Control for Water Industry (CCWI), Amsterdam, pp. 1–10 (2016)
17. Magiera, E., Froelich, W.: Application of Bayesian networks to the forecasting of daily water demand. In: Intelligent Decision Technologies 2015: Proceedings of the 7th KES International Conference on Intelligent Decision Technologies (KES-IDT 2015), pp. 385–393. Springer International Publishing (2015)

18. Mellios, N., Kofinas, D., Papageorgiou, E., Laspidou, C.: A multivariate analysis of the daily water demand of Skiathos Island, Greece, implementing the artificial neuro-fuzzy inference system (anfis). In: E-proceedings of the 36th IAHR World Congress, 28 June – 3 July, 2015, The Hague, the Netherlands, pp. 1–8 (2015)
19. Nardo, A.D., Alcocer-Yamanaka, V.H., Altucci, C., Battaglia, R., Bernini, R., Bodini, S., Bortone, I., Bourguett-Ortiz, V.J., Cammissa, A., Capasso, S., Cascetta, F., Cocco, M., D'acunto, M., Ventura, B.D., Martino, F.D., Mauro, A.D., Natale, M.D., Doveri, M., Mansouri, B.E., Funari, R., Gesuele, F., Greco, R., Iovino, P., Koenig, R., Korakis, T., Laspidou, C.S., Lupi, L., Maietta, M., Musmarra, D., Paleari, O., Santonastaso, G.F., Savic, D., Scozzari, A., Soldovieri, F., Smorra, F., Tuccinardi, F.P., Tzatchkov, V.G., Vamvakeridou-Lyroudia, L.S., Velotta, R., Venticinque, S., Vetrano, B.: New perspectives for smart water network monitoring, partitioning and protection with innovative on-line measuring sensors. In: E-proceedings of the 36th IAHR World Congress, 28 June – 3 July, 2015, The Hague, the Netherlands, pp. 1–10 (2015)
20. Perren, K., Yang, L., He, J., Yang, S.H., Shan, Y.: Incorporating persuasion into a decision support system: the case of the water user classification function. In: 2016 22nd International Conference on Automation and Computing (ICAC), pp. 429–434 (2016)
21. Perren, K., Yang, L.: Psychosocial and behavioural factors associated with intention to save water around the home: a greek case study. Procedia Eng. **119**, 1447–1454 (2015)
22. Piasecka, J., Samborska, K., Ulańczyk, R., Magiera, E., Froelich, W.: Projekt ISS ewatus a zrównoważone zużycie wody (in polish). Gospodarka Wodna (2016)
23. Safa El-Jamal, A.C., Wang, Z.: A holistic dissemination strategy to deliver water conservation messages through gamification and social networks. In: Conference Water Efficency Network (WATEF), pp. 1–10 (2016)
24. Salmeron, J.L., Froelich, W.: Dynamic optimization of fuzzy cognitive maps for time series forecasting. Knowl.-Based Syst. **105**, 29–37 (2016)
25. Salmeron, J.L., Froelich, W., Papageorgiou, E.I.: Forecasting daily water demand using fuzzy cognitive maps. In: Time Series Analysis and Forecasting, pp. 329–340. Springer International Publishing (2016)
26. Shan, Y., Yang, L., Perren, K., Zhang, Y.: Household water consumption: insight from a survey in Greece and Poland. Procedia Eng. **119**, 1409–1418 (2015)
27. Ulańczyk, R., Pecka, T., Skotak, K., Samborska, K., Kliś, C., Suschka, J., Laspidou, C., Kokkinos, K., Froelich, W., Bragiel, T., A.Batóg, Selvik, J.R.: Systemy wspomagania dla gospodarki wodno-ściekowej w obliczu zmian klimatu i innych zmian w środowisku (in polish). In: XIV Konferencja Gospodarka wodno-ściekowa na terenach niezurbanizowanych, Kielce, Poland (2016)
28. Ulańczyk, R., Pecka, T., Skotak, K., Samborska, K., Suschka, J., Laspidou, C., Kokkinos, K., Froelich, W., Brągiel, T., Batóg, A.: Systemy wspomagania dla gospodarki wodno-ściekowej (in polish). Wodociagi i Kanalizacja **1**(155), 30–33 (2017)
29. Ulańczyk, R., Samborska, K., Froelich, W., Magiera, E., Laspidou, C., Salmeron, J.: A fuzzy-stochastic modelling approach for urban water supply systems – ISS ewatus project's concept. In: E-proceedings of the 36th IAHR World Congress, 28 June – 3 July, 2015, The Hague, the Netherlands, pp. 1–10 (2015)
30. Wang, Z., Capiluppi, A.: A social-centred gamification approach to improve household water use efficiency. In: 2015 7th International Conference on Games and Virtual Worlds for Serious Applications (VS-Games), pp. 1–4 (2015)

31. Wang, Z., Capiluppi, A.: A specialised social network software architecture for efficient household water use management. In: Weyns, D., Mirandola, R., Crnkovic, I. (eds.) Software Architecture: 9th European Conference, ECSA 2015, Dubrovnik/Cavtat, Croatia, 7–11 September 2015. Proceedings, pp. 146–153 (2015)
32. Yang, L., Yang, S.H., Magiera, E., Froelich, W., Jach, T., Laspidou, C.: Domestic water consumption monitoring and behaviour intervention by employing the internet of things technologies. In: 8th International Conference on Advances in Information Technology, IAIT 2016, December 2016, Macau, China, pp. 1–10. Procedia Computer Science (2016)

Interval-Valued Intuitionistic Fuzzy Cognitive Maps for Supplier Selection

Petr Hajek[(✉)] and Ondrej Prochazka

Faculty of Economics and Administration,
Institute of System Engineering and Informatics, University of Pardubice,
Studentska 84, 532 10 Pardubice, Czech Republic
petr.hajek@upce.cz, st47576@student.upce.cz

Abstract. Fuzzy cognitive maps (FCMs) are used to aid decision-making in complex highly nonlinear problems dealing with uncertainty. In FCMs, decision concepts are linked in order to represent causal relationships. In business decision-making, it is difficult to precisely estimate the strengths of the relationships. To address this issue, we propose interval-valued intuitionistic FCMs (IVIFCMs). A multi-criteria decision making method is introduced in which the concepts of IVIFCM represent criteria and the edges represent interaction effects among the criteria. To identify the best alternative, the steady state solution of the IVIFCM is compared with the desired values of the concepts. The proposed method provides us with an effective tool for multi-attribute decision making in complex decision-making problems with a strong uncertainty.

Keywords: Fuzzy cognitive map · Interval-valued intuitionistic fuzzy set · Supplier selection · Multi-criteria decision making

1 Introduction

Fuzzy cognitive maps (FCMs) [1] can be defined as fuzzy signed digraphs in which nodes represent descriptive concepts and edges correspond to causal relationships between concepts. Concepts are expressed in terms of fuzzy sets and the cumulative impact of causal concepts is transformed by a nonlinear activation function. FCMs' capacity to incorporate a high level of uncertainty in decision-making has made them appealing for supporting business decisions [2–4]. However, many business decisions are taken under a strong uncertainty in dynamic and unstructured environments. Thus, determining the precise values of both concepts and causal relationships can be difficult. To overcome this problem, several generalizations of FCMs have recently been proposed such as intuitionistic FCMs [5, 6] and interval-valued FCMs [7–9]. The generalizations have attracted increasing attention mainly owing to the additional freedom in assigning the membership degrees to concepts and causal relationships. In addition, information granularity in FCMs [10] have recently been examined.

Although several generalizations of FCMs have been introduced based on corresponding generalizations of fuzzy sets, note that intuitionistic fuzzy sets, interval-valued fuzzy sets, grey sets and vague sets represent equipollent generalizations of fuzzy sets and *L*-fuzzy sets, respectively. As a result, the main differences between the

I. Czarnowski et al. (eds.), *Intelligent Decision Technologies 2017*,
Smart Innovation, Systems and Technologies 72, DOI 10.1007/978-3-319-59421-7_19

generalizations can be found in their motivation and application domains. For example, the concepts in intuitionistic fuzzy sets are approached by separately envisaging positive (membership degree) and negative instances (non-membership degree), typically in medical diagnosis [5]. In contrast, interval-valued fuzzy sets use intervals to express uncertainty related to the context or to the lack of model accuracy. Business and engineering are typical domains [9]. Interval-valued intuitionistic fuzzy sets combine the advantages of both approaches so that membership (and non-membership) degree are represented by intervals rather than real numbers [11].

Interval-valued intuitionistic fuzzy sets have been widely used in business decision making due to the increasing complexity of business environment and the lack of knowledge about the problem domain [12]. Specifically, the literature on supplier selection (for reviews, see [13, 14]) suggests that evaluating suppliers under hesitation fuzzy environments represents a very promising direction in this domain. In such an environment, a decision maker may provide his/her preferences over alternatives with interval-valued intuitionistic fuzzy values. This provides additional freedom in the evaluation of suppliers compared with fuzzy decision environment [13]. The process of ranking alternatives under such situations has recently received much attention and become an interesting and important research topic in supplier selection [15–21]. So far, however, far too little attention has been paid to complex, highly nonlinear problems with interaction effects among the decision-making criteria. To overcome this problem, we introduce a multi-criteria decision making method based on interval-valued intuitionistic FCMs (IVIFCMs). Here, the concepts of IVIFCM represent criteria and the interaction effects among the criteria are expressed using the directions and weights of the edges. To rank alternatives, the steady state solution of the IVIFCM is compared with the desired values of the concepts using a weighted normalized Euclidean distance between interval-valued intuitionistic fuzzy sets. On a case study of supplier selection, we demonstrate that the proposed method can be used as an effective decision-support tool in complex problems with a high level of uncertainty.

The remainder of this paper has been organized in the following way. Section 2 briefly reviews related literature on supplier selection. Section 3 provides theoretical background on FCMs and interval-valued intuitionistic fuzzy sets. Section 4 introduces IVIFCMs, including operators used in the inference process. In Sect. 5, we develop a multi-criteria decision making method based on IVIFCMs and demonstrate its effectiveness on a supplier selection problem. Section 6 concludes this paper and discusses possible future research directions.

2 Related Literature on Supplier Selection

The problem of supplier selection has received considerable attention for its importance in logistic and supply chain management. As indicated by recent surveys [13], recent research tends to deal with practical supplier selection problems via uncertainty hybrid approaches, particularly by fuzzy hybrid methods roughly categorized into basic concept of fuzzy logic, fuzzy sets, intuitionistic fuzzy sets, and interval-valued intuitionistic fuzzy sets.

In supplier selection problems under the interval-valued intuitionistic fuzzy environment, both assessments of alternatives on attributes and attribute weights are provided as interval-valued intuitionistic fuzzy sets. For example, the notion of relative closeness was extended by [15] to interval values and fractional programming models based on TOPSIS method was developed to determine a relative closeness interval. A fuzzy TOPSIS approach to interval-valued intuitionistic fuzzy group decision making was proposed by [16] to maximize consensus between experts from the perspectives of both the ranking and the magnitude of decision data. Another fuzzy TOPSIS based approach was proposed by [17] using the normalized Hamming distance to calculate the distance between interval-valued intuitionistic fuzzy sets. Similarly, novel entropy measures have been developed to calculate the distances [18]. Furthermore, new score and accuracy functions were introduced for ranking interval-valued intuitionistic fuzzy sets by taking into account of the decision makers' attitudinal character [19].

Another core decision-making technique extended to take account of the decision makers' interval-valued intuitionistic fuzzy set assessment information was Electre I [20]. A rule-based group decision model was proposed by [21] to exploit the extracted if-then rules for prediction of preference orders of all potential suppliers.

Despite this interest in supplier selection using interval-valued intuitionistic fuzzy sets, previous studies failed to take into account the dynamics of interaction effects among the decision-making criteria. Moreover, related comparative studies have shown that the ranks of alternatives are strongly affected by the choice of aggregation operators [18].

3 Theoretical Background

3.1 Fuzzy Cognitive Maps

In FCMs, causal relationships are represented by fuzzy weights w_{ji}. These are used to assess the edge from concept j to concept i. In other words, concepts j and i in FCMs are connected by edges representing the positive (negative) relationships between the concepts. Similarly, the fuzzy value of node c_i^k is assigned to the i-th concept, where k denotes the index of iteration. For the known values of concepts c_i^k and weights w_{ji}, inference can be performed in FCMs. The concept value c_i^{k+1} for the next iteration $k+1$ can be calculated as:

$$c_i^{k+1} = f\Big(c_i^k + \sum_{\substack{j=1 \\ j \neq i}}^{N} c_j^k \times w_{ji}\Big), \tag{1}$$

where i and j refer to the i-th and j-th concept, respectively, and N concepts are included in the whole FCM. Note that the new value of a concept is calculated based on all edges connected to that concept. The nonlinear activation function f such as sigmoid function is used to transform the linear values of concepts.

3.2 Interval-Valued Fuzzy Sets

An interval-valued fuzzy set is defined by an interval-valued membership function, where the membership degree of an element $x \in X$ in an interval-valued fuzzy set A is an interval [22]. An interval-valued fuzzy set A on X can be expressed as

$$A = \{\langle x, M_A(x)\rangle | x \in X\}, \tag{2}$$

where the interval function $M_A : X \rightarrow D\,[0,\ 1]$ such that $x \rightarrow M_A(x) = \left[\mu_A^L(x),\ \mu_A^U(x)\right]$ determines the lower and upper bounds, respectively, of the interval $M_A(x)$, $0 \leq \mu_A^L(x) \leq 1, 0 \leq \mu_A^U(x) \leq 1$. The length of the interval $M_A(x) = \left[\mu_A^L(x),\ \mu_A^U(x)\right]$ is called the degree of uncertainty of x and is defined as

$$\pi_A(x) = \mu_A^U(x) - \mu_A^L(x). \tag{3}$$

3.3 Intuitionistic Fuzzy Sets

The concept of intuitionistic fuzzy sets was introduced by [23]. Intuitionistic fuzzy sets can be defined as follows. Let x be an element of a set X. An intuitionistic fuzzy set A in a finite set of the universe of discourse X is an object having the form

$$A = \{\langle x, \mu_A(x), \nu_A(x)\rangle | x \in X\}, \tag{4}$$

where $\mu_A(x)$ is the membership degree and $\nu_A(x)$ is the non-membership degree of element x belonging to the intuitionistic fuzzy set A, $0 \leq \mu_A(x) \leq 1, 0 \leq \nu_A(x) \leq 1$, and $0 \leq \mu_A(x) + \nu_A(x) \leq 1$. Hesitation degree $\pi_A(x)$ of element x belonging to the intuitionistic fuzzy set A is defined as

$$\pi_A(x) = 1 - \mu_A(x) - \nu_A(x). \tag{5}$$

The hesitation degree represents the indeterminacy degree of the membership of element x, and is equivalent to the degree of uncertainty in Eq. (3).

3.4 Interval-Valued Intuitionistic Fuzzy Sets

Interval-valued intuitionistic fuzzy sets were developed by [11] to express a stronger uncertainty. An interval-valued intuitionistic fuzzy set A on X can be represented as follows

$$A = \left\{\left\langle x, [\mu_A^L(x), \mu_A^U(x)], [\nu_A^L(x), \nu_A^U(x)]\right\rangle | x \in X\right\}, \tag{6}$$

where $\left[\mu_A^L(x),\ \mu_A^U(x)\right]$ denotes the interval membership degree of element x belonging to the interval-valued intuitionistic fuzzy set, $\left[\nu_A^L(x),\ \nu_A^U(x)\right]$ denotes the interval non-membership degree of element x belonging to the interval-valued intuitionistic fuzzy set, $0 \leq \mu_A^L(x) \leq \mu_A^U(x) \leq 1$, $0 \leq v_A^L(x) \leq v_A^U(x) \leq 1$, and $0 \leq \mu_A^U(x) + v_A^U(x) \leq 1$. The interval hesitation degree $\pi_A(x)$ of element x belonging to the interval-valued intuitionistic fuzzy set A can be defined as

$$\pi_A(x) = [1 - \mu_A^U(x) - v_A^U(x), 1 - \mu_A^L(x) - v_A^L(x)]. \tag{7}$$

Note that interval-valued intuitionistic fuzzy sets have a physical interpretation [24], with the vote for resolution between $\mu_A^L(x)$ and $\mu_A^U(x)$ in favour, between $v_A^L(x)$ and $v_A^U(x)$ against and between $\pi_A^L(x)$ and $\pi_A^U(x)$ abstentions.

4 Interval-Valued Intuitionistic Fuzzy Cognitive Maps

To reformulate inference in conventional FCMs, the addition and multiplication operators for interval-valued intuitionistic fuzzy set [25] have to be defined as follows

$$A \oplus B = \left\{ \left\langle \begin{array}{l} x, [\mu_A^L(x) + \mu_B^L(x) - \mu_A^L(x) \cdot \mu_B^L(x), \\ \mu_A^U(x) + \mu_B^U(x) - \mu_A^U(x) \cdot \mu_B^U(x)], \\ [v_A^L(x) \cdot v_B^L(x), v_A^U(x) \cdot v_B^U(x)] \end{array} \right\rangle | x \in X \right\}, \tag{8}$$

$$A \otimes B = \left\{ \left\langle \begin{array}{l} x, [\mu_A^L(x) \cdot \mu_B^L(x), \mu_A^U(x) \cdot \mu_B^U(x)], \\ [v_A^L(x) + v_B^L(x) - v_A^L(x) \cdot v_B^L(x), \\ v_A^U(x) + v_B^U(x) - v_A^U(x) \cdot v_B^U(x)] \end{array} \right\rangle | x \in X \right\}. \tag{9}$$

For negative influence in IVIFCMs, it is also necessary to define the negation operator for IVIFSs as follows

$$\neg A = \left\{ \left\langle x, [v_A^L(x), \mu_A^L(x)], [v_A^U(x), \mu_A^U(x)] \right\rangle | x \in X \right\}. \tag{10}$$

Applying the addition and multiplication operators for interval-valued intuitionistic fuzzy set from Eqs. (8) and (9), the inference in IVIFCMs can be defined as

$$\begin{aligned} c_i^{k+1} &= \{[\mu_A^L(c), \mu_A^U(c)], [v_A^L(c), v_A^U(c)]\}_i^{k+1} \\ &= f(\{[\mu_A^L(c), \mu_A^U(c)], [v_A^L(c), v_A^U(c)]\}_i^k \\ &\quad \oplus (\bigoplus_{\substack{j=1 \\ j \neq i}}^{N} (\{[\mu_A^L(c), \mu_A^U(c)], [v_A^L(c), v_A^U(c)]\}_j^k \\ &\quad \otimes \{[\mu_A^L(w), \mu_A^U(w)], [v_A^L(w), v_A^U(w)]\}_{ji}))) \end{aligned} \tag{11}$$

5 Supplier Selection Using IVIFCM

The topology for IVIFCMs was adopted from the case study on supplier selection using FCMs [26]. In the case study, a manufacturing company selects a suitable material supplier to purchase the key components of new products. A decision-making group was formed, including experts from each strategic decision area. The criteria for assessing the risk suppliers were based on the literature in the field of supplier selection. Three alternatives s_1, s_2 and s_3 remained after a preliminary screening.

Table 1. Description of supplier selection criteria

Category	Criterion	Description
Quality risk of the product	c_1	Acceptation rate of the product
	c_2	On-time delivery rate
	c_3	Product qualification ratio
	c_4	Remedy for quality problems
Service risk	c_5	Response to changes
	c_6	Technological and R&D support
	c_7	Ease of communication
Supplier's profile risk	c_8	Financial status
	c_9	Customer base
	c_{10}	Performance history
	c_{11}	Production facility and capacity
Long-term cooperation risk	c_{12}	Supplier's delivery ratio
	c_{13}	Management level
	c_{14}	Technological capability

Step 1: The committee of decision-makers identified the 14 evaluation criteria for supplier selection as presented in Table 1.

Step 2: The committee used interval-valued intuitionistic fuzzy sets to evaluate three alternatives (s_1, s_2, s_3) with respect to the criteria, as shown in Table 2.

Step 3: The experts were asked to indicate the strengths of influence among the criteria using interval-valued intuitionistic fuzzy sets (Fig. 1).

Step 4: The desired values of the criteria were defined by the experts (Table 3).

Step 5: Using the initial values of the criteria (concepts) from Table 2 and the strengths of influence (weights) from Fig. 1, the IVIFCMs were simulated for each alternative in order to reach a steady state. Hyperbolic tangent functions were used in the experiments as activation functions. The final interval-valued intuitionistic fuzzy sets of the concepts (after 10 iterations) are presented in Table 4. When compared with the initial values, the average interval hesitation degree $\pi_A(x)$ increased in time, which corresponds to the growing level of uncertainty about the suppliers in the future.

Step 6: The criteria weight vector was defined by the committee of decision-makers as follows [26]:

Table 2. Initial values of criteria for suppliers

Suppl.	c_1	c_2	c_3	c_4	c_5	c_6	c_7
s_1	[0.20,0.30]	[0.63,0.76]	[0.38,0.46]	[0.81,0.90]	[0.40,0.45]	[0.90,0.95]	[0.40,0.50]
	[0.20,0.22]	[0.10,0.15]	[0.51,0.54]	[0.03,0.08]	[0.17,0.55]	[0.01,0.05]	[0.30,0.37]
s_2	[0.80,0.87]	[0.40,0.55]	[0.75,0.80]	[0.60,0.67]	[0.60,0.70]	[0.70,0.75]	[0.72,0.80]
	[0.10,0.12]	[0.30,0.35]	[0.10,0.15]	[0.15,0.20]	[0.10,0.15]	[0.17,0.20]	[0.15,0.20]
s_3	[0.50,0.60]	[0.80,0.90]	[0.32,0.37]	[0.13,0.20]	[0.35,0.40]	[0.90,0.95]	[0.50,0.67]
	[0.25,0.30]	[0.05,0.10]	[0.50,0.55]	[0.70,0.75]	[0.35,0.40]	[0.00,0.05]	[0.15,0.25]
	c_8	c_9	c_{10}	c_{11}	c_{12}	c_{13}	c_{14}
s_1	[0.50,0.55]	[0.80,0.85]	[0.68,0.75]	[0.55,0.59]	[0.70,0.80]	[0.72,0.90]	[0.30,0.40]
	[0.23,0.42]	[0.05,0.10]	[0.10,0.15]	[0.10,0.11]	[0.15,0.18]	[0.03,0.10]	[0.10,0.29]
s_2	[0.70,0.80]	[0.50,0.55]	[0.75,0.87]	[0.75,0.85]	[0.55,0.62]	[0.40,0.45]	[0.50,0.55]
	[0.15,0.18]	[0.20,0.30]	[0.05,0.05]	[0.05,0.10]	[0.22,0.30]	[0.20,0.30]	[0.25,0.30]
s_3	[0.35,0.50]	[0.95,1.00]	[0.65,0.72]	[0.90,0.95]	[0.87,0.90]	[0.80,0.87]	[0.75,0.78]
	[0.45,0.47]	[0.00,0.00]	[0.20,0.23]	[0.00,0.05]	[0.02,0.07]	[0.10,0.12]	[0.16,0.21]

$$\boldsymbol{W} = (0.0251, 0.0722, 0.1423, 0.0252, 0.0890, 0.0216, 0.0728, \\ 0.1079, 0.0153, 0.0456, 0.1066, 0.0346, 0.0621, 0.1797)^{\mathrm{T}} \tag{12}$$

Step 7: The output values obtained by the IVIFCMs in Table 4 was compared with the target values of the criteria from Table 3. To calculate the distance

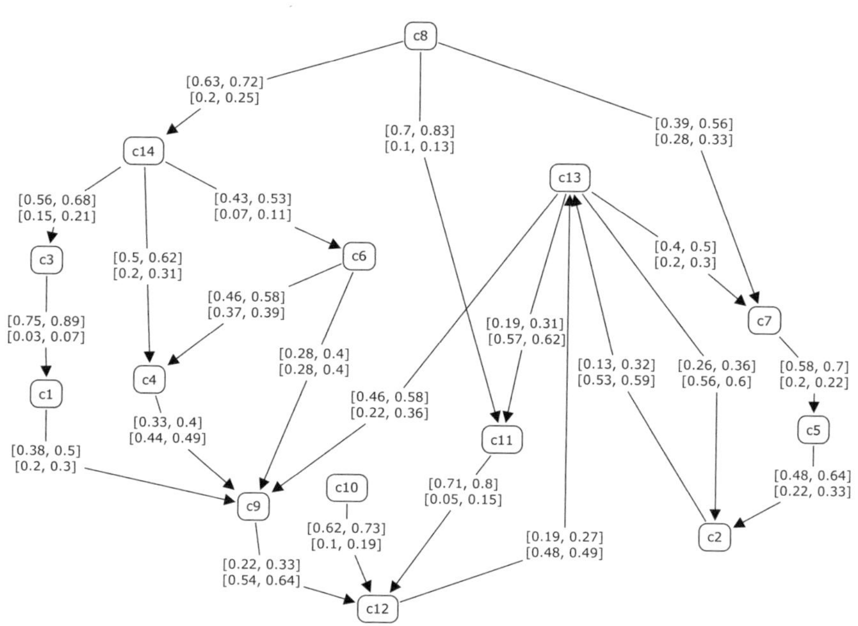

Fig. 1. Topology of the IVIFCM for supplier selection

Table 3. Desired (target) values of criteria

	c_1	c_2	c_3	c_4	c_5	c_6	c_7
$\hat{x}$	[0.95,0.02] [0.00,0.00]	[0.95,1.00] [0.00,0.00]	[0.99,1.00] [0.00,0.00]	[0.96,1.00] [0.00,0.00]	[0.95,1.00] [0.00,0.00]	[0.92,1.00] [0.00,0.00]	[0.91,1.00] [0.00,0.00]
	c_8	c_9	c_{10}	c_{11}	c_{12}	c_{13}	c_{14}
$\hat{x}$	[0.92,1.00] [0.00,0.00]	[0.93,1.00] [0.00,0.00]	[0.97,1.00] [0.00,0.00]	[0.94,1.00] [0.00,0.00]	[0.80,1.00] [0.00,0.00]	[0.87,1.00] [0.00,0.00]	[0.95,1.00] [0.00,0.00]

between the output and target values, we used the weighted normalized Euclidean distance between interval-valued intuitionistic fuzzy sets [27]:

$$D(A,B) = \sqrt{\frac{1}{2}\sum\nolimits_{i=1}^{n} \begin{matrix} w_i[\left|\mu_A^L(c_i) - \mu_B^L(c_i)\right|^2 + \left|\mu_A^U(c_i) - \mu_B^U(c_i)\right|^2 \\ + \left|v_A^L(c_i) - v_B^L(c_i)\right|^2 + \left|v_A^U(c_i) - v_B^U(c_i)\right|^2] \end{matrix}}. \quad (13)$$

Using (13), we obtained the distances for suppliers s_1, s_2 and s_3 as follows: $D_1 = 0.391$, $D_2 = 0.375$ and $D_3 = 0.406$. According to the final scores, the most desirable supplier was $s_2, s_2 \succ s_1 \succ s_3$.

To demonstrate the differences to previous studies, we compared the results with three methods: (1) fuzzy TOPSIS interval-valued intuitionistic fuzzy decision making (IVIF-TOPSIS) [16], (2) method based on interval-valued intuitionistic fuzzy weighted average operator and fuzzy ranking method (IVIFWA+FR) [24], and (3) method based on interval-valued FCM (IVFCM) [9]. First, the relative closeness indices (*CI*) of suppliers s_1, s_2 and s_3 with respect to the positive ideal solution were calculated as follows: $CI_1 = 0.417$, $CI_2 = 0.629$ and $CI_3 = 0.395$, this is the results obtained by the IVIF-TOPSIS are in agreement with the method proposed in this study $s_2 \succ s_1 \succ s_3$. Similarly to our method, the IVIF-TOPSIS enables a comparison with ideal solution, but interaction effects among the decision-making criteria cannot be taken into account. Second, the ranking values (RV) of the IVIFWA+FR were calculated, with RV(s_1) = 0.244, RV(s_2) = 0.441 and RV(s_3) = 0.315, i.e. $s_2 \succ s_3 \succ s_1$. The difference to our method was most likely due to the impossibility to compare the solution with its ideal counterpart. Moreover, the interaction effects could not be incorporated. Finally, the distances for suppliers s_1, s_2 and s_3 were obtained by the IVFCM as follows: $D_1 = 0.375$, $D_2 = 0.290$ and $D_3 = 0.344$, i.e. $s_2 \succ s_3 \succ s_1$. To calculate the distances between the values and the ideal solution, we used weighted normalized Euclidean distance between interval-valued intuitionistic fuzzy sets. The results are again different from those observed by the method proposed in this study, mainly due to the limited capability of the IVFCM to model the underlying uncertainty (only interval-valued fuzzy sets are taken into account).

Table 4. Final values of criteria for suppliers obtained by IVIFCMs

Supplier	c_1	c_2	c_3	c_4	c_5	c_6	c_7
s_1	[0.68,0.71] [0.00,0.00]	[0.67,0.71] [0.00,0.00]	[0.64,0.67] [0.00,0.00]	[0.69,0.71] [0.00,0.00]	[0.65,0.72] [0.00,0.00]	[0.61,0.64] [0.00,0.00]	[0.64,1.00] [0.00,0.00]
s_2	[0.68,0.71] [0.00,0.00]	[0.67,0.71] [0.00,0.00]	[0.65,0.67] [0.00,0.00]	[0.69,0.71] [0.00,0.00]	[0.66,0.72] [0.00,0.00]	[0.61,0.64] [0.00,0.00]	[0.65,1.00] [0.00,0.00]
s_3	[0.68,0.71] [0.00,0.00]	[0.67,0.71] [0.00,0.00]	[0.64,0.67] [0.00,0.00]	[0.68,0.71] [0.00,0.00]	[0.65,0.72] [0.00,0.00]	[0.61,0.64] [0.00,0.00]	[0.64,1.00] [0.00,0.00]
	c_8	c_9	c_{10}	c_{11}	c_{12}	c_{13}	c_{14}
s_1	[0.31,0.33] [0.20,0.29]	[0.72,0.74] [0.00,0.00]	[0.35,0.36] [0.10,0.14]	[0.64,0.67] [0.10,0.11]	[0.72,0.73] [0.00,0.00]	[0.59,0.66] [0.00,0.00]	[0.59,0.61] [0.00,0.00]
s_2	[0.35,0.36] [0.14,0.16]	[0.72,0.74] [0.00,0.00]	[0.36,0.37] [0.05,0.05]	[0.65,0.68] [0.05,0.10]	[0.72,0.73] [0.22,0.30]	[0.59,0.66] [0.00,0.00]	[0.60,0.62] [0.00,0.00]
s_3	[0.27,0.31] [0.30,0.31]	[0.72,0.74] [0.00,0.00]	[0.34,0.35] [0.18,0.20]	[0.62,0.67] [0.00,0.05]	[0.71,0.73] [0.02,0.07]	[0.59,0.66] [0.00,0.00]	[0.57,0.61] [0.00,0.00]

6 Conclusion

In this study, we propose a novel generalization of FCM that copes with complex relationships among decision concepts under a highly uncertain supplier selection environment. Instead of traditional fuzzy sets theory, IVIFCMs are based on a more general concept of interval-valued intuitionistic fuzzy sets, providing an effective tool to deal with strong uncertainty in the values of criteria and their causal relationships. In supplier selection domain, the physical interpretation of interval-valued intuitionistic fuzzy sets is also appropriate, allowing vote for resolution in favour, against and abstention, respectively. In contrast to previous methods, the proposed model utilized dependence and interactions among criteria.

Although the convergence of the concepts strongly depends on the topology of IVIFCMs, we observed that the final values of criteria for suppliers obtained by IVIFCMs differed mainly in supplier's profile risk. This suggests that supplier's profile risk is considered the most important category of criteria in the IVIFCM model.

Finally, several important limitations need to be considered. First, the proposed inference mechanism for the IVIFCM does not evolve dynamically. Future research should therefore concentrate on the investigation of adaptive IVIFCMs [28]. Second, sensitivity analysis on IVIFCM topology should be undertaken to investigate its evolving behaviour. Finally, more complex decision-making problems should be examined to investigate the scalability of IVIFCMs.

Acknowledgments. This article was supported by the scientific research project of the Czech Sciences Foundation Grant No.: 16-19590S and by the grant No. SGS_2017_017 of the Student Grant Competition.

References

1. Kosko, B.: Fuzzy cognitive maps. Int. J. Man Mach. Stud. **24**, 65–75 (1986)
2. Papageorgiou, E., Salmeron, J.L.: A review of fuzzy cognitive maps research during the last decade. IEEE Trans. Fuzzy Syst. **21**(1), 66–79 (2013)
3. Groumpos, P.P.: Modelling business and management systems using fuzzy cognitive maps: a critical overview. IFAC-PapersOnLine **48**(24), 207–212 (2015)
4. Prochazka, O., Hajek, P.: Modelling knowledge management processes using fuzzy cognitive maps. LNBIP, vol. 224, pp. 41–50 (2015)
5. Papageorgiou, E., Iakovidis, D.K.: Intuitionistic fuzzy cognitive maps. IEEE Trans. Fuzzy Syst. **21**(2), 342–354 (2013)
6. Prochazka, O., Hajek, P.: Intuitionistic fuzzy cognitive maps for corporate performance modeling. In: 33rd International Conference on Mathematical Methods in Economics, Cheb, pp. 683–688 (2015)
7. Papageorgiou, E., Stylios, C., Groumpos, P.: Introducing interval analysis in fuzzy cognitive map framework. LNCS, vol. 3955, pp. 571–575 (2006)
8. Salmeron, J.L.: Modelling grey uncertainty with fuzzy grey cognitive maps. Expert Syst. Appl. **37**(12), 7581–7588 (2010)
9. Hajek, P., Prochazka O.: Interval-valued fuzzy cognitive maps for supporting business decisions. In: IEEE World Congress on Computational Intelligence, Vancouver, pp. 531–536 (2016)
10. Pedrycz, W., Homenda, W.: From fuzzy cognitive maps to granular cognitive maps. IEEE Trans. Fuzzy Syst. **22**(4), 859–869 (2014)
11. Atanassov, K., Gargov, G.: Interval valued intuitionistic fuzzy sets. Fuzzy Sets Syst. **31**(3), 343–349 (1989)
12. Xu, Z.: A method based on distance measure for interval-valued intuitionistic fuzzy group decision making. Inf. Sci. **180**(1), 181–190 (2010)
13. Chai, J., Liu, J.N., Ngai, E.W.: Application of decision-making techniques in supplier selection: a systematic review of literature. Expert Syst. Appl. **40**(10), 3872–3885 (2013)
14. Ho, W., Xu, X., Dey, P.K.: Multi-criteria decision making approaches for supplier evaluation and selection: a literature review. Eur. J. Oper. Res. **202**(1), 16–24 (2010)
15. Wang, Z., Li, K.W., Xu, J.: A mathematical programming approach to multi-attribute decision making with interval-valued intuitionistic fuzzy assessment information. Expert Syst. Appl. **38**(10), 12462–12469 (2011)
16. Zhang, X., Xu, Z.: Soft computing based on maximizing consensus and fuzzy TOPSIS approach to interval-valued intuitionistic fuzzy group decision making. Appl. Soft Comput. **26**, 42–56 (2015)
17. Izadikhah, M.: Group decision making process for supplier selection with TOPSIS method under interval-valued intuitionistic fuzzy numbers. Adv. Fuzzy Syst. **2**, 1–14 (2012)
18. Chen, T.Y., Wang, H.P., Lu, Y.Y.: A multicriteria group decision-making approach based on interval-valued intuitionistic fuzzy sets: a comparative perspective. Expert Syst. Appl. **38** (6), 7647–7658 (2011)
19. Wu, J., Chiclana, F.: A risk attitudinal ranking method for interval-valued intuitionistic fuzzy numbers based on novel attitudinal expected score and accuracy functions. Appl. Soft Comput. **22**, 272–286 (2014)
20. Xu, J., Shen, F.: A new outranking choice method for group decision making under Atanassov's interval-valued intuitionistic fuzzy environment. Knowl. Based Syst. **70**, 177–188 (2014)

21. Chai, J., Liu, J.N., Xu, Z.: A rule-based group decision model for warehouse evaluation under interval-valued intuitionistic fuzzy environments. Expert Syst. Appl. **40**(6), 1959–1970 (2013)
22. Zadeh, L.A.: The concept of a linguistic variable and its application to approximate reasoning – I. Inf. Sci. **8**(3), 199–249 (1975)
23. Atanassov, K.T.: Intuitionistic fuzzy sets. Fuzzy Sets Syst. **20**(1), 87–96 (1986)
24. Chen, S.M., Lee, L.W., Liu, H.C., Yang, S.W.: Multiattribute decision making based on interval-valued intuitionistic fuzzy values. Expert Syst. Appl. **39**(12), 10343–10351 (2012)
25. Atanassov, K.T.: Operators over interval valued intuitionistic fuzzy sets. Fuzzy Sets Syst. **64** (2), 159–174 (1994)
26. Xiao, Z., Chen, W., Li, L.: An integrated FCM and fuzzy soft set for supplier selection problem based on risk evaluation. Appl. Math. Model. **36**(4), 1444–1454 (2012)
27. Burillo, P., Bustince, H.: Entropy on intuitionistic fuzzy sets and on interval-valued fuzzy sets. Fuzzy Sets Syst. **78**(3), 305–316 (1996)
28. Froelich, W., Salmeron, J.L.: Evolutionary learning of fuzzy grey cognitive maps for the forecasting of multivariate, interval-valued time series. Int. J. Approximate Reasoning **55**(6), 1319–1335 (2014)

MLEM2 Rule Induction Algorithm with Multiple Scanning Discretization

Patrick G. Clark[1], Cheng Gao[1], and Jerzy W. Grzymala-Busse[1,2](✉)

[1] Department of Electrical Engineering and Computer Science, University of Kansas, Lawrence, KS 66045, USA
patrick.g.clark@gmail.com, {cheng.gao,jerzy}@ku.edu

[2] Department of Expert Systems and Artificial Intelligence, University of Information Technology and Management, 35-225 Rzeszow, Poland

Abstract. In this paper we show experimental results on the MLEM2 rule induction algorithm and the Multiple Scanning discretization algorithm. The MLEM2 algorithm of rule induction has its own mechanisms to handle missing attribute values and numerical data. We compare, in terms of an error rate, two setups: MLEM2 used for rule induction directly from incomplete and numerical data and MLEM2 inducing rule sets from data sets previously discretized by Multiple Scanning and then converted to be incomplete. In both setups certain and possible rule sets were induced. For certain rule sets, the former setup was more successful for two data sets, while the latter setup was more successful for four data sets, for eight data sets the difference was not significant (Wilcoxon test, 5% significance level). Similarly, for possible rule sets the former setup was more successful for two data sets, while the latter setup was more successful for three data sets. Thus we may conclude that there is not significant difference between both setups and that we may use MLEM2 for rule induction directly from incomplete and numerical data.

Keywords: Incomplete data · MLEM2 rule induction algorithm · Multiple discretization algorithm · Concept lower and upper approximations

1 Introduction

LEM2 (Learning from Examples Module, Version 2) is a basic component of the LERS (Learning from Examples based on Rough Sets) data mining system [9,11]. MLEM2 (Modified LEM2) is a rule induction algorithm that can induce rules directly from incomplete and numerical data sets. Missing attribute sets are handled by rough set theory [11–13] while any numerical attribute is converted to a set of binary attributes defined by all possible cutpoints [10]. On the other hand, a highly successful Multiple Scanning strategy to discretization, based on entropy, was introduced in [15]. In general, entropy based discretization is considered to be one of the best [1–8,19–22]. The Multiple Scanning discretization method is significantly better, in terms of an error rate, than well-known discretization methods such as globalized versions of Equal Interval Width and

I. Czarnowski et al. (eds.), *Intelligent Decision Technologies 2017*,
Smart Innovation, Systems and Technologies 72, DOI 10.1007/978-3-319-59421-7_20

Equal Frequency per Interval [16]. This result was obtained using rule induction methodology. Using independent methodology of decision tree generation, it was shown that the Multiple Scanning discretization is better not only than both globalized discretization methods but also C4.5 [26].

In our experiments, for each concept lower and upper approximations were computed, using a rough set approach to rule induction [23–25]. Rules induced from lower approximations are called certain, rules induced from upper approximations are called possible. Note that for incomplete data sets three types of approximations may be used: singleton, subset and concept [11]. We used concept lower and upper approximations. In this paper we use only one interpretation of missing attribute values, called lost [11].

Our main objective is to compare, in terms of an error rate, two setups: MLEM2 applied for rule induction directly from incomplete and numerical data and MLEM2 inducing rule sets from data sets previously discretized by the Multiple Scanning method and then converted to be incomplete. We wanted to compare the internal discretization included with the MLEM2 rule induction algorithm with the Multiple Scanning discretization applied to numerical data and then rule induction using MLEM2. As follows from our results, there is no significant difference between both setups. Thus, we may save time and skip the Multiple Discretization and induce rule sets directly from incomplete and numerical data using MLEM2.

2 Multiple Scanning Discretization

An example of numerical data set is presented in Table 1. A *concept* is a set of all cases with the same decision value. In Table 1 there are three concepts, {1, 2}, {3, 4, 5} and {6, 7, 8}. The set of all cases will be denoted by U.

Let a be a numerical attribute, let x be the smallest value of a and let y be the largest value of a. A discretization is based on computing a partition of $[x, y]$ into k intervals,

$$\{[a_{i_0}, a_{i_1}), [a_{i_1}, a_{i_2}), ..., [a_{i_{k-2}}, a_{i_{k-1}}), [a_{i_{k-1}}, a_{i_k}]\},$$

where $a_{i_0} = x$, $a_{i_k} = y$, and $a_{i_l} < a_{i_{l+1}}$ for $l = 0, 1, ..., k-1$. The numbers a_{i_1}, a_{i_2},..., $a_{i_{k-1}}$ are called *cut-points*. We will denote such intervals by

$$a_{i_0}..a_{i_1}, a_{i_1}..a_{i_2}, ..., a_{i_{k-2}}..a_{i_{k-1}}, a_{i_{k-1}}..a_{i_k}.$$

An example of a data set with numerical attributes is presented in Table 1. In this table all cases are described by variables called *attributes* and one variable called a *decision*. The set of all attributes is denoted by A. The decision is denoted by d. The set of all cases is denoted by U. In Table 1 the attributes are *Length*, *Width* and *Height* while the decision is *Price*. Moreover, $U = \{1, 2, 3, 4, 5, 6, 7, 8\}$.

For any subset B of the set A of all attributes, an *indiscernibility* relation $IND(B)$ is defined, for any $x, y \in U$, in the following way

$$(x, y) \in IND(B) \text{ if and only if } a(x) = a(y) \text{ for any } a \in B,$$

Table 1. A numerical data set

	Attributes			Decision
Case	Length	Width	Height	Price
1	4.1	1.7	1.4	Low
2	4.1	1.8	1.4	Low
3	4.5	1.7	1.4	Medium
4	4.1	1.9	1.5	Medium
5	4.3	1.9	1.5	Medium
6	4.5	1.8	1.5	High
7	4.3	1.8	1.6	High
8	4.5	1.9	1.6	High

where $a(x)$ denotes the value of the attribute $a \in A$ for the case $x \in U$. The relation $IND(B)$ is an equivalence relation. The equivalence classes of $IND(B)$ are denoted by $[x]_B$.

A partition on U constructed from all B-elementary sets of $IND(B)$ is denoted by B^*. Sets from d^* are called *concepts*. For example, for Table 1, if $B = \{Length\}$, $B^* = \{\{1, 2, 4\}, \{3, 5, 7\}, \{6, 7\}\}$ and $\{d\}^* = \{\{1, 2\}, \{3, 4, 5\}, \{6, 7, 8\}\}$. A data set is consistent if $A^* \leq \{d\}^*$, i.e., if for each set X from A^* there exists set Y from $\{d\}^*$ such that $X \subseteq Y$. For the data set from Table 1, each set from A^* is a singleton, so the data set from Table 1 is consistent.

During discretization the domain of the attribute is sorted. Potential cutpoints are selected as means of two consecutive numbers from the sorted attribute domain. For example, for *Length* there are two potential cutpoints: 4.2 and 4.4.

In Multiple Scanning discretization, the set of all attributes is scanned k times, where the number k id given by the user. In our experiments k was equal to three. During the first scan, for each attribute the best cutpoint is selected. Multiple Scanning discretization is based on the conditional entropy $H_S(a|q)$.

Let S be a subset of the set U. An entropy $H_S(a)$ of an attribute a, with values a_1, a_2,..., a_n is defined as follows

$$-\sum_{i=1}^{n} p(a_i) \cdot \log p(a_i),$$

where $p(a_i)$ is a probability (relative frequency) of value a_i in the set S, logarithms are binary, and $i = 0, 1, ..., n$.

Let a be an attribute and q be a cutpoint that splits the set S into two subsets, S_1 and S_2. The conditional entropy $H_S(a|q)$ is defined as follows

$$\frac{|S_1|}{|U|} H_{S_1}(a) + \frac{|S_2|}{|U|} H_{S_2}(a),$$

where $|X|$ denotes the cardinality of the set X. The cut-point q for which the conditional entropy $H_S(a|q)$ has the smallest value is selected as the best cutpoint. For the potential cutpoint 4.2 of the attribute *Length*,

Table 2. A partially discretized data set

	Attributes			Decision
Case	Length[1]	Width[1]	Height[1]	Price
1	4.1..4.2	1.7..1.85	1.4..1.55	Low
2	4.1..4.2	1.7..1.85	1.4..1.55	Low
3	4.2..4.5	1.7..1.85	1.4..1.55	Medium
4	4.1..4.2	1.85..1.9	1.4..1.55	Medium
5	4.2..4.5	1.85..1.9	1.4..1.55	Medium
6	4.2..4.5	1.7..1.85	1.4..1.55	High
7	4.2..4.5	1.85..1.9	1.55..1.6	High
8	4.2..4.5	1.7..1.85	1.55..1.6	High

$$H_U(Length|4.2) = \frac{3}{8}(-\frac{1}{3} \cdot \log\frac{1}{3} - \frac{2}{3} \cdot \log\frac{2}{3}) + \frac{5}{8}(-\frac{2}{5} \cdot \log\frac{2}{5} - \frac{3}{5} \cdot \log\frac{3}{5}) \approx 0.951.$$

Similarly, $H_U(Length|4.4) = 1.295$. Thus, for *Length* the cutpoint 4.2 is selected. Additionally, the best cutpoints are 1.85 for *Width* and 1.55 for *Height*. After the first scan, the discretized data set for the data set from Table 1 is presented on Table 2.

Let a^i denote the attribute a discretized by i scans. Let A^i be the set of all attributes discretized by i scans. In the example of Table 2, $(A^1)^* = \{\{1, 2\}, \{3, 6\}, \{4, 5\}, \{7\}, \{8\}\}$. Obviously, $(A^1)^* \not\leq \{d\}^*$.

In general, the process of discretization is completed if after i scans $(A^i)^* \leq \{d\}^*$. If this condition fails, we have to identify all blocks $X \in (A^1)^*$ such that for such X no Y exists in $\{d\}^*$ with $X \subseteq Y$. For each attribute, the best cutpoint among all such sets X should be selected. If after k steps $(A^k)^* \not\leq \{d\}^*$, the remaining discretization should be conducted using other methods, e.g., *Dominant Attribute* [16–18].

Table 3. Completely discretized data set

	Attributes			Decision
Case	Length[2]	Width[2]	Height[2]	Price
1	4.1..4.2	1.7..1.75	1.4..1.55	Low
2	4.1..4.2	1.75..1.85	1.4..1.55	Low
3	4.2..4.5	1.7..1.75	1.4..1.55	Medium
4	4.1..4.2	1.85..1.9	1.4..1.55	Medium
5	4.2..4.5	1.85..1.9	1.4..1.55	Medium
6	4.2..4.5	1.75..1.85	1.4..1.55	High
7	4.2..4.5	1.75..1.85	1.55..1.6	High
8	4.2..4.5	1.85..1.9	1.55..1.6	High

In our example, $(A^1)^* \not\leq \{d\}^*$, an offending set X from $(A^1)^*$ is $\{3, 6\}$. We cannot use attribute *Length*, for remaining two attributes, $H_{\{3,6\}}(Width|1.75) = H_{\{3,6\}}(Height|1.55)$, so we use the cutpoint 1.75 for the attribute *Width.* Our discretized data set is presented on Table 3.

3 Incomplete Data Sets

An example of incomplete data set is presented in Table 4. For complete data sets, for an attribute-value pair $(a,\ v)$, a *block* of $(a,\ v)$, denoted by $[(a,\ v)]$, is the following set

$$[(a,\ v)] = \{x|x \in U, a(x) = v\}.$$

Table 4. Incomplete data set

	Attributes			Decision
Case	Length	Width	Height	Price
1	4.1..4.2	?	1.4..1.55	Low
2	4.1..4.2	1.75..1.85	1.4..1.55	Low
3	?	1.7..1.75	?	Medium
4	4.1..4.2	1.85..1.9	1.4..1.55	Medium
5	4.2..4.5	1.85..1.9	?	Medium
6	?	1.75..1.85	1.4..1.55	High
7	4.2..4.5	?	1.55..1.6	High
8	4.2..4.5	1.85..1.9	1.55..1.6	High

For incomplete decision tables the definition of a block of an attribute-value pair must be modified in the following way [11,14]: If for an attribute a and a case x, if $a(x) = ?$, the case x should not be included in any blocks $[(a, v)]$ for all values v of attribute a.

For a case $x \in U$ the *characteristic set* $K_B(x)$ is defined as the intersection of the sets $K(x, a)$, for all $a \in B$, where B is a subset of the set A of all attributes and the set $K(x, a)$ is defined in the following way:

- If $a(x)$ is specified, then $K(x, a)$ is the block $[(a, a(x))]$ of attribute a and its value $a(x)$,
- If $a(x) = ?$ then the set $K(x, a) = U$.

For the data set from Table 4, the set of blocks of attribute-value pairs is

$[(Length, 4.1..4.2)] = \{1, 2, 4\}$,
$[(Length, 4.2..4.5)] = \{5, 7, 8\}$,
$[(Width, 1.7..1.75)] = \{3\}$,
$[(Width, 1.75..1.85)] = \{2, 6\}$,

$[(Width, 1.85..1.9)] = \{4, 5, 8\}$,
$[(Height, 1.4..1.55)] = \{1, 2, 4, 6\}$,
$[(Height, 1.55..1.6)] = \{7, 8\}$.

The corresponding characteristic sets are

$K_A(1) = \{1, 2, 4\}$,
$K_A(2) = \{2\}$,
$K_A(3) = \{3\}$,
$K_A(4) = \{4\}$,
$K_A(5) = \{5, 8\}$,
$K_A(6) = \{2, 6\}$,
$K_A(7) = \{7, 8\}$,
$K_A(8) = \{8\}$.

4 Approximations

For incomplete data sets there exist a number of different definitions of approximations. In this paper we will use only *concept* approximations.

The B-*lower approximation* of X, denoted by $\underline{appr}(X)$, is defined as follows

$$\cup \ \{K_B(x) \mid x \in X, K_B(x) \subseteq X\}. \tag{1}$$

Such lower approximations were introduced in [11,13].

The B-*upper approximation* of X, denoted by $\overline{appr}(X)$, is defined as follows

$$\cup \{K_B(x) \mid x \in X, K_B(x) \cap X \neq \emptyset\} = \cup \{K_B(x) \mid x \in X\}.$$

These approximations were studied in [11,13]. For Table 4, lower and upper approximations for the concept [(Price, low)] are

$$\underline{appr}(\{1, 2\}) = \{2\} \text{ and } \overline{appr}(\{1, 2\} = \{1, 2, 4\}).$$

5 Experiments

Our experiments were conducted on 14 well-known numerical data sets from the University of California at Irvine *Machine Learning Repository*. Every data set was converted into six incomplete and numerical data sets by incremental and random replacement of numerical values by question marks, representing lost values, with 5%, 10%, 15%, 20%, 25% and 30% of missing attribute values. For each data sets discretized by Multiple Scanning, a set of six incomplete data sets was created by replacing intervals by question marks in the same places where question marks were located in the original numerical data sets converted to incomplete.

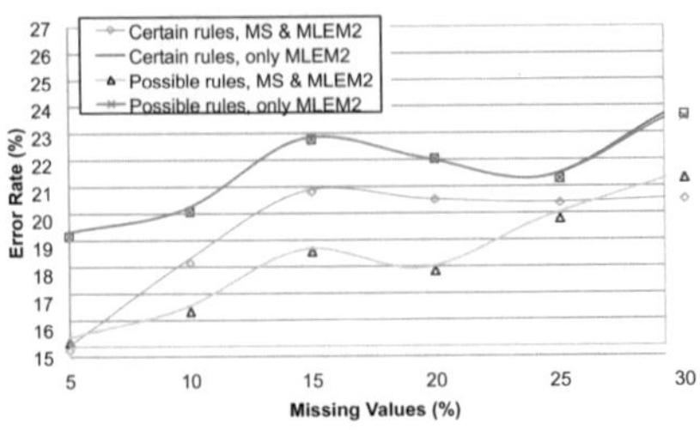

Fig. 1. Error rates for the *australian* data set

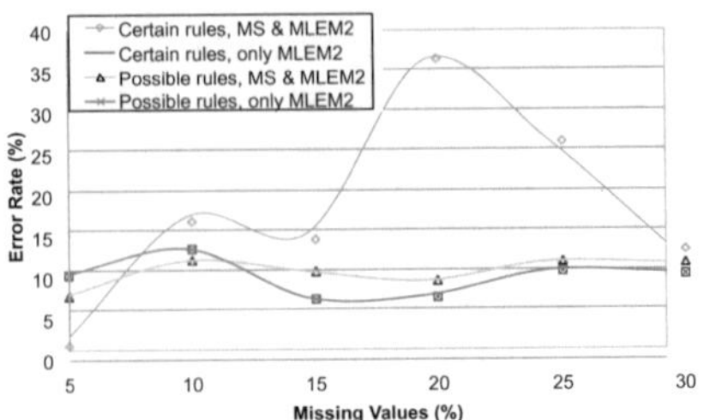

Fig. 2. Error rates for the *bankruptcy* data set

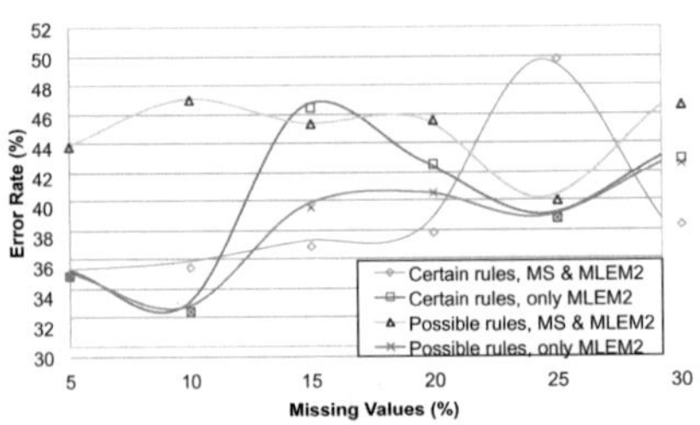

Fig. 3. Error rates for the *bupa* data set

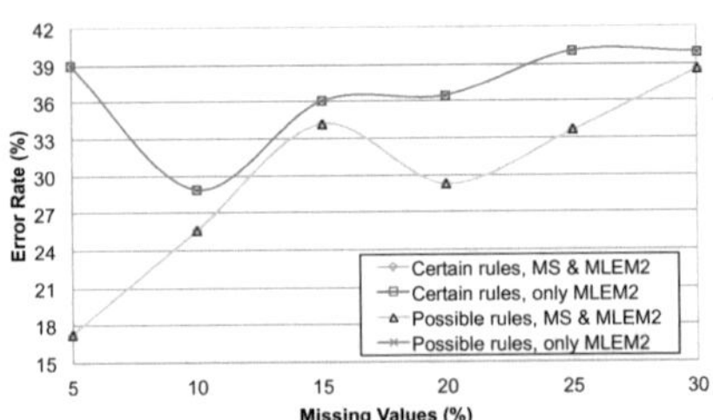

Fig. 4. Error rates for the *connectionist bench* data set

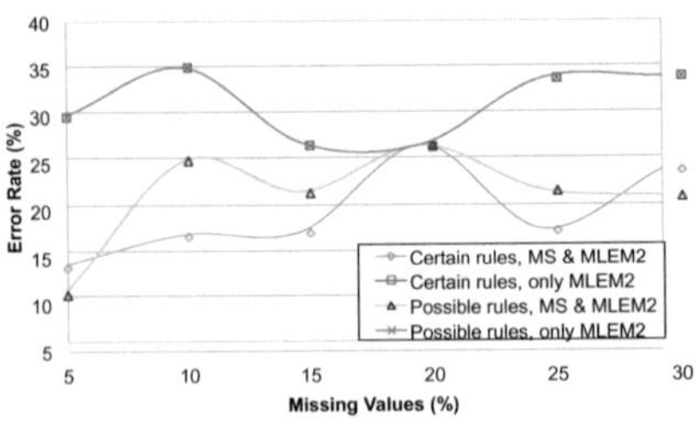

Fig. 5. Error rates for the *echocardiogram* data set

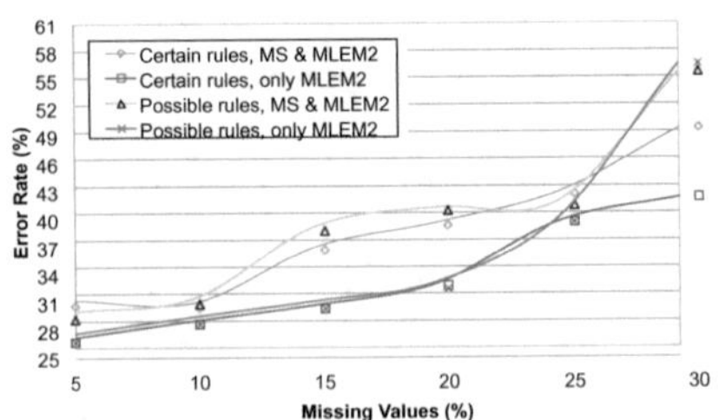

Fig. 6. Error rates for the *ecoli* data set

Results of our experiments are presented on Figs. 1, 2, 3, 4, 5, 6, 7, 8, 9, 10, 11, 12, 13 and 14. For certain rule sets, MLEM2 was more successful for two data sets, while the MLEM2, preceded by Multiple Scanning discretization, was more successful for four data sets, for eight data sets the difference was not significant (Wilcoxon test, 5% significance level). Similarly, for possible rule sets MLEM2 was more successful for two data sets, while MLEM2 preceded by Multiple Scanning was more successful for three data sets.

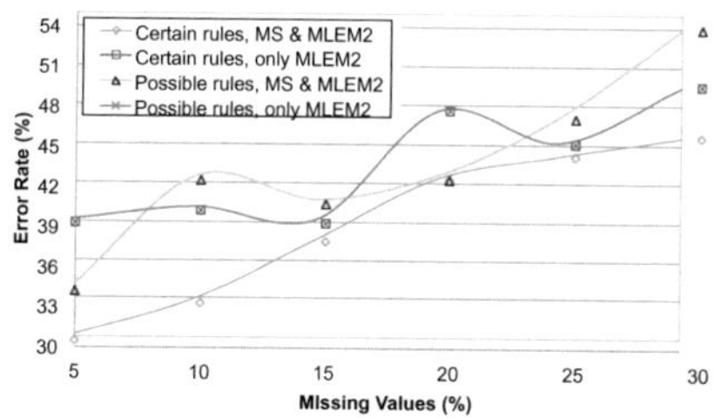

Fig. 7. Error rates for the *glass* data set

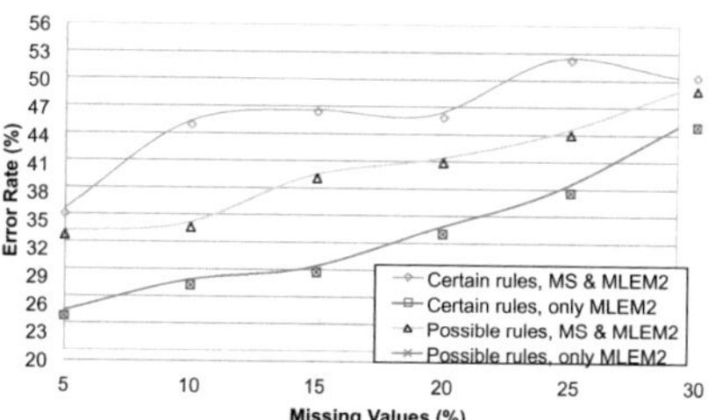

Fig. 8. Error rates for the *image segmentation* data set

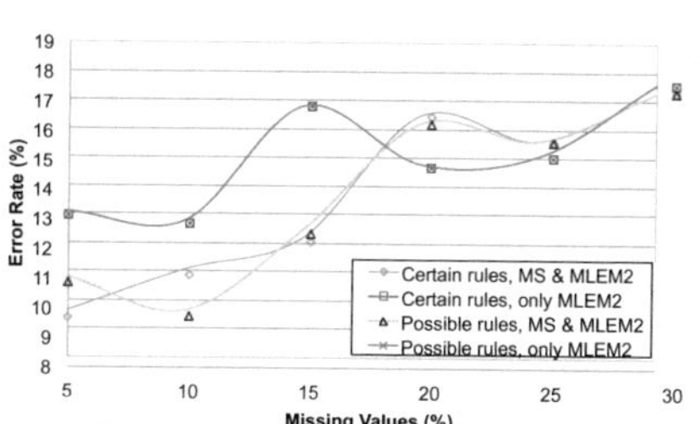

Fig. 9. Error rates for the *ionosphere* data set

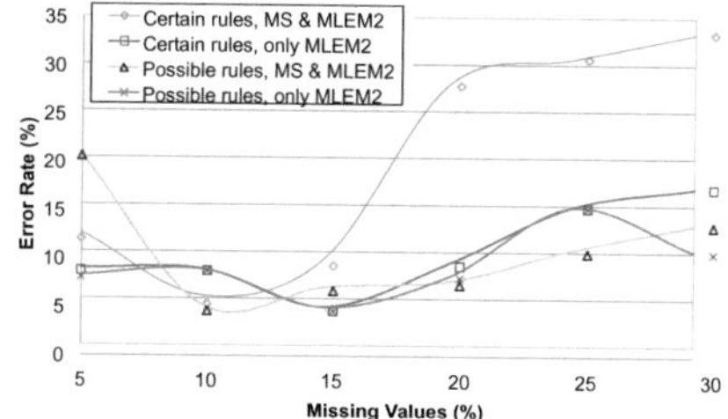

Fig. 10. Error rates for the *iris* data set

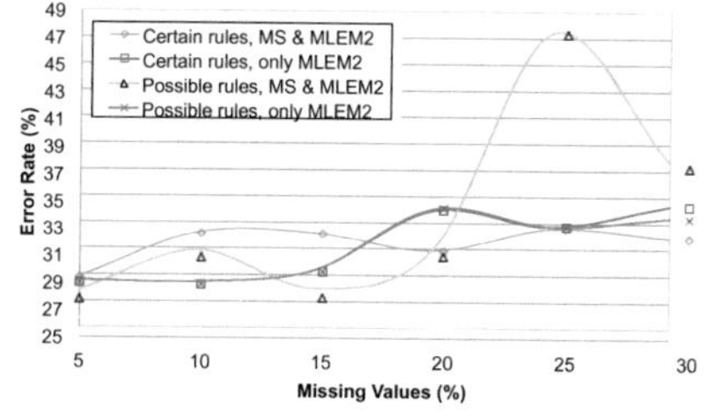

Fig. 11. Error rates for the *pima* data set

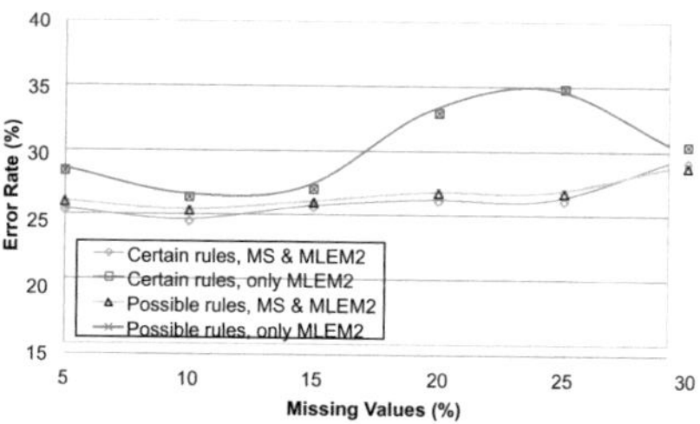

Fig. 12. Error rates for the *wave* data set

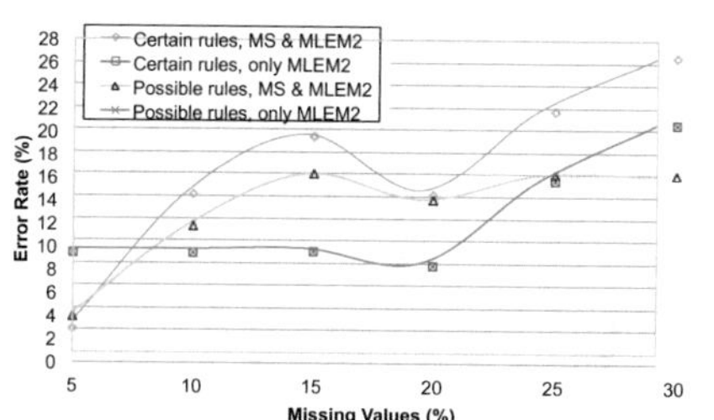

Fig. 13. Error rates for the *wine recognition* data set

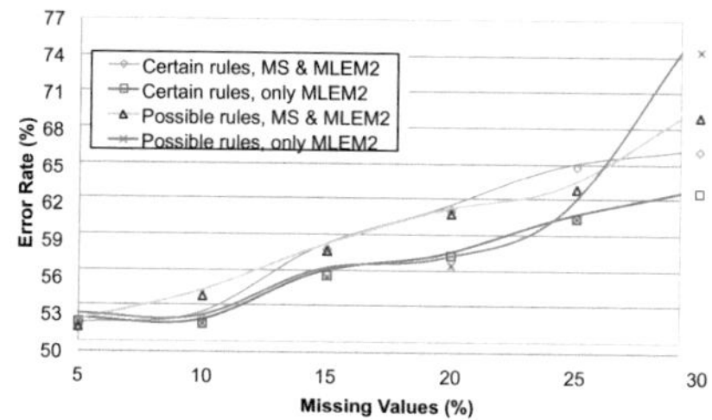

Fig. 14. Error rates for the *yeast* data set

6 Conclusions

Our main objective is to compare, in terms of an error rate, two setups: MLEM2 applied for rule induction directly from incomplete and numerical data and MLEM2 inducing rule sets from data sets previously discretized by Multiple Scanning and then converted to be incomplete. Our conclusion is that there is not significant difference between both setups. Rule sets may be induced directly from incomplete and numerical data using MLEM2 as well.

References

1. Blajdo, P., Grzymala-Busse, J.W., Hippe, Z.S., Knap, M., Mroczek, T., Piatek, L.: A comparison of six approaches to discretization—a rough set perspective. In: Proceedings of the Rough Sets and Knowledge Technology Conference, pp. 31–38 (2008)
2. Chan, C.C., Batur, C., Srinivasan, A.: Determination of quantization intervals in rule based model for dynamic. In: Proceedings of the IEEE Conference on Systems, Man, and Cybernetics, pp. 1719–1723 (1991)
3. Chmielewski, M.R., Grzymala-Busse, J.W.: Global discretization of continuous attributes as preprocessing for machine learning. Int. J. Approximate Reasoning **15**(4), 319–331 (1996)
4. Clarke, E.J., Barton, B.A.: Entropy and MDL discretization of continuous variables for bayesian belief networks. Int. J. Intell. Syst. **15**, 61–92 (2000)
5. Elomaa, T., Rousu, J.: General and efficient multisplitting of numerical attributes. Mach. Learn. **36**, 201–244 (1999)
6. Elomaa, T., Rousu, J.: Efficient multisplitting revisited: optima-preserving elimination of partition candidates. Data Min. Knowl. Disc. **8**, 97–126 (2004)
7. Fayyad, U.M., Irani, K.B.: On the handling of continuous-valued attributes in decision tree generation. Mach. Learn. **8**, 87–102 (1992)
8. Fayyad, U.M., Irani, K.B.: Multi-interval discretization of continuous-valued attributes for classification learning. In: Proceedings of the Thirteenth International Conference on Artificial Intelligence, pp. 1022–1027 (1993)
9. Grzymala-Busse, J.W.: A new version of the rule induction system LERS. Fundamenta Informaticae **31**, 27–39 (1997)
10. Grzymala-Busse, J.W.: MLEM2—discretization during rule induction. In: Proceedings of the International Conference on Intelligent Information Processing and WEB Mining Systems, pp. 499–508 (2003)
11. Grzymala-Busse, J.W.: Rough set strategies to data with missing attribute values. In: Notes of the Workshop on Foundations and New Directions of Data Mining, in Conjunction with the Third International Conference on Data Mining, pp. 56–63 (2003)
12. Grzymala-Busse, J.W.: Characteristic relations for incomplete data: a generalization of the indiscernibility relation. In: Proceedings of the Fourth International Conference on Rough Sets and Current Trends in Computing, pp. 244–253 (2004)
13. Grzymala-Busse, J.W.: Data with missing attribute values: generalization of indiscernibility relation and rule induction. Trans. Rough Sets **1**, 78–95 (2004)
14. Grzymala-Busse, J.W.: Three approaches to missing attribute values—a rough set perspective. In: Proceedings of the Workshop on Foundation of Data Mining, in Conjunction with the Fourth IEEE International Conference on Data Mining, pp. 55–62 (2004)

15. Grzymala-Busse, J.W.: A multiple scanning strategy for entropy based discretization. In: Proceedings of the 18th International Symposium on Methodologies for Intelligent Systems, pp. 25–34 (2009)
16. Grzymala-Busse, J.W.: Discretization based on entropy and multiple scanning. Entropy **15**, 1486–1502 (2013)
17. Grzymala-Busse, J.W., Mroczek, T.: A comparison of two approaches to discretization: multiple scanning and c4.5. In: Proceedings of the 6th International Conference on Pattern Recognition and Machine Learning, pp. 44–53 (2015)
18. Grzymala-Busse, J.W., Mroczek, T.: A comparison of four approaches to discretization based on entropy. Entropy **18**, 1–11 (2016)
19. Kohavi, R., Sahami, M.: Error-based and entropy-based discretization of continuous features. In: Proceedings of the Second International Conference on Knowledge Discovery and Data Mining, pp. 114–119 (1996)
20. Kotsiantis, S., Kanellopoulos, D.: Discretization techniques: a recent survey. GESTS Int. Trans. Comput. Sci. Eng. **32**(1), 47–58 (2006)
21. Liu, H., Hussain, F., Tan, C.L., Dash, M.: Discretization: an enabling technique. Data Min. Knowl. Disc. **6**, 393–423 (2002)
22. Nguyen, H.S., Nguyen, S.H.: Discretization methods in data mining. In: Polkowski, L., Skowron, A. (eds.) Rough Sets in Knowledge Discovery 1: Methodology and Applications, pp. 451–482. Physica-Verlag, Heidelberg (1998)
23. Pawlak, Z.: Rough sets. Int. J. Comput. Inform. Sci. **11**, 341–356 (1982)
24. Pawlak, Z.: Rough Sets. Theoretical Aspects of Reasoning about Data. Kluwer Academic Publishers, Dordrecht (1991)
25. Pawlak, Z., Grzymala-Busse, J.W., Slowinski, R., Ziarko, W.: Rough sets. Commun. ACM **38**, 89–95 (1995)
26. Quinlan, J.R.: C4.5: Programs for Machine Learning. Morgan Kaufmann Publishers, San Mateo (1993)

Heuristic Method of Air Defense Planning for an Area Object with the Use of Very Short Range Air Defense

Tadeusz Pietkiewicz(✉), Adam Kawalec, and Bronisław Wajszczyk

Faculty of Electronics, Institute of Radioelectronics, Military University of Technology, Gen. S. Kaliskiego 2 Street, 00-908 Warsaw, Poland
{tadeusz.pietkiewicz,adam.kawalec, bronislaw.wajszczyk}@wat.edu.pl

Abstract. This paper presents a heuristic method of planning the deployment of very short-range anti-air missile and artillery sets (VSHORAD) around the protected area object. The function dependent on the distance between the earliest feasible points of destroying targets and the center of the protected area object is taken as an objective function. This is a different indicator from those commonly used in the literature based on the likelihood of a defense zone penetration by the means of air attack (MAA), the kill probability of the MAA and the probability of area object losses. The model constraints result directly from the restrictions imposed by the real air defense system and the nature of the area objects. The presented sub-optimal heuristic method has been implemented. In the final part of the paper computational results have been presented.

Keywords: Optimal planning · Air defense system · Area objects protection · Deployment of anti-air missile and artillery sets of very short-range (VSHORAD) sets · Binary programing · Heuristic method

1 Introduction

Modern means of air attack allow penetration of protected territory and the destruction of important industrial or defense installations. One of the most important means of air defense are anti-aircraft missile and artillery sets (AAMAS). They offer an opportunity to cover the strategically important point or area objects. These sets enable the creation of a short range air defense (SHORAD) or a very short range air defense (VSHORAD) system around objects of a strategic nature. This system is usually composed of the following elements:

- anti-aircraft missile and artillery sets,
- short range radar,
- control post connected with a superior air defense command system.

Systems based on AAMAS can be effectively used to repel air raids, if the sets are properly deployed at the planning stage. A lot of factors should be considered in this process. This includes terrain characteristics that may alter the exposure zone of the

I. Czarnowski et al. (eds.), *Intelligent Decision Technologies 2017*,
Smart Innovation, Systems and Technologies 72, DOI 10.1007/978-3-319-59421-7_21

local radar and kill zones of anti-aircraft missile and artillery sets. In the process of planning the AAMA sets deployment we are dealing with a large indeterminacy of many factors, including the type and means of an air attack, the type of weapons used by the enemy, the composition of the air raid, the trajectory of air attack means, the air attack means tactics, the enemy's knowledge about the area object and many others. Hence, we think that one can assume simplified models of the AAMA sets kill zones and the local radar exposure zone.

In this research we assumed that the planning phase of the deployment of the AAMA sets is preceded by a stage of reconnaissance on the terrain. The purpose of this reconnaissance mission is to identify of potential positions, which can be used to deploy the AAMA sets. The suitability of specific sites depend on favorable terrain conditions and the necessary technical equipment.

In summarizing, the problem of the AAMA sets deployment in air defense system of the area object consists in the optimal deployment of a fixed number of the AAMA sets around an area or around a set of points defining a defended object, taking into account the characteristics of the surrounding terrain and features of kill zones of the AAMA sets.

2 An Optimization Problem Formulation for an Area Object Air Defense Planning

In another paper at this conference titled "An optimization problem of Air Defense Planning for an Area Object" [4] a formulation of the problem of optimal the AAMA sets deployment for an area object air defense has been presented.

Selection of the best deployment points, in which a given number of AAMA sets will be distributed, means choosing pairs: the number of the deployment point from the feasible number set and the number of AAMA set type in such a way as to maximize a criterion (evaluation) function, and the number of selected AAMA sets does not exceed the maximum number of AAMA sets of specific types.

A formulation of the problem of optimal pair selection: deployment point and associated type of AAMA set is presented below.

Let $\boldsymbol{V} = [v_{jt}]_{J \times T}$ be a binary matrix of decision variables with the following interpretation:

$$v_{jt} = \begin{cases} 0 & - \text{ a AAMA set of the } t\text{ - th type is not located in } j \text{ - th point ,} \\ 1 & - \text{ a AAMA set of the } t\text{ - th type is located in } j \text{ - th point.} \end{cases} \quad (1)$$

Solution of optimization problem consist in finding such a matrix $\boldsymbol{V}^* = [v^*_{jt}]_{J \times T}$ to maximize the function $f(\boldsymbol{V})$

$$f(\boldsymbol{V}^*) = \sum_{\substack{j \in \boldsymbol{D} \\ d^o(j,t) \neq -\infty}} \sum_{t=1}^{T} v^*_{jt}\, d^o(j,t) = \max_{\boldsymbol{V} \in \boldsymbol{\Omega}} \sum_{\substack{j \in \boldsymbol{D} \\ d^o(j,t) \neq -\infty}} \sum_{t=1}^{T} v_{jt}\, d^o(j,t) \quad (2)$$

with constraints

$$\boldsymbol{\Omega} = \left\{ \boldsymbol{V} = [v_{jt}]_{J \times T} : v_{jt} \in \{0, 1\} \wedge \forall j \in \boldsymbol{D} \sum_{t=1}^{T} v_{jt} \leq 1 \wedge \forall t \in \{1, \ldots, T\} \sum_{j \in \boldsymbol{D}} v_{jt} \leq K_t \right\}, \tag{3}$$

where

$$\boldsymbol{D} = \left\{ D_j = (x_j^d, y_j^d),\ j = 1, \ldots, J \right\} \in \bar{\boldsymbol{D}} \tag{4}$$

is a set of potential location points, which are taken into account during plan elaborating;

T - the number of AAMA set types;

K_t - the maximum number of AAMA sets of t-th type (t = 1,…,T) that can be used to deploy.

The function $d^o(j, t)$ is an assessment function of pair selection: the deployment point and type of AAMA set

$$d^o(j,t) = \begin{cases} \sum\limits_{\substack{n=1 \\ z_{jnt} \neq -\infty \\ p_n > 0}}^{N} s_t p_n z_{jnt} & \exists n = 1, \ldots, N\ z_{jnt} \neq -\infty \\ -\infty & \forall n = 1, \ldots, N\ z_{jnt} = -\infty \end{cases} \quad j \in \boldsymbol{D}, t \in \{1, \ldots, T\}; \tag{5}$$

where

N - the number of possible courses of the air raid,

p_n - the estimated probability of the raid from n-th direction,

s_t - the average efficiency of AAMA set of t-th type

z_{jnt} - value which describes the possibility of n-th air target kill by AAMA set of t-th type from j-th deployment point:

$$z_{jnt} = \begin{cases} -\infty & \text{when there is no possibility of } n \text{ - th air target kill} \\ & \text{by AAMA set of } t \text{ - th type from } j \text{ - th deployment point} \\ 1 & \text{when there is a possibility of } n \text{ - th air target kill} \\ & \text{by AAMA set of } t \text{ - th type from } j \text{ - th deployment point} \end{cases}, \tag{6}$$

The method of calculating the value of z_{jnt} is presented in the paper "An optimization problem of Air Defense Planning for an Area Object".

The problem (2) with constraints (3) belongs to the class of the binary programming tasks. We propose an heuristic algorithm to solve this task.

3 AAMA Sets Deployment Algorithm

The problem (2) solution with constraints (3) provides the following algorithm.

1. Calculation of the center protected area $S = (x_s, y_s)$.
2. For each air raid direction n conversion of azimuth angles α_A^n (expressed in degrees) for angles β_n expressed in radians.
3. Calculation of unit vectors $\boldsymbol{w}_n$ in the air raid direction.
4. Calculation of feasible deployment points from which one can kill air targets in all raid directions.
5. Calculation of the kill points coordinates for those deployment points for which killing is possible ($z_{jnt} = 1$).
6. Calculation of the vectors connecting the center of the protected area $\boldsymbol{S}$ with all points $O_{jt}^{n,1} = (x_{jt}^{o,n,1}, y_{jt}^{o,n,1})$ and $O_{jt}^{n,2} = (x_{jt}^{o,n,2}, y_{jt}^{o,n,2})$ and their scalar products with unit vector of n-th direction $\boldsymbol{w}^n$.
7. Calculation of the value of the evaluation function for feasible pairs: a number of deployment point and a number of AAMA set type.
8. Choosing the best deployment points and AAMA set types.

Calculation of the matrix $\boldsymbol{V}^* = [v_{jt}^*]_{J\times T}$ for optimization task with constraints $\boldsymbol{\Omega}$ can be iteratively realized in $LK = \min\{K_1 + \ldots + K_T, J\}$ iterations.

1. Iteration 1
   1a. $\boldsymbol{D}^* := \boldsymbol{D}$;
   1b. Find such $j_1^* \in \boldsymbol{D}^*$ and $t_1^* \in \{1, \ldots, T\}$ that

$$d^o(j_1^*, t_1^*) = \max_{\substack{j \in \boldsymbol{D}^* \\ t \in \{1,\ldots,T\} \wedge K_t > 0 \\ d^o(j,\ t) \neq -\infty}} d^o(j, t)$$

   1c. $v_{j_1^* t_1^*} := 1$;
   1d. $\boldsymbol{D}^* := \boldsymbol{D}^* \backslash \{j_1^*\}$;
   1e. $K_{t_1^*} := K_{t_1^*} - 1$.
2. Iteration i-th
   2a. Find such $j_i^* \in \boldsymbol{D}^*$ and $t_i^* \in \{1, \ldots, T\}$ that

$$d^o(j_i^*, t_i^*) = \max_{\substack{j \in \boldsymbol{D}^* \\ t \in \{1,\ldots,T\} \wedge K_t > 0 \\ d^o(j,t) \neq -\infty}} d^o(j, t)$$

   2b. $v_{j_i^* t_i^*} := 1$;
   2c. $\boldsymbol{D}^* := \boldsymbol{D}^* \backslash \{j_i^*\}$;
   2d. $K_{t_i^*} := K_{t_i^*} - 1$.

3. Iteration *LK* (final)

   3a. Find such $j^*_{LK} \in \boldsymbol{D}$ and $t^*_{LK} \in \{1, \ldots, T\}$ that

$$d^o(j^*_{LK}, t^*_{LK}) = \max_{\substack{j \in \boldsymbol{D}^* \\ t \in \{1,\ldots,T\} \wedge K_t > 0 \\ d^o(j,\, t) \neq -\infty}} d^o(j, t)$$

   3b. $v_{j^*_{LK} t^*_{LK}} := 1;$

(a)

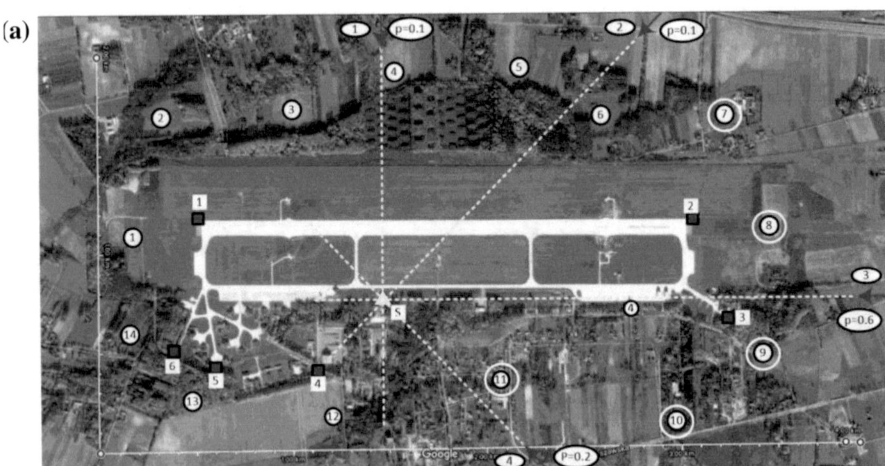

(b)

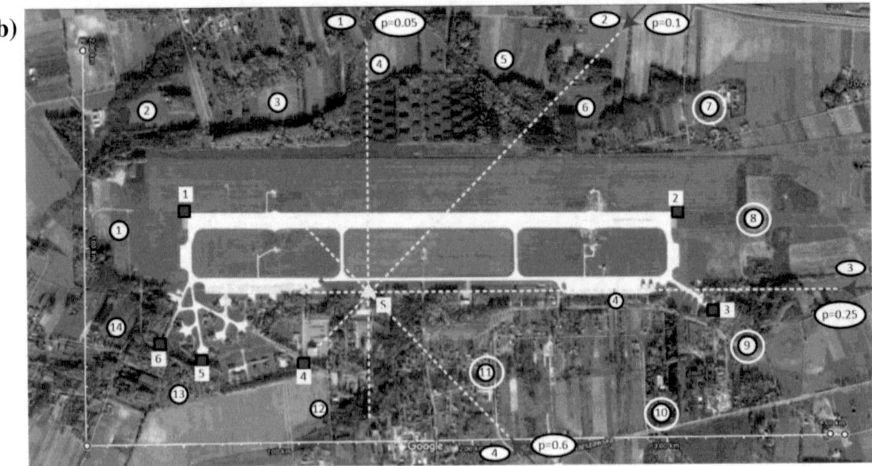

Fig. 1. An illustration of optimal deployment of AAMA sets around the area object of the airport type for different variants of the air raid. The air raid variants differ in probability of the target appearance for each direction.

(c)

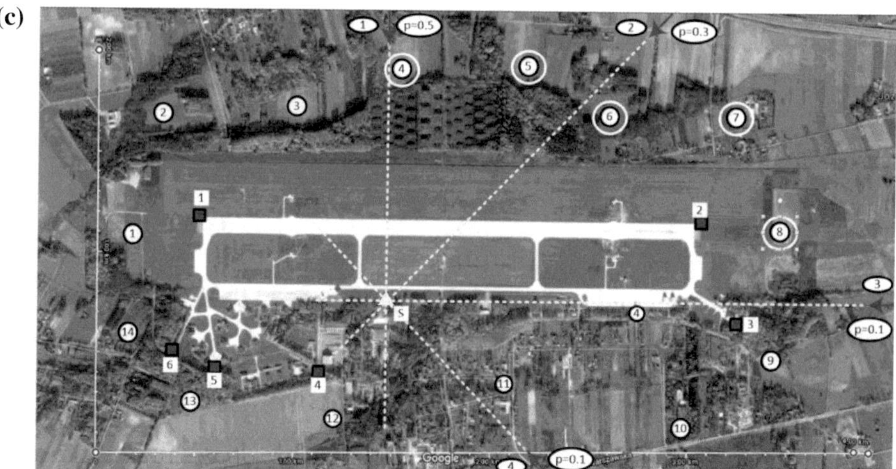

(d)

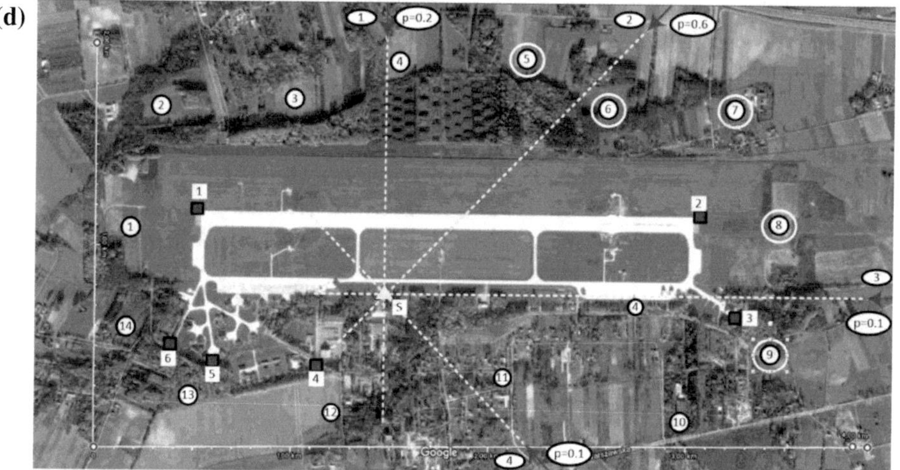

Fig. 1. (continued)

4 Calculation Examples

This section presents the results of applying the algorithm to the plan air defense of certain military airports (image taken from the portal "Google map") with the use of one type of AAMA sets with the range of 6 km.

In these examples, the following assumptions have been taken into account:

1. The protected area object is described by a polygon with five vertices shown in Figs. 1a–d using the squares.
2. The number of potential AAMA sets deployment points equals 14.

Table 1. Air raid configuration and calculation results.

Air raid attributes		Air raid configuration			
		Z1	Z2	Z3	Z4
Direction azimuths α_i, (i = 1, …,4) [degrees]	α_1	180	180	180	180
	α_2	225	225	225	225
	α_3	270	270	270	270
	α_4	315	315	315	315
Direction probabilities p_i, (i = 1,…,4)	p_1	0.5	0.2	0.1	0.05
	p_2	0.3	0.6	0.1	0.1
	p_3	0.1	0.1	0.6	0.25
	p_4	0.1	0.1	0.2	0.6
Numbers of calculated deployment points		7, 5, 6, 4, 8	7, 8, 6, 5, 9	9, 10, 8, 7, 11	9, 10, 8, 7, 11

3. The location of potential AAMA sets deployment points are shown in Figs. 1a–d using small circles.
4. The maximum number of AAMA sets is 5.
5. The air raid can be divided into four directions, and the probability and the azimuths of which five different raid configurations are shown in Table 1.

The selected deployment points are surrounded in Figs. 1a–d using larger white circles, and their numbers are shown in Table 1 in order of decreasing assessment function values (5). A phenomenon of moving the selected deployment point set with a change of the number of the most likely air raid direction can be observed in Figs. 1a–d.

5 Conclusion

This paper has presented the heuristic method of planning the anti-aircraft missile and artillery sets deployment in an air defense system of important industrial or defense area objects. A new approach to building the evaluation criterion function for AAMA set deployment has been proposed. A sub-optimal heuristic method has been suggested as a method to solve the problem of the deployment of the AAMA sets.

The presented model can be developed. In particular, one should enter the allowed distance between the AAMA sets and combat potential distribution within the air raid. As a method to solve the problem of the optimal deployment of anti-aircraft missile and artillery sets one can use, dynamic programming [1], adaptive dynamic programming [2] or genetic programming [3].

References

1. Ghode, D., Prasad, U.R., Guruprasad, K.: Missile battery placement for air defense: a dynamic programming approach. Appl. Math. Model. **17**(9), 450–458 (1993). doi:10.1016/0307-904X(93)90086-V

2. Pant, M., Deep, K.: Building a better air defence system using genetic algorithms. In: Gabrys, B., Howlett, R.J., Jain, L.C. (eds.) Knowledge-Based Intelligent Information and Engineering Systems, KES 2006. LNCS, vol. 4251, pp. 951–959. Springer, Heidelberg (2006). doi:10.1007/11892960_114
3. Ahner, D.K., Parson, C.R.: Optimal multi-stage allocation of weapons to targets using adaptive dynamic programming. Optim. Lett. **9**(8), 1689–1701 (2015). doi:10.1007/s11590-014-0823-x
4. Pietkiewicz, T., Kawalec, A., Wajszczyk, B.: An optimization problem of air defense planning for an area object. In: Intelligent Decision Technologies 2017. Smart Innovation, Systems and Technologies. Proceedings of the 9th KES International Conference on Intelligent Decision Technologies (KES-IDT 2017). Springer

An Optimization Problem of Air Defense Planning for an Area Object

Tadeusz Pietkiewicz(✉), Adam Kawalec, and Bronisław Wajszczyk

Faculty of Electronics, Institute of Radioelectronics, Military University of Technology, 00-908 Warsaw gen. S. Kaliskiego 2 Street, Poland
{tadeusz.pietkiewicz, adam.kawalec, bronislaw.wajszczyk}@wat.edu.pl

Abstract. This paper presents an optimization problem of planning the deployment of very short-range anti-air missile and artillery sets (VSHORAD) around the protected area object. The function dependent on the distance between the earliest feasible points of destroying targets and the center of the protected area object is taken as an objective function. This is a different indicator from those commonly used in the literature based on the likelihood of a defense zone penetration by the means of air attack (MAA), the kill probability of the MAA and the probability of area object losses. The model constraints result directly from the restrictions imposed by the real air defense system and the nature of the area objects. In the final part of the paper an optimization problem formulation as a task of binary programming is presented.

Keywords: Optimal planning · Air defense system · Area objects protection · Deployment of anti-air missile and artillery sets of very short-range (VSHORAD) sets

1 Introduction

Modern means of air attack allow penetration of protected territory and the destruction of important industrial or defense installations. One of the most important means of air defense are anti-aircraft missile and artillery sets (AAMAS). They offer an opportunity to cover the strategically important point or area objects. These sets enable the creation of a short range air defense (SHORAD) or a very short range air defense (VSHORAD) system around objects of a strategic nature. This system is usually composed of the following elements:

- anti-aircraft missile and artillery sets,
- short range radar,
- control post connected with a superior air defense command system.

We assumed that the control post is connected to the air defense command system of the country according to the network-centric warfare principle. This means that the control post receives:

- recognized air picture of the region and on the approaches,
- plans of a higher command post (superior).

I. Czarnowski et al. (eds.), *Intelligent Decision Technologies 2017*,
Smart Innovation, Systems and Technologies 72, DOI 10.1007/978-3-319-59421-7_22

Systems based on AAMAS can be effectively used to repel air raids, if the sets are properly deployed at the planning stage. A lot of factors should be considered in this process. This includes terrain characteristics that may alter the exposure zone of the local radar and kill zones of anti-aircraft missile and artillery sets. In the bibliography one can notice a clear aspire to accurately take into account as many factors as possible during the planning process [3]. The authors believe that this approach is fully justified in another problem, namely the weapon target assignment (WTA) problem in real time during the repelling of air raids [5, 6]. In the process of planning the AAMA sets deployment we are dealing with a large indeterminacy of many factors, including the type and means of an air attack, the type of weapons used by the enemy, the composition of the air raid, the trajectory of air attack means, the air attack means tactics, the enemy's knowledge about the area object and many others. Hence, we think that one can assume simplified models of the AAMA sets kill zones and the local radar exposure zone.

In this research we assumed that the planning phase of the deployment of the AAMA sets is preceded by a stage of reconnaissance on the terrain. The purpose of this reconnaissance mission is to identify of potential positions, which can be used to deploy the AAMA sets.

2 Model of Protected Object

In this paper we assumed that the protected object is an area object, of which a defense or economy value is spread over a certain area or at certain points in an area.

We assumed that the protected area object $\boldsymbol{\Psi}$ is limited to a polygon.

Let

$\boldsymbol{B}$ - a set of vertices of protected area object polygon, where

$$\boldsymbol{B} = \left\{B_i = (x_i^b, y_i^b) \in \mathscr{R}^2,\ i = 1,\ldots,I\right\} \tag{1}$$

We define a set $\boldsymbol{\Psi}$ as a convex combination of vertices from the set $\boldsymbol{B}$:

$$\boldsymbol{\Psi} = \left\{(x,y) \in \mathscr{R}^2 : (x,y) = \sum_{i=1}^{I} \alpha_i (x_i^b, y_i^b),\ \alpha_i \in [0,1],\ \sum_{i=1}^{I} \alpha_i = 1,\ (x_i^b, y_i^b) \in \boldsymbol{B}\right\} \tag{2}$$

A value function for protected area objects can be introduced. In the case of planar distribution one can define the function of the value density $w : \mathscr{R}^2 \rightarrow \mathscr{R}$ as follows:

$$w(x,y) = \begin{cases} = 0 \text{ for } (x,y) \notin \boldsymbol{\Psi} \\ \geq 0 \text{ for } (x,y) \in \boldsymbol{\Psi} \end{cases} \tag{3}$$

In the case of point value distribution within the object one should specify the elementary point protected objects belonging to the area object. Let $\boldsymbol{A} = \{1,\ldots,A\}$

denote a set of numbers of elementary objects included in the area object with coordinates (x_a, y_a) and values w_a.

We determine the center of the area object (x_s, y_s) as a center of object value. In the continuous case we have [7]:

$$x_s = \frac{\int_{-\infty}^{\infty}\int_{-\infty}^{\infty} xw(x,y)dxdy}{\int_{-\infty}^{\infty}\int_{-\infty}^{\infty} w(x,y)dxdy}, \quad y_s = \frac{\int_{-\infty}^{\infty}\int_{-\infty}^{\infty} yw(x,y)dxdy}{\int_{-\infty}^{\infty}\int_{-\infty}^{\infty} w(x,y)dxdy}. \tag{4}$$

In the discrete case we have [7]:

$$x_s = \frac{\sum_{a=1}^{A} x_a w_a}{\sum_{a=1}^{A} x_a}, \quad y_s = \frac{\sum_{a=1}^{A} y_a w_a}{\sum_{a=1}^{A} x_a}. \tag{5}$$

3 Model of Area Object Air Defense System

3.1 Model of AAMA Sets

For practical reasons, we assumed that AAMA sets can have different operational properties. There are three types of AAMA sets. The first two are basic: anti-aircraft artillery set and anti-aircraft missile set. The third is a complex that results from joining two basic types of sets.

Each type of set has the following attributes:

- the range of the set d_t, where t is the number of AAMA set type ($t = 1,\ldots,T$), it is a radius of the kill zone at a height equal to half the maximum height of the zone,
- the average efficiency of AAMA set type s_t, where t is the number of AAMA set type ($t = 1,\ldots,T$); the average efficiency of AAMA set type is defined as an average kill probability under different operational conditions.

The above-mentioned characteristics are average. In the bibliography of the AAMA sets deployment and weapon target assignment problems [1–8] a number of specific characteristics are taken into account.

The impact range and efficiency of the third type of AAMA sets depends on the corresponding attributes of the first and second type of AAMA sets in accordance with the following formulas:

$$d_3 = \max\{d_1, d_2\}, \quad s_3 = 1 - (1 - s_1)(1 - s_2). \tag{6}$$

Moreover, we introduce limitations K_t on the maximum number of AAMA sets of t-th type ($t = 1,\ldots,T$) that can be used to deploy.

3.2 AAMA Sets Deployment Model

The AAMA set may be arranged in one of a predetermined set location points.

At one location point only one AAMA set may be arranged. We define a set of all possible location points,

$$\bar{\boldsymbol{D}} = \left\{ \bar{D}_j = (\bar{x}_j^d, \bar{y}_j^d) \in \mathscr{R}^2,\ j = 1, \ldots, \bar{\boldsymbol{J}} \right\}. \tag{7}$$

Furthermore, we assumed that the air defense plan is developed taking into account the forecasted scenario of enemy action (air raid), for which it is possible to reduce the set of all possible location points. Therefore we introduce a set of selected potential location points, which are taken into account during plan elaborating,

$$\boldsymbol{D} = \left\{ D_j = (x_j^d, y_j^d),\ j = 1, \ldots, J \right\} \in \bar{\boldsymbol{D}}. \tag{8}$$

3.3 Air Raid Model

The content of assumptions for the cover plan (AfCP) should be submitted from the superior command post and should include assumptions regarding the forecasted enemy air raid. Each direction of the air raid is characterized by azimuth α_n and the estimated probability p_n of the raid from this direction $(0 \le p_n \le 1, \sum_{n=1}^{N} p_n = 1)$, where N is the number of possible courses of the raid.

The above air raid model assumes some simplifications - all the directions of the raid will pass through the center of the area object defined by (4) or (5).

AfCP can also specify the maximum number K_t of AAMA sets of t-th type (t = 1,..,T) that can be used during deployment planning.

Due to the fact that in the content of combat documents the angles of azimuth α_n are expressed in degrees measured from the north in a clockwise direction, it is necessary to convert them into angles β_n expressed in radians and calculated relative to the horizontal axis (OX) of the local Cartesian coordinate system in an counter-clockwise direction for each direction of the air raid n.

4 Calculation of Potential Points of Air Targets Kill

In this chapter, we determine feasible points of AAMA sets deployment, from which forecasted air targets kill is possible, and the coordinates of the kill points. We used the models of area objects, AAMA sets and forecasted air raids.

4.1 Calculation of the Feasible Deployment Point

The unit vectors of the air raid directions $\boldsymbol{w}_n = (w_x^n, w_y^n)$(marked in Fig. 1 with a bold line) are helpful in determining the physical possibilities of air targets kill. One can notice that:

$$w_x^n = \cos \beta_n, \quad w_y^n = \sin \beta_n. \tag{9}$$

Moreover, there is

$$\|\boldsymbol{w}_n\| = \sqrt{\left(w_x^n\right)^2 + \left(w_y^n\right)^2} = 1. \tag{10}$$

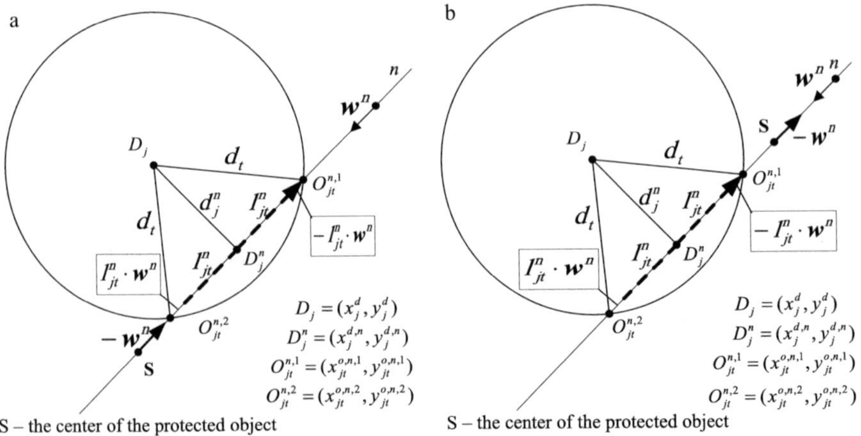

Fig. 1. An illustration of the concept of determining the air target kill point

In order to determine the feasible points of AAMA deployment for all air raid directions (Fig. 1), we use the equation of a line passing through the center of the protected object $\boldsymbol{S} = (x_s, y_s)$ and parallel to the vector $\boldsymbol{w}_n = (w_x^n, w_y^n)$:

$$\frac{x - x_s}{w_x^n} = \frac{y - y_s}{w_y^n}, \tag{11}$$

from where we get

$$w_y^n x - w_x^n y - w_y^n x_s + w_x^n y_s = 0. \tag{12}$$

The coefficients of the general equation of the line (12) are the following:

$$A_n = w_y^n \quad B_n = -w_x^n \quad C_n = -w_y^n x_s + w_x^n y_s. \tag{13}$$

The distance d_j^n the point $D_j = (x_j^d, y_j^d)$ from the n-th direction is determined by the following formula:

$$d_j^n = \frac{\left| w_y^n x_j^d - w_x^n y_j^d - w_y^n x_s + w_x^n y_s \right|}{\sqrt{\left(w_x^n\right)^2 + \left(w_y^n\right)^2}}. \tag{14}$$

Because there is (10) we finally get:

$$d_j^n = \left| w_y^n x_j^d - w_x^n y_j^d - w_y^n x_s + w_x^n y_s \right|. \tag{15}$$

If $d_j^n \le d_t$ (d_t is the range of an AAMA set of the t-th type), then an AAMA set of this type can kill an air target on the n-th air raid direction from the location point D_j. If $d_j^n > d_t$, then an AAMA set of this type cannot kill an air target on the n-th air raid direction from the location point D_j.

The d_j^n values allow to determine the values of the matrix $\mathbf{Z}$ which describe the possibility of an air target kill by all types of AAMA sets from all deployment points:

$$\mathbf{Z} = [z_{jnt}]_{J \times N \times T}, \quad z_{jnt} \in \mathcal{R}, \tag{16}$$

$$z_{jnt} = \begin{cases} -\infty & \text{when} \quad d_j^n > d_t \\ 1 & \text{when} \quad d_j^n \le d_t \end{cases}, \tag{17}$$

where symbol $-\infty$ indicates the smallest negative number possible to save in the given type of variable.

4.2 Calculation of the Coordinates of Air Target Kill Points

This section specifies a method of determining the coordinates of air target kill points when it is possible $(z_{jnt} = 1)$.

The concept of problem solving is shown in Fig. 1a and b. Segment $\overline{O_j^{n,1} O_j^{n,2}}$ is a set of points in which the n-th air target can be killed.

The first step is to determine the coordinates of the point $D_j^n = (x_j^{d,n}, y_j^{d,n})$ which is the orthogonal projection of point $D_j = (x_j^d, y_j^d)$ on the line n. If the point $D_j = (x_j^d, y_j^d)$ lies on the line n, then its coordinates satisfy the equation of the line

$$A_n x_j^d + B_n y_j^d + C_n = 0, \tag{18}$$

and the point $D_j^n = (x_j^{d,n}, y_j^{d,n})$ coincides with the point $D_j = (x_j^d, y_j^d)$.

If the point $D_j = (x_j^d, y_j^d)$ does not lie on the line n, then its coordinates do not satisfy the equation of the line (18).

Vector $\overrightarrow{D_jD_j^n}$ must be perpendicular to the vector $\boldsymbol{w}^n$, thus a scalar product:

$$\left(\overrightarrow{D_jD_j^n}|\boldsymbol{w}_n\right) = 0 \tag{19}$$

and after joining the equation of line n (18) we get a system of equations

$$\begin{cases} (x_j^{d,n} - x_j^d)w_x^n + (y_j^{d,n} - y_j^d)w_y^n = 0 \\ A_n x_j^{d,n} + B_n y_j^{d,n} + C_n = 0 \end{cases}, \tag{20}$$

which is transformed to the following formulas

$$\begin{cases} w_x^n x_j^{d,n} + w_y^n y_j^{d,n} = w_x^n x_j^d + w_y^n y_j^d \\ A_n x_j^{d,n} + B_n y_j^{d,n} = -C_n \end{cases}. \tag{21}$$

We obtain the determinants from this system of equations

$$W = \begin{vmatrix} w_x^n & w_y^n \\ A_n & B_n \end{vmatrix}, \quad W_1 = \begin{vmatrix} w_x^n x_j^d + w_y^n y_j^d & w_y^n \\ -C_n & B_n \end{vmatrix}, \quad W_2 = \begin{vmatrix} w_x^n & w_x^n x_j^d + w_y^n y_j^d \\ A_n & -C_n \end{vmatrix} \tag{22}$$

and the following solution

$$x_j^{d,n} = \frac{W_1}{W}, \quad y_j^{d,n} = \frac{W_2}{W}. \tag{23}$$

Taking into account (10) one can show that the value of the determinant W is not equal to zero:

$$W = \begin{vmatrix} w_x^n & w_y^n \\ A_n & B_n \end{vmatrix} = w_x^n B_n - w_y^n A_n = -\left[\left(w_x^n\right)^2 + \left(w_y^n\right)^2\right] = -1. \tag{24}$$

Finally, the solution takes the form:

$$x_j^{d,n} = -W_1, \quad y_j^{d,n} = -W_2. \tag{25}$$

Considering (9) and (22) we finally obtain from formulas (13) and (25):

$$x_j^{d,n} = -\left(w_y^n\right)^2\left(x_j^d - x_s\right) + w_y^n w_x^n\left(y_j^d - y_s\right) + x_j^d \tag{26}$$

$$y_j^{d,n} = -\left(w_x^n\right)^2\left(y_j^d - y_s\right) + w_y^n w_x^n\left(x_j^d - x_s\right) + y_j^d \tag{27}$$

Next, we determine the length l_j^n of vectors $\overrightarrow{D_j^n O_{jt}^{n,1}}$ and $\overrightarrow{D_j^n O_{jt}^{n,2}}$.

$$l_{jt}^{n}=\sqrt{d_t^2-\left\|\overrightarrow{D_jD_j^n}\right\|^2}=\sqrt{d_t^2-\left(\left(x_j^{d,n}-x_j^d\right)^2+\left(y_j^{d,n}-y_j^d\right)^2\right)}=\sqrt{d_t^2-\left(d_j^n\right)^2}. \tag{28}$$

The coordinates of the ends of the section, in which the air target can be destroyed, we can calculate as the vector end coordinates $\overrightarrow{D_j^nO_{jt}^{n,1}}$ and $\overrightarrow{D_j^nO_{jt}^{n,2}}$

$$\overrightarrow{D_j^nO_{jt}^{n,1}}=\left[l_{jt}^n\cdot\cos\beta_n,l_{jt}^n\cdot\sin\beta_n\right]\quad\overrightarrow{D_j^nO_{jt}^{n,2}}=\left[-l_{jt}^n\cdot\cos\beta_n,-l_{jt}^n\cdot\sin\beta_n\right]. \tag{29}$$

Hence, we get the coordinates of points $O_{jt}^{n,1}=(x_{jt}^{o,n,1},y_{jt}^{o,n,1})$ and $O_{jt}^{n,2}=(x_{jt}^{o,n,2},y_{jt}^{o,n,2})$

$$x_{jt}^{o,n,1}=x_j^{d,n}+l_{jt}^n\cdot\cos\beta_n,\quad y_{jt}^{o,n,1}=y_j^{d,n}+l_{jt}^n\cdot\sin\beta_n \tag{30}$$

$$x_{jt}^{o,n,2}=x_j^{d,n}-l_{jt}^n\cdot\cos\beta_n,\quad y_{jt}^{o,n,2}=y_j^{d,n}-l_{jt}^n\cdot\sin\beta_n \tag{31}$$

One should choose one point from the set of two points $O_{jt}^{n,1}=(x_{jt}^{o,n,1},y_{jt}^{o,n,1})$ and $O_{jt}^{n,2}=(x_{jt}^{o,n,2},y_{jt}^{o,n,2})$, that lies closer to the point of forecasted air target detection. The vectors connecting the center of the protected area $\boldsymbol{S}$ with all points $O_{jt}^{n,1}=(x_{jt}^{o,n,1},y_{jt}^{o,n,1})$ and $O_{jt}^{n,2}=(x_{jt}^{o,n,2},y_{jt}^{o,n,2})$ have the following form:

$$\overrightarrow{SO_{jt}^{n,1}}=\left[x_{jt}^{o,n,1}-x_s,y_{jt}^{o,n,1}-y_s\right]\quad\overrightarrow{SO_{jt}^{n,2}}=\left[x_{jt}^{o,n,2}-x_s,y_{jt}^{o,n,2}-y_s\right], \tag{32}$$

and the scalar products of these vectors with the vector $-\boldsymbol{w}_n=-(w_x^n,w_y^n)$ are described by formulas:

$$\begin{aligned}d_{jt}^{o,n,1}&=\left(\overrightarrow{SO_{jt}^{n,1}}|\boldsymbol{w}_n\right)=-[(x_{jt}^{o,n,1}-x_s)\cos\beta_n+(y_{jt}^{o,n,1}-y_s)\sin\beta_n]\\ d_{jt}^{o,n,2}&=\left(\overrightarrow{SO_{jt}^{n,2}}|\boldsymbol{w}_n\right)=-[(x_{jt}^{o,n,2}-x_s)\cos\beta_n+(y_{jt}^{o,n,2}-y_s)\sin\beta_n]\end{aligned} \tag{33}$$

Taking into account points $O_{jt}^{n,1}=(x_{jt}^{o,n,1},y_{jt}^{o,n,1})$ and $O_{jt}^{n,2}=(x_{jt}^{o,n,2},y_{jt}^{o,n,2})$ we choose the point for which the value of the scalar product (35) is greater and this value is entered to the matrix $\boldsymbol{Z}$:

$$z_{jnt}:=\max\left\{d_{jt}^{o,n,1},\,d_{jt}^{o,n,2}\right\},\,j=1,\ldots,J,\quad n=1,\ldots,N,\quad t=1,\ldots,T \tag{34}$$

5 Selection of the Best AAMA Deployment Points

5.1 The Objective Function for Evaluation of Feasible Pairs: The Number of Deployment Point and the Number of AAMA Set

Selection of the best deployment points, in which a given number of AAMA sets will be distributed, means choosing pairs: the number of the deployment point from the feasible number set and the number of AAMA set type in such a way as to maximize a criterion (evaluation) function, and the number of selected AAMA sets does not exceed the maximum number of AAMA sets of specific types.

Criterion functions of the AAMA sets deployment [1–4] and weapon target assignment [5, 6] presented in the bibliography use models taking into account both the probability of the air target kill and the likelihood of the destruction of the object and defense system. We propose a different approach to the construction of the criterion function. It was assumed that the criterion function should take into account the following factors: the distance between the possible kill points and the center of protected area, the effectiveness of AAMA sets and the expected distribution of probabilities assigned to air raid directions.

These postulates meet the following assessment function of pair selection: the deployment point and type of AAMA set

$$d^o(j,t) = \begin{cases} \sum\limits_{\substack{n=1 \\ z_{jnt} \neq -\infty \wedge p_n > 0}}^{N} s_t \quad p_n z_{jnt} & \exists n = 1, \ldots, N \; z_{jnt} \neq -\infty \\ -\infty & \forall n = 1, \ldots, N \; z_{jnt} = -\infty \end{cases} \quad j \in \boldsymbol{D}, t \in \{1, \ldots, T\}. \tag{35}$$

5.2 Selection of the Best Deployment Points and the Best AAMA Types

A formulation of the problem of optimal pair selection: deployment point and associated type of AAMA set is presented below.

Let $\boldsymbol{V} = [v_{jt}]_{J \times T}$ be a binary matrix of decision variables with the following interpretation:

$$v_{jt} = \begin{cases} 0 & - \text{ a AAMA set of the } t - \text{th type is not located in } j - \text{th point,} \\ 1 & - \text{ a AAMA set of the } t - \text{th type is located in } j - \text{th point.} \end{cases} \tag{36}$$

Solution of optimization problem consist in finding such a matrix $\boldsymbol{V}^* = [v^*_{jt}]_{J \times T}$ to maximize the function $f(\boldsymbol{V})$

$$f(\boldsymbol{V}^*) = \sum_{\substack{j \in \boldsymbol{D} \\ d^o(j,t) \neq -\infty}} \sum_{t=1}^{T} v^*_{jt} \, d^o(j,t) = \max_{\boldsymbol{V} \in \Omega} \sum_{\substack{j \in \boldsymbol{D} \\ d^o(j,t) \neq -\infty}} \sum_{t=1}^{T} v_{jt} \, d^o(j,t) \tag{37}$$

with constraints

$$\Omega = \left\{ \boldsymbol{V} = [v_{jt}]_{J \times T} : v_{jt} \in \{0,1\} \wedge \forall j \in \boldsymbol{D} \sum_{t=1}^{T} v_{jt} \le 1 \wedge \forall t \in \{1, \ldots, T\} \sum_{j \in \boldsymbol{D}} v_{jt} \le K_t \right\}. \tag{38}$$

6 Conclusion

This paper has presented an optimization problem of planning the anti-aircraft missile and artillery sets deployment in an air defense system of area objects. A new approach to building the evaluation criterion function for AAMA set deployment has been proposed. Models of protected area objects, air defense systems equipped with AAMA sets, air raids and a decisive evaluation system have been presented.

A method to solve the problem of the optimal deployment of anti-aircraft missile and artillery sets is presented in another paper at this conference titled "Heuristic Method of Air Defense Planning for an Area Object with the Use of Very Short Range Air Defense" [8].

References

1. Chen, C., Chen, J., Zhang, C.: Deployment optimization for air defense base on artificial potential field. In: 2011 8th Asian Control Conference (ASCC 2011), pp. 812–816. IEEE Press (2011)
2. Liu, L., Li, X, Liu, R., Yan, J.: Mathematic model of key-point anti-air position ring-deployment and optimization. In: 2012 2nd International Conference on Computer Science and Network Technology, pp. 166–171. IEEE Press (2012). doi:10.1109/ICCSNT.2012.6525913
3. Ghode, D., Prasad, U.R., Guruprasad, K.: Missile battery placement for air defense: a dynamic programming approach. Appl. Math. Model. **17**(9), 450–458 (1993). doi:10.1016/0307-904X(93)90086-V
4. Pant, M., Deep, K.: Building a better air defence system using genetic algorithms. In: Gabrys, B., Howlett, R.J., Jain, L.C. (eds.) Knowledge-Based Intelligent Information and Engineering Systems, KES 2006. LNCS, vol. 4251, pp. 951–959. Springer, Heidelberg (2006). doi:10.1007/11892960_114
5. Chen, J., Xin, B., Peng, Z.H., et al.: Evolutionary decision-makings for the dynamic weapon-target assignment problem. Sci. China Ser. F-Inf. Sci. **52**(11), 2006–2018 (2009). doi:10.1007/s11432-009-0190-x
6. Bogdanowicz, Z.R., Tolano, A., Patel, K., Coleman, N.P.: Optimization of weapon-target pairings based on kill probabilities. IEEE Trans. Cybern. **43**(6), 1835–1844 (2013). doi:10.1109/TSMCB.2012.2231673
7. Kasprzak, W.: Image and speech signals recognition. Oficyna Wydawnicza Politechniki Warszawskiej, Warszawa (2009). (in Polish)
8. Pietkiewicz, T., Kawalec, A., Wajszczyk, B.: Heuristic method of air defense planning for an area object with the use of very short range air defense. In: Proceedings of the 9th KES International Conference on Intelligent Decision Technologies (KES-IDT 2017) Intelligent Decision Technologies 2017. SIST. Springer (2017)

Forecasting Social Security Revenues in Jordan Using Fuzzy Cognitive Maps

Ahmad Zyad Alghzawi[1(✉)], Gonzalo Nápoles[1], George Sammour[2], and Koen Vanhoof[1]

[1] Department of Business Informatics, Hasselt University, Hasselt, Belgium
ahmad.alghzawi@uhasselt.be
[2] Department of Management Information System, Princess Sumaya University for Technology, Amman, Jordan

Abstract. In recent years, Fuzzy Cognitive Maps (FCMs) have become a convenient knowledge-based tool for economic modeling. Perhaps, the most attractive feature of these cognitive networks relies on their transparency when performing the reasoning process. For example, in the context of time series forecasting, an FCM-based model allows predicting the next outcomes while expressing the underlying behavior behind the investigated system. In this paper, we investigate the forecasting of social security revenues in Jordan using these neural networks. More specifically, we build an FCM forecasting model to predict the social security revenues in Jordan based on historical records comprising the last 120 months. It should be remarked that we include expert knowledge related to the sign of each weights, whereas the intensity in computed by a supervised learning procedure. This allows empirically exploring a sensitive issue in such models: the trade-off between interpretability and accuracy.

Index Terms: Fuzzy cognitive maps · Time series prediction · Economic modeling

1 Introduction

Fuzzy Cognitive Maps (FCMs) have been developed as a knowledge-based tool to model and analyze complex systems using causal relations [1]. From the structural perspective, an FCM can be defined as a fuzzy digraph that describes the underlying behavior of an intelligent system in terms of concepts (i.e., objects, states, variables or entities) and causal relations. As a matter of fact, FCMs are a kind of recurrent neural networks that support backward connections that sometimes form cycles in the causal graph [2]. This implies that concepts in the information network can be understood as neural processing entities with inference capabilities.

Although FCMs inherited many aspects from neural systems, there are important differences with regards to other Artificial Neural Network (ANNs). More explicitly, ANNs regularly perform like black-boxes, where both neurons and connections do not have any clear meaning for the problem domain, or the decision process cannot easily be explained [3]. In contrast, all neurons in an FCM have a precise meaning for the

I. Czarnowski et al. (eds.), *Intelligent Decision Technologies 2017*,
Smart Innovation, Systems and Technologies 72, DOI 10.1007/978-3-319-59421-7_23

physical system being modeled and correspond to specific variables that form part of the solution. It is worth mentioning that an FCM does not comprise hidden neurons since these entities could not be interpreted nor help at explaining why a solution is suitable for a given problem instance. This suggests that the representation capability of FCM-based systems is superior compared to ANN models.

In recent years, the use of FCMs in time series forecasting has been noticeable due to the transparency of FCM-based models. For example, the approaches proposed in [4–8] rely on fuzzy information granules to forecast the time series with high accuracy. In these works, all values of the learning part of the time series are clustered through the *fuzzy c-means* algorithm [9], where the number of clusters is a user-defined parameter. In this way, at each time iteration, the value of the time series belongs to each cluster with some membership degree. It is assumed that every cluster (i.e., concept) plays the role of a fuzzy set. Besides, they often use a heuristic search method to estimate the weight matrix attached to the system.

The algorithms proposed in [10–13] adopt a low-level approach where neurons denote attributes instead of comprising information granules. However, all of these methods suffer from the same drawback: there is no guarantee that the produced weight set involves a realistic interpretation for the physical system, even is the model is able to achieve good prediction rates. This suggests that the modeled system cannot be interpreted, although the FCM reasoning is still transparent.

In this paper, we investigate this complex issue using a real case study concerning to the Jordanian Social Security System (JSS). More specifically, this paper comprises two key contributions. On the one hand, we build a FCM-based system to forecast the social security revenues in Jordan in a comprehensible way. Using a cognitive model allows experts to forecast the revenues in next years and understand the underlying behavior behind such predictions. On the other hand, we include expert knowledge to investigate the trade-off between interpretability and accuracy. This is equivalent to (i) learn a first FCM model without any restriction, (ii) learn a second FCM model using knowledge coming from experts, and (iii) compare the accuracy.

The remaining of this paper is organized as follow. Section 2 introduces the Fuzzy Cognitive Maps. Section 3 presents the method for constructing the FCM structure. Section 4 outlines the results and their ensuing discussion, while Sect. 5 provides the conclusions and discusses future research direction.

2 Fuzzy Cognitive Maps

Cognitive mapping has become a convenient knowledge-based tool for modeling and simulation [1]. In point of fact, FCMs can be understood as recurrent neural networks with learning capabilities, consisting of concepts and weighted arcs. Concepts denote entities, variables, entities and are equivalent to neurons in neural models; whereas weights associated to connections denote the *causality* among such nodes. Each link takes values in the range $[-1, 1]$, denoting the causality degree between two concepts as a result of the quantification of a fuzzy linguistic variable, which is often assigned by experts during the modeling phase [14]. The activation value of neurons is also fuzzy in nature and regularly takes values in the range $[0, 1]$. The higher the activation value of a neuron, the stronger its influence over the investigated system, offering to decision-makers an overall picture of the systems behavior.

Without loss of generality, we can entirely define the semantics behind an FCM using a 4-tuple (C, W, A, f) where $C = \{C_1, C_2, \ldots, C_M\}$ denotes a set of M neural processing entities, $W : (C_i, C_j) \rightarrow w_{ij}$ is a function that associates a causal weight $w_{ij} \in [-1, 1]$ to each pair of neurons (C_i, C_l). Similarly, $A : (C_i) \rightarrow A_i$ is a function that associates the activation degree $A_i \in \mathbb{R}$ to the C_i neuron at each iteration-step moment $t(t = 1, 2, \ldots, T)$. Finally, a transformation function $f : \mathbb{R} \rightarrow [0, 1]$ is used to keep the neurons' activation value in the allowed interval. Equation (1) portrays the inference mechanism attached to an FCM-based system, using the $A^{(0)}$ vector as the initial activation. This neural procedure is repeated until either a fixed-point attractor is discovered or a maximal number of iterations is reached.

$$A_i^{(t+1)} = f\left(\sum_{j=1}^{M} w_{ji} A_j^{(t)} + A_i^{(t)}\right), i \neq j \tag{1}$$

The most used threshold functions are: the bivalent function, the trivalent function, and the sigmoid variants. It should be stated that authors will be focused on Sigmoid FCMs, instead of discrete ones. It is inspired by the benchmarking analysis discussed in reference [15] where results revealed that the sigmoid function outperformed the other functions by the same decision model. Therefore, the adequate selection of this transfer function becomes crucial for the system behavior. From [2] some important observations were formalized and summarized as follows:

- Binary and trivalent FCMs cannot represent the degree of an increase or a decrease of a concept. Such discrete maps always converge to a fixed-point attractor or limit cycle since FCMs are deterministic models.
- Sigmoid FCMs, by allowing neuron's activation level, can also represent the neuron's activation degree. They are suitable for qualitative and quantitative tasks, however, may additionally show chaotic behaviors.

Perhaps the most attractive feature of FCMs relies on their transparency, that is, the network's capability to explain the decision process. This features can be achieved by relying on activation degrees and causal relations between neurons. There are three types of causal relations between two neural entities.

- If $w_{ij} > 0$ then an increase (decrement) in the cause neuron C_i will produce an increment (decrement) of the effect neuron C_j with intensity $|w_{ij}|$.
- If $w_{ij} < 0$ then an increase (decrement) in the cause neuron C_i will produce an decrement (increment) of the effect neuron C_j with intensity $|w_{ij}|$.
- If $w_{ij} = 0$ denotes the absence of relation between C_i and C_j.

FCMs can be constructed either using the knowledge coming from domain experts or using a learning method. In the next section, we describe an evolutionary procedure to derive the network structure in a supervised fashion, i.e. by minimizing the global dissimilarity between the expected outputs and the predicted ones.

3 Fuzzy Cognitive Maps Learning

Let us assume that $Z^{(t)} = \left[Z_1^{(t)}, Z_2^{(t)}, \ldots, Z_M^{(t)}\right]$ is the desired system response for the $Z^{(t-1)}$ activation vector, $A^{(t)} = \left[A_1^{(t)}, A_2^{(t)}, \ldots, A_M^{(t)}\right]$ is the FCM output for the $Z^{(t-1)}$ initial vector, while T is the number of the learning records. Equation (2) displays an error function used in the context of time series forecasting, where W represents the candidate weight matrix, M is the number of neurons, while t indexes the iteration-steps (i.e., the learning records). In short, this learning scheme attempt minimizing the dissimilarity between the expected outputs and the predicted ones.

$$E(W) = \frac{1}{(T-1)M} \sum_{t=1}^{T-1} \sum_{i=1}^{M} \left[Z_i^{(t)} - A_i^{(t)}\right]^2 \tag{2}$$

In this supervised learning model, a continuous search method (i.e., Particle Swarm Optimization, Genetic Algorithms) generates the weights matrices to be evaluated by the algorithm. Equation (3) shows the structure of the weight set.

$$W = \left[w_{12}, \ldots, w_{1M}, w_{21}, w_{23}, \ldots, w_{2M}, \ldots, w_{MM}\right] \tag{3}$$

In this research, we adopt the *Real-Coded Genetic Algorithm* (RCGA) as standard continuous optimizer. The RCGA [16] is an evolutionary search method that codifies genes directly as real numbers and can be used to optimize parametrical problems for continuous variables. Therefore, each chromosome involves a vector of floating point numbers that involves a candidate solution. The size of the chromosomes is identical to the length of the vector, which is the solution to the optimization problem. In this way, each gene represents a variable of the problem, i.e. a weight component. Genetic operators only have to observe the fact that the values of the genes remain within the interval established by the variables they represent.

Each chromosome in the population is evaluated on the basis of a fitness function according to the error function. Due to the fact that *Genetic Algorithms* are frequently expressed like maximization-type problems, the previous error function is expressed in terms of a fitness function, which is formalizes as follows:

$$F(W) = \frac{1}{a * E(W) + 1} \tag{4}$$

where $\alpha > 0$ is a user-specified parameter, W is the candidate weight set computed by the RCGA optimizer, while $E(W)$ is the error function.

During the optimization, parents are selected and new population of chromosomes are generated with some probability. In this research, we use the well-known roulette wheel method as standard selection operator. For each chromosome in the population, the probability of directly copying this solution in the next (improved) population can be calculated according to the following equation:

$$P_i = F_i(W) / \sum_{K=1}^{P} F_k(W) \tag{5}$$

where $F_i(W)$ represents the fitness of the i th candidate solution (i.e., weight matrix) and P denotes the number of chromosomes in the population.

After performing the crossover procedure, a mutation operator modifies elements of the selected solution with a specific probability. The use of mutation prevents the premature convergence of the genetic algorithm to suboptimal solutions [16]. In this research, we adopt the Mühlenbein's operator [17] as the standard mutation method. On the other hand, in order to guarantee the survival of the best chromosome in the artificial population, an elitist strategy was applied.

The learning process stops either when a maximum number of iterations is reached or alternatively when the following condition is satisfied:

$$F_{best}(W) > F_{max} \tag{6}$$

where $F_{best}(W)$ denotes the fitness function value related with the best chromosome found so far in the population, while F_{max} is a parameter.

4 Forecasting Social Security Revenues in Jordan

In this section, we introduce a study case concerning to the forecasting social security revenues in Jordan. The aim of this study is focused on predicting the revenue values and understanding the underlying interrelation between concepts; the latter is the main motivation to use cognitive mapping models. The dataset contains the social security revenues in Jordan for the last 120 months (from 2006 until 2015) and comprises the eight fields (i.e., map concepts) that are listed below:

- C_1 - Aging subscription
- C_2 - Work related injuries
- C_3 - Maternity insurance
- C_4 - Years earlier service
- C_5 - Optional subscriptions
- C_7 - Different revenue
- C_8 - Stamps

The available data records were normalized to the [0, 1] range in order to activate the neurons with values inside the activation interval. On the other hand, the dataset was randomly divided into two disjoint subsets: the learning set (114 records) and the test (6 records). Aiming at evaluating the quality of the FCM-base forecasting model, we adopt the *Mean squared Error* (MSE) and the *Root Mean Squared Error* (RMSE). Equations (7) and (8) formalizes both quality measures.

$$MSE = \frac{1}{M(T-1)}\sum_{t=1}^{T-1}\sum_{i=1}^{M}\left[Z_i^{(t)} - A_i^{(t)}\right]^2 \tag{7}$$

$$RMSE = \sqrt{\frac{1}{M(T-1)}\sum_{t=1}^{T-1}\sum_{i=1}^{M}\left[Z_i^{(t)} - A_i^{(t)}\right]^2} \tag{8}$$

Due to the stochastic nature of the heuristic search methods, we perform 20 trials and average the quality measures. Moreover, we use the following parametric setting in all the simulations: 50 chromosomes as in the artificial population, 200 generations, the α parameter is set to 10, the crossover probability is set to 0.8, the mutation probability is set to 0.1, while the F_{max} parameter is set to 0.999.

In addition to explore two different configurations. In the first one, genes can take values in the [−1, 1] range, whereas in the second one we include expert knowledge concerning to the sign of causal weights. Due to the nature of concepts use to model this real-world problem, the experts determined three rules:

- $[R_1]$ $w_{i7} \in [-1, 0)$, $w_{7i} \in [-1, 0), \forall i = \{1, 2, \ldots, 6\}$
- $[R_2]$ $w_{ij} \in (0, 1]$, $\forall j = \{1, 2, \ldots, 6\}, \forall i = \{1, 2, \ldots, 6\}$ and $i \neq j$.
- $[R_3]$ $w_{ii} = 0$, $\forall i = \{1, 2, \ldots, 7\}$ since a concept cannot influence itself.

The goal is not to increase performance but to preserve the coherence in the causal cognitive network. Weights freely estimated in the [−1, 1] range may actually produce improved prediction rates due to the attached freedom degree. However, there is no way to ensure that weights produced by a heuristic search method comprises a causal meaning for the problem under investigation. Hence, we introduce expert knowledge in an attempt of establishing a proper trade-off between the accuracy in the forecasted revenues values and the network interpretability.

The average error measures for the first configuration (i.e., without any restriction) are $MSE_1 = 0.0431$ and $RMSE_1 = 0.207$, whereas for the second configuration the error measures are $MSE_2 = 0.1992$ and $RMSE_2 = 0.4463$. The reader can observe that, for the case study under investigation, promoting the FCM interpretability leads to higher forecasting errors. This is often the price we have to pay in order to produce truly interpretable causal cognitive models. However, whether such forecasted values are acceptable for this real-world problem is questionable.

But why the FCM forecasting model produces such results if, in principle, a neuron naturally produces a time series? The response relies on the convergence properties of the causal network. In most FCM-based systems, ensuring convergence is mandatory, otherwise making reliable decisions is not possible [3, 18]. Nevertheless, in the time series context, convergence is not desirable since it decreases the network's capability of computing both short and long-term predictions. Figure 1 portrays the expected and forecasted values for both configurations, where the horizontal axis denotes each time step, while the vertical axis shows the activation values. (Figure 2)

From the above figures we can confirm our hypothesis: the convergence properties of the FCM-based system affects the quality of forecasted revenues. In the case of the

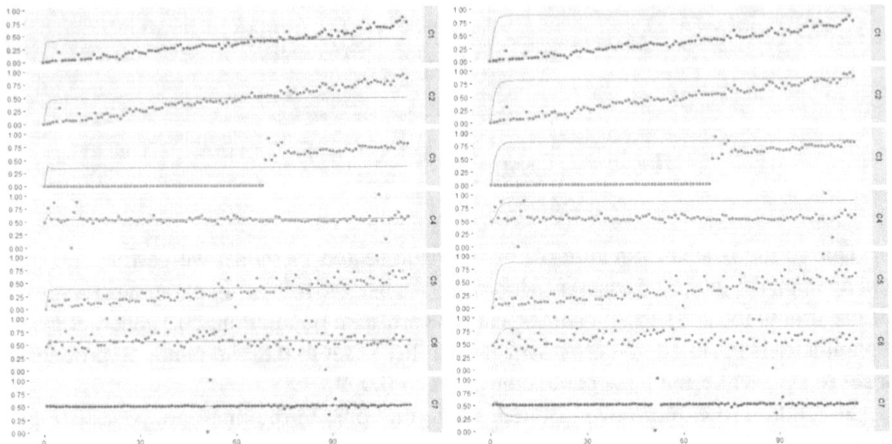

Fig. 1. Forecasted revenue values using the current activation value to predict the next time series point according to (a) the unrestricted model, (b) the restricted model.

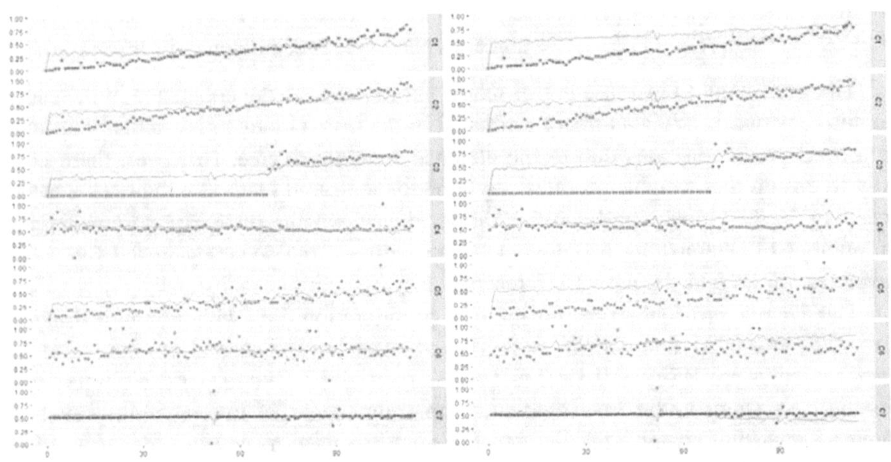

Fig. 2. Forecasted revenue values using the real activation value to predict the next time series point according to (a) the unrestricted model, (b) the restricted model.

second configuration, the error increases as a result of decreasing the freedom degree during the search process. But this outcome is expected.

Aiming at alleviating the convergence issue, we could use a different rule to update the activation values of map entities. For instance, we can define a set of independent neurons such that their current activation values are not computed from their previous ones but from their real values stored in the dataset. Figure 1 shows the expected and forecasted values using this approach for both configurations.

In this case, the average error measures computed by the unrestricted configuration are $MSE_1 = 0.026$ and $RMSE_1 = 0.166$, whereas for the restricted configuration the error measures are $MSE_2 = 0.064$ and $RMSE_2 = 0.254$. The reader can observe that this modified variant leads to lower error rates as both models fit better the time series associated to each variable. Therefore, the model will produce forecasted values with lower convergence features, but performing multiple-ahead predictions is no longer possible since the network make its predictions based on the real value stored in the dataset. In spite of this fact, the results showed that encouraging the coherence in the network implies to slightly sacrifice the accuracy.

5 Conclusions

In this paper, we proposed an FCM-based forecasting model to investigate the factors affecting revenues on the social security in Jordan. The proposed model uses a *Real-Code Genetic Algorithm* to learn the map structure together with well-known methods for FCM learning. As a first step, we allowed the algorithm to indicate a relationship value between the neurons to fluctuate between -1 and 1 without any constraint. But the model occasionally computed a positive relation where it should be a negative one and vice versa, therefore producing a forecasting model that does not have a coherent meaning for the real-world problem under investigation.

In order to produce meaningful weights, we included domain knowledge related to the sign of each causal relation. The numerical simulations have shown that using the current value to forecast the revenue values leads to higher error rates since the model converges to an equilibrium attractor. As an alternative, we compute the next revenue value from the (real) previous value stored in the dataset. This strategy leads to lower error rates in both scenarios while rescinding the network's capability of performing multiple-ahead predictions. Besides, we noticed that the error increased when adding the domain constraints but this outcome is actually expected. Promoting the network's transparency is a key, otherwise the FCM becomes a black-box and there is no reason to employ other (perhaps more accurate) forecasting models. The feature research will be focused on increasing the prediction rates while retaining the network capability of naturally performing multiple-ahead predictions.

Acknowledgments. We would like to thank our colleague Frank Vanhoenshoven for his valuable support and constructive comments.

References

1. Kosko, B.: Fuzzy cognitive maps. Int. J. Man-Mach. Stud. **24**, 65–75 (1986)
2. Tsadiras, A.K.: Comparing the inference capabilities of binary, trivalent and sigmoid fuzzy cognitive maps. Inf. Sci. **178**, 3880–3894 (2008)
3. Nápoles, G., Papageorgiou, E.I., Bello, R., Vanhoof, K.: On the convergence of sigmoid fuzzy cognitive maps. Inf. Sci. **349–350**, 154–171 (2016)
4. Pedrycz, W.: The design of cognitive maps: A study in synergy of granular computing and evolutionary optimization. Expert Syst. Appl. **37**, 7288–7294 (2010)

5. Pedrycz, W., Jastrzebska, A., Homenda, W.: Design of fuzzy cognitive maps for modeling time series. IEEE Trans. Fuzzy Syst. **24**, 120–130 (2016)
6. Lu, W., Yanga, J., Liua, X., Pedrycz, W.: The modeling and prediction of time series based on synergy of high-order fuzzy cognitive map and fuzzy c-means clustering. Knowl.-Based Syst. **70**, 242–255 (2014)
7. Froelich, W., Pedrycz, W.: Fuzzy cognitive maps in the modeling of granular time series. Knowl.-Based Syst. **115**, 110–122 (2017)
8. Homenda, W., Jastrzebska, A., Pedrycz, W.: Modeling time series with fuzzy cognitive maps. In: FUZZ-IEEE 2014, Beijing, China, pp. 2055–2062 (2014)
9. Bezdek, J.C., Ehrlich, R., Full, W.: FCM: The fuzzy c-means clustering algorithm. Comput. Geosci. **10**, 191–203 (1984)
10. Froelich, W., Salmeron, J.L.: Evolutionary learning of fuzzy grey cognitive maps for the forecasting of multivariate, interval-valued time series. Int. J. Approximate Reasoning **55**, 1319–1335 (2014)
11. Poczęta, K., Yastrebov, A.: Monitoring and prediction of time series based on fuzzy cognitive maps with multi-step gradient methods. In: Progress in Automation, Robotics and Measuring Techniques, pp. 197–206. Springer (2015)
12. Papageorgiou, E.I., Poczęta, K., Yastrebov, A., Laspidou, C.: Fuzzy cognitive maps and multi-step gradient methods for prediction: applications to electricity consumption and stock exchange returns. In: Intelligent Decision Technologies, pp. 501–511. Springer (2015)
13. Salmeron, J.L., Froelich, W.: Dynamic optimization of fuzzy cognitive maps for time series forecasting. Knowl.-Based Syst. **105**, 29–37 (2016)
14. Nápoles, G., et al.: Learning and convergence of fuzzy cognitive maps used in pattern recognition. Neural Process. Lett. **43**, 1–14 (2016)
15. Bueno, S., Salmeron, J.L.: Benchmarking main activation functions in fuzzy cognitive maps. Expert Syst. Appl. **36**, 5221–5229 (2009)
16. Herrera, F., Lozano, M., Verdegay, J.L.: Tackling real-coded genetic algorithms: operators and tools for behavioral analysis. Artif. Intell. Rev. **12**, 265–319 (1998)
17. Scrucca, L.: GA: A Package for Genetic Algorithms in R. J. Stat. Softw. **54**, 1–37 (2013)
18. Nápoles, G., Bello, R., Vanhoof, K.: How to improve the convergence on sigmoid fuzzy cognitive maps? Intell. Data Anal. **18**, S77–S88 (2014)

Fuzzy Cognitive Maps Employing ARIMA Components for Time Series Forecasting

Frank Vanhoenshoven(✉), Gonzalo Nápoles, Samantha Bielen, and Koen Vanhoof

Faculty of Business Economics, Hasselt University, Agoralaan, 3590 Diepenbeek, Belgium
frank.vanhoenshoven@uhasselt.be

Abstract. In this paper, we address some shortcomings of Fuzzy Cognitive Maps (FCMs) in the context of time series prediction. The transparent and comprehensive nature of FCMs provides several advantages that are appreciated for decision-maker. In spite of this fact, FCMs also have some features that are hard to match with time series prediction, resulting in a prediction power that is probably not as extensive as other techniques can boast. By introducing some ideas from ARIMA models, this paper aims at overcoming some of these concerns. The proposed model is evaluated on a real-world case study, captured in a dataset of crime registrations in the Belgian province of Antwerp. The results have shown that our proposal is capable of predicting multiple steps ahead in an entire system of fluctuating time series. However, these enhancements come at the cost of a lower prediction accuracy and less transparency than standard FCM models can achieve. Therefore, further research is required to provide a comprehensive solution.

1 Introduction

Fuzzy Cognitive Maps (FCMs), proposed by Kosko [11], comprise a very suitable tool for modeling and simulation purposes. Without loss of generality, we can define FCMs as *recurrent neural networks* that produce a state vector at each iteration. From the structural perspective, FCM-based models comprise a set of neural processing entities called concepts (or simply neurons) and a set of causal relations defining a weight matrix. One of the advantages of cognitive mapping relies on the interpretability attached to the causal network, where both neurons and weights have a precise meaning for the problem domain.

The reader can find several interesting applications of these cognitive neural networks in [18], while the survey in [19] is devoted to their extensions. Meanwhile, due to the relevance of the weight matrix in designing the cognitive model, several supervised learning methods have been introduced to compute the weights characterizing the system under investigation (e.g., [1,16,28,30,31]). Estimating a high-quality weight matrix is a key issue towards building the forecasting model; otherwise, the system interpretation will not be meaningful even if the FCM reasoning is still transparent.

I. Czarnowski et al. (eds.), *Intelligent Decision Technologies 2017*,
Smart Innovation, Systems and Technologies 72, DOI 10.1007/978-3-319-59421-7_24

One field where FCMs have found great applicability is related to the time series prediction [9,10,12,13,24,25,27]. In [26] the authors focused on the learning mechanism and causality estimation from the perspective of time series. Multivariate, interval-valued time series are addressed using fuzzy grey cognitive maps in [8], where near-continuous time series are transformed into intervals of increasing or decreasing values. Modifications of the learning algorithm to better deal with time series prediction are proposed in [7]. Likewise, FCMs and time lags in time series prediction are considered in [20]. After this, not much attention seems to be paid to the issue of lagging effects in time series.

The power of FCMs does not lay in superior forecasting in time series analysis. Indeed, looking only at fitness measures for fitted and predicted data, specialized techniques like even the well-known ARIMA methods [3] are able to outperform an FCM. However, the weight matrix, which is the center of an FCM-based model, provides a degree of transparency that cannot be easily achieved with other forecasting models. As a further advantage, a discovered weight matrix, constrained or not, can be presented to experts for either validation of the model or as input for supporting the policy making process.

Regrettably, FCM-based forecasting models tend to converge to a fixed-point when performing the inference process. While this feature is highly attractive in simulation and pattern classification situations [17], its effect on time series scenarios is less desirable. More explicitly, if the network converges to a fixed-point, then the forecasting model will not fit the time series, unless the data points match with the fixed-point, which is rather unlikely.

In this paper, we investigate the use of FCMs in time series prediction in the context of a real case study. Particularly, our paper introduces two contributions, moving from theory to practice. The first contribution is related to the study of four neural updating rules and their impact over the accuracy achieved by the cognitive forecasting model. As a second contribution, we adopt the proposed models to predict the future influx of a prosecutor's office in a Belgian province using both demographic and economic features. The numerical simulations have shown that selecting a specific neural updating rule may improve the model accuracy; however, further investigation is required.

This paper follows the principles and structure of Design Science, as proposed by Peffers et al. in [22]. Section 2 introduces the *Fuzzy Cognitive Maps*, while Sect. 3 describes the theoretical problem to be confronted. Section 4 presents some alternatives to cope with the identified problem. Section 5 is devoted to the numerical analysis by using a real-world case study, whereas Sect. 6 discusses the results and outlines the future research directions.

2 Fuzzy Cognitive Maps

The semantics behind an FCM-based system can be completely defined using a 4-tuple $(\mathcal{C}, \mathcal{W}, \mathcal{A}, f)$ where $\mathcal{C} = \{C_1, C_2, C_3, \ldots, C_M\}$ is the set of M neural processing entities, $\mathcal{W} : \mathcal{C} \times \mathcal{C} \rightarrow [-1, 1]$ is a function that attaches a causal value w_{ij} to each pair of neurons (C_i, C_j) and the value w_{ij} denotes the direction and intensity of the edge connecting the cause C_i and the effect C_j. This function describes a causal weight matrix that defines the behavior of the investigated system [11]. On the other hand, the function $\mathcal{A} : \mathcal{C} \rightarrow A_i^{(t)}$ computes the activation degree $A_i \in \Re$ of each neuron C_i at the discrete-time step t, where $t = \{1, 2, \ldots, T\}$. Finally, the transfer function $f : \Re \rightarrow I$ retains the activation degree of each neuron confined into the activation space.

Causal relations have an associated numerical value in the $[-1, 1]$ range that govern the intensity of the relationships between the concepts defining the system semantics. Let w_{ij} be the weight associated with the connection between two neurons C_i and C_j. Generally speaking, the interpretation of causal influences between two neural entities can be summarized as follows:

- If $w_{ij} > 0$, this indicates that an increment (decrement) in the neuron C_i will produce an increment (decrement) on C_j with intensity $|w_ij|$.
- If $w_{ij} < 0$, this indicates that an increment (decrement) in the neuron C_i will produce a decrement (increment) on C_j with intensity $|w_ij|$.
- If $w_{ij} = 0$ (or very close to 0), this denotes the absence of a causal relationship from C_i upon C_j.

Equation 1 shows the updating rule attached to an FCM-based system, where $A^{(t)}$ is the vector of activation values at the tth iteration step, and W is the weight matrix defining the causal relations between concepts. The $f(x)$ function is introduced to normalize the values and is often defined as a sigmoid function (see Eq. 2). It is worth mentioning that, in this model, a neuron cannot influence itself, so the main diagonal of W only contains zeros.

$$\boldsymbol{A}^{(t+1)} = f(\boldsymbol{W} \times \boldsymbol{A}^{(t)}) \tag{1}$$

$$f(x) = \frac{a}{1 + exp(-\lambda x)} - b \tag{2}$$

The features that turn FCMs into an attractive artefact for scientists and decision makers are multiple. (1) Their visual and transparent structure enables domain experts to introduce their knowledge into the model. (2) By applying learning techniques to an FCM, a weight matrix can be discovered. This weight matrix provides insights into the internal mechanism of the analyzed system, thus contributing to the decision-making process. (3) By modeling complex structures of dependencies between several concepts, the FCM is not constraint to focus on a single variable, but can simulate a system as a whole.

3 Problem Statement

The output of a neural processing entity in an FCM-based network produces a time-series-like sequence. In most FCM models, convergence is a desired (sometimes mandatory) feature when iterating the causal model. Understanding and encouraging convergence features have been studied actively in recent papers by Nápoles et al. [15,17]. In contrast, time series typically display a more chaotic behavior.

More explicitly, when an iteration t in an FCM is equated with a time observation k in a time series, measures have to be taken to reconcile the conflicting requirements coming to the convergence features.

As an alternative we can adopt a modified updating rule that uses different strategies to avoid the convergence issue. In this paper, we describe and compare different models and study their effect on the outcome.

As a practical example, we could compute a specific weight matrix $W^{(t)}$ to predict the next value [13]. Besides the increased complexity towards learning the weight sets, and an inherently increased capacity for over-fitting, the multiplex nature of the weight sets decrease their transparency in decision-making scenarios. On top of it, predictions for future values can only be obtained if those future weight matrices are known as well.

A form of one-step-prediction is also used in [21]. Instead of using predicted values $A^{(t)}$ to calculate $A^{(t+1)}$, this adaptation uses the real values stored in the row data records. This approach seems adequate to analyze historical data; however, but it cannot solve the convergence issue when forecasting future values using the learned model (i.e., with no actual values available).

This paper proposes a modified strategy that addresses the aforementioned issues while maintaining the advantages of using cognitive mapping. Concretely, our model should satisfy the following requirements:

1. The prediction of chaotic time series with controlled chaos features.
2. The multiple-step-ahead predictions.
3. To preserve high prediction rates.
4. To preserve the transparency of the weight matrix.

4 Proposed Solutions

This paper proposes a modification of the FCM updating rule by introducing ARIMA-inspired components. Regardless the transfer function [4], Eq. 3 is selected as benchmark. In contrast with Eq. 1, it is allowed for neurons to influence themselves in the next iteration. In other words, the main diagonal of the weight matrix W can take values in the $[-1, 1]$ range.

$$\boldsymbol{A}^{(t+1)} = f(\boldsymbol{A}^{(t)} + \boldsymbol{W} \times \boldsymbol{A}^{(t)}) \tag{3}$$

Previous proposals advocate for the time series analysis using one-step ahead predictions [21,23]. In these approaches, $A_i^{(t+1)}$ is not calculated based on the calculated $A_i^{(t)}$, but on the real, *actual* record $R_i^{(t)}$. In a more generic form, this could be translated into Eq. 4, where part of the concepts are considered independent variables. Let $\boldsymbol{C}^{-}$ be the set of independent variables and $C_i \in \boldsymbol{C}^{-}$, then $R_i^{(t)}$ the actual value in the data set will be used to calculate the activation value in the next iteration. These variables are providing independent input before each iteration. For the remaining variables, $C_i \in \boldsymbol{C}^{*}$, the $A_i^{(t)}$ calculated activation values will be used to evolve the model.

$$\boldsymbol{A}^{(t+1)} = f(\boldsymbol{A}^{(t)} + \boldsymbol{W}_{c^*} \times \boldsymbol{A}^{(t)} + \boldsymbol{W}_{c^-} \times \boldsymbol{R}^{(t)}) \tag{4}$$

Note that the model assumes that an iteration t coincides with an observation at time k in the time series. In this interpretation, the evolution through the iterations is implicitly considered as an evolution over time.

In our solution, we propose the introduction of ARIMA-inspired components to strengthen the applicability of FCM in the domain of time series analysis. Firstly, instead of using $A_i^{(t)}$ as a base for calculating $A_i^{(t+1)}$, a moving average will be introduced. On top of that, to avoid a converging time series, a weight amplification method is proposed. Just like ARIMA uses the auto-correlation parameters, the FCM would use correlation between concepts to increase or decrease the effect of the attributed weights.

$$\boldsymbol{A}^{(t+1)} = \hat{f}(\boldsymbol{MAC}_{window}^{(t)} + \hat{g}(\boldsymbol{W}) \times \boldsymbol{A}^{(t)}) \tag{5}$$

The modified reasoning rule is depicted in Eq. 5. $MAC_{window}^{(t)}$ denotes the moving average of the concept values for a time period $[t - window : t]$. The amplification of the weight effects is defined as $\hat{g}(x)$ in Eq. 6, where ρ_t^{window} is a matrix of Pearson Coefficients [29].

$$\hat{g}(W) = \rho_t^{window} \times W \tag{6}$$

Aiming at confining the activation values into the allowed interval, we adopt a modified version of the sigmoid function 2 (see Eq. 7). The parameters of this function are set to $a = 1.2$, $b = 0.2$ and $\lambda = 1$.

$$\hat{f}(x) = \begin{cases} \dfrac{1.2}{1 + exp(-x)} - 0.2, & \text{if } \hat{f}(x) \geq 0 \\ 0, & \text{otherwise} \end{cases} \tag{7}$$

5 Demonstration

5.1 Dataset

The proposed models are applied to support the prosecutors in the Belgian criminal justice system. Accurately forecasting criminal activity and the demand on the criminal justice system enables authorities to proactively plan resources, which is particularly important because there has been variation in the intensity of crime over the last years, both over time and geographic units.

We follow the traditional law and economics literature on criminal behavior, which goes back to [2]. The size of the police force (POL, C_4) is a critical component of the price of committing crime because a larger police force is associated with a higher probability of apprehension and therefore an increased price of committing crime [5,6,14]. Furthermore, socio-economic (SOC) and demographic (DEM) variables[1] that are expected to affect crime are included in the form of the proportion of young, single males (C_5), poverty (C_6) and unemployment (C_7) are included as well. Equation 8 defines crime (CR) as a function of

[1] https://bestat.economie.fgov.be/bestat/.

police (POL), socio-economic (SOC) and demographic (DEM) characteristics. AS such the crime function is defined as in 8.

$$CR = f(CR, POL, SOC, DEM) \tag{8}$$

In the dataset adopted for simulation purposes, crime consists of the number of drug (C_1), theft (C_2) and violence (C_3) related cases in the Belgian province of Antwerp. The data set contains 31 four-monthly observations, starting in the year 2006. Compared to other time series used in literature, the small nature of the data set provides an added level of complexity.

5.2 Methodology

The results from the experiments are obtained using a 80-20 division between training and test. This results in the 25 first observations to be used to learn the model, while the last six are used for validation.

All available data is normalized linearly in the $[0,1]$ range, with 0 representing the effective theoretical minimum of 0, and 1 representing the maximum observed value for the entire Belgian area, plus a small margin. Although this paper does not intend to make long-term predictions, this margin gives the models a larger range of error to operate in.

The results compare predictions for different FCM-related techniques and are obtained after 20 independent runs for each configuration. Model *M1* uses Eq. 3. In Eq. 4 used in *M2*, activation values are used for concepts, C_1, C_2 and C_3, while actual values are used for the other concept. *M3* uses actual values for *all* concepts, also using 4. The model introducing the ARIMA concepts is *M4*, using Eq. 5, with only activation values to evolve the model.

A real-valued Genetic Algorithm is used to determine the weight matrices for each of the different settings. The algorithm is implemented in the package GA, available for R[2]. The size of the population is 49, with a maximum number of iterations equal to 500. The crossover probability is 0.7, while mutation probability is 0.2. All values in the weight matrix are restricted to the $[-1,1]$ interval. The goal function to be optimized is a standardized form of the mean absolute error (MAE) and can be seen in Eq. 9.

$$fitness = \left(1 + \frac{\sum_{c=1}^{C^*} \sum_{t=1}^{T} |A_c^i - R_d^i|}{C^* \times T}\right)^{-1} \tag{9}$$

5.3 Results

Figure 1 displays the values computed by each forecasting method, using the weights that yield the lowest MAE on the test set. Predictions are shown for C_1, C_2 and C_3, which are the concepts of interest in the dataset. The line graph

[2] https://cran.r-project.org/web/packages/GA/index.html.

depicts the predicted values according to the model, while the data points display the actual values from the data set.

From this numerical simulation we can notice that *M1* converges quite early in the time series. Convergence is not achieved in *M2* and *M3*, each showing varying degrees of fluctuations in the predictions. However, both need actual values from a data set to do a next-step prediction. Lacking this input data, they are not capable of making predictions, as can be seen for the last observations in the time series. The *M4* method does not suffer from a converged series while also being capable of predicting multiple steps ahead.

According to the prediction accuracy - as MAE - of 20 independent runs in the boxplots in Table 1, *M4* causes a drop in both accuracy and robustness compared to *M1* and *M3*. This is noticeable on both the training and the test set. As expected, the one-step ahead prediction of *M3* produces the highest accuracy results. Furthermore, the robustness of *M1* seems noteworthy. Its boxplot is dwarfed by the large range of *M2*, which occasionally outperforms *M1*, but does not seem to produce stable results.

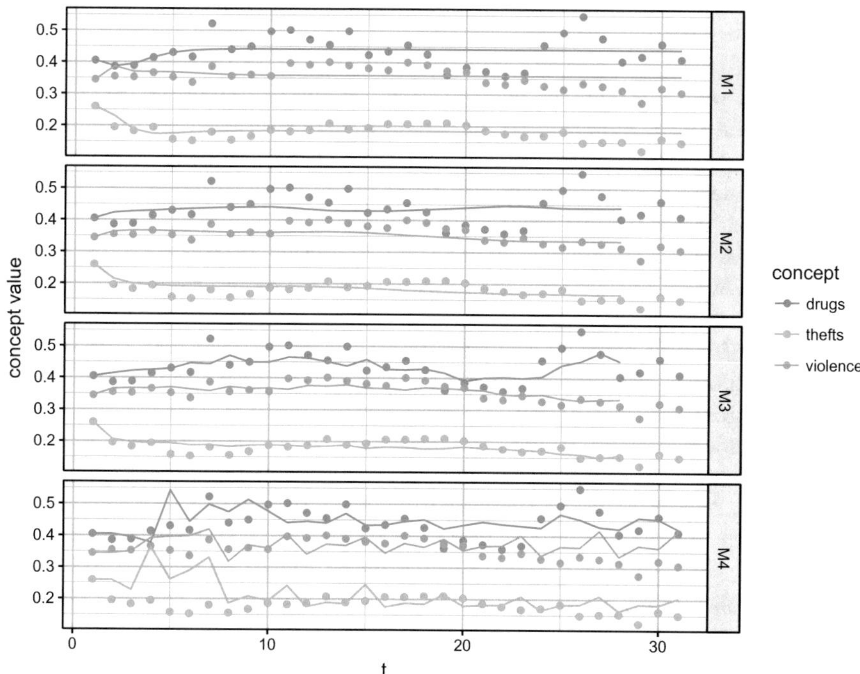

Fig. 1. Predictions for the models with the best validation fitness for each method.

The relatively worse MAE results attached to the *M4* method can be explained by the fact that its model is quite complex, compared to the others, with the weights not having a direct effect but rather being modified by

Table 1. Mean absolute error and standard deviations.

Method	Test avg.	Test std.	Training avg.	Training std.
$M1$	0.0449	±0.0012	0.0227	±0.0005
$M2$	0.0830	±0.0660	0.0702	±0.0661
$M3$	0.0534	±0.0577	0.0464	±0.0533
$M4$	0.0867	±0.0293	0.0413	±0.0084

correlations of previous predictions. As such, there is a higher degree of freedom as well as less predictability in the behavior the model.

The facts that the weights are amplified based on correlations between previous predictions, the model has become less transparent. The weight matrix can only be interpreted by combining it with those correlations, which are dynamically changing throughout the iterations. As a result, the impact of the causal relations cannot be completely understood from the weight matrix.

6 Conclusions

In this paper, we try to remedy some of the inherent drawbacks of FCMs when applied to time series analysis. By introducing concepts familiar from ARIMA models, the goal is to enable an FCM to (1) predict a fluctuating time series, (2) predict multiple steps ahead in the future, (3) maintain the prediction accuracy and (4) maintain the transparency of the weight matrix.

The results seem to suggest that the introduction of the ARIMA components does allow the FCM to predict multiple steps ahead in the future, without converging to a single set of values.

However, this has come at the expense of decreased accuracy scores. More importantly though, the transparency of the weight matrix has been blurred by the weight amplification function.

Despite some promising results, the method needs some optimization in order to achieve all goals set out in this paper.

Other unsolved issues regarding time series prediction remain a topic for future research. First of all, in the FCM model proposed by Kosko, an iteration has no meaning. An FCM is a type of recurrent neural network, representing a set of complex interactions within a system. When using the technique for analysis and predictions of time series, it is thus not self-evident that an iteration corresponds to a time interval.

Clarification of the relation between an observation at time k and an activation value at iteration t, will also provide insights into the issue of lagging effects in a multivariate time series. Ideally, the model should be able to deal with effects that have varying lags in a $[0, T]$ range, in which case both contemporary and historic causality can be taken into account.

Addressing iterations and time intervals also inevitably leads to the topic of convergence. Convergence is sought after in the area of cognitive mapping, as it represents a stable state of the system. Without that stable state, it is not obvious to extract decision-making rules from the results. In contrast, in a time series, convergence is less likely to be desirable. This paper already exposes a possible path towards a solution. However, further research is needed to develop a model that exhibits all strengths of an FCM, while also boasting the strong time series prediction capabilities.

References

1. Baykasoglu, A., Durmusoglu, Z.D.U., Kaplanoglu, V.: Training fuzzy cognitive maps via extended great deluge algorithm with applications. Comput. Ind. **62**(2), 187–195 (2011)
2. Becker, G.S.: Crime and punishment: an economic approach. J. Polit. Econ. **76**(2), 169–217 (1968)
3. Box, G.E.P., Jenkins, G.M., Reinsel, G.C., Ljung, G.M.: Time Series Analysis: Forecasting and Control. Wiley, New York (2015). google-Books-ID: lCy9BgAAQBAJ
4. Bueno, S., Salmeron, J.L.: Benchmarking main activation functions in fuzzy cognitive maps. Expert Syst. Appl. **36**(3, (Part 1)), 5221–5229 (2009)
5. Corman, H., Mocan, H.N.: A time-series analysis of crime, deterrence, and drug abuse in new york city. Am. Econ. Rev. **90**(3), 584–604 (2000)
6. Corman, H., Mocan, N.: Carrots, sticks, and broken windows. J. Law Econ. **48**(1), 235–266 (2005)
7. Froelich, W., Papageorgiou, E.I.: Extended evolutionary learning of fuzzy cognitive maps for the prediction of multivariate time-series. In: Papageorgiou, E.I. (ed.) Fuzzy Cognitive Maps for Applied Sciences and Engineering. Intelligent Systems Reference Library, vol. 54, pp. 121–131. Springer, Heidelberg (2014)
8. Froelich, W., Salmeron, J.L.: Evolutionary learning of fuzzy grey cognitive maps for the forecasting of multivariate, interval-valued time series. Int. J. Approximate Reasoning **55**(6), 1319–1335 (2014)
9. Homenda, W., Jastrzebska, A., Pedrycz, W.: Modeling time series with fuzzy cognitive maps. In: 2014 IEEE International Conference on Fuzzy Systems (FUZZ-IEEE), pp. 2055–2062, July 2014
10. Homenda, W., Jastrzebska, A., Pedrycz, W.: Time series modeling with fuzzy cognitive maps: simplification strategies. In: Computer Information Systems and Industrial Management, pp. 409–420. Springer, Heidelberg (2014)
11. Kosko, B.: Fuzzy cognitive maps. Int. J. Man Mach. Stud. **24**(1), 65–75 (1986)
12. Lu, W., Yang, J., Liu, X.: The linguistic forecasting of time series based on fuzzy cognitive maps, pp. 649–654. IEEE (2013)
13. Lu, W., Yang, J., Liu, X., Pedrycz, W.: The modeling and prediction of time series based on synergy of high-order fuzzy cognitive map and fuzzy c-means clustering. Knowl.-Based Syst. **70**, 242–255 (2014)
14. Mocan, H.N., Corman, H.: An economic analysis of drug use and crime. J. Drug Issues **28**(3), 613–629 (1998)
15. Nápoles, G., Bello, R., Vanhoof, K.: How to improve the convergence on sigmoid Fuzzy Cognitive Maps? Intell. Data Anal. **18**(6S), S77–S88 (2014)
16. Nápoles, G., Grau, I., Prez-Garca, R., Bello, R.: Learning of fuzzy cognitive maps for simulation and knowledge discovery. In: Bello, R. (ed.) Studies on Knowledge Discovery, Knowledge Management and Decision Making, pp. 27–36. Atlantis Press, Paris (2013)

17. Nápoles, G., Papageorgiou, E., Bello, R., Vanhoof, K.: On the convergence of sigmoid Fuzzy Cognitive Maps. Inf. Sci. **349**, 154–171 (2016)
18. Papageorgiou, E.I.: Review study on fuzzy cognitive maps and their applications during the last decade. In: Glykas, M. (ed.) Business Process Management. Studies in Computational Intelligence, vol. 444, pp. 281–298. Springer, Heidelberg (2013)
19. Papageorgiou, E.I., Salmeron, J.L.: Methods and algorithms for fuzzy cognitive map-based modeling. In: Papageorgiou, E.I. (ed.) Fuzzy Cognitive Maps for Applied Sciences and Engineering. Intelligent Systems Reference Library, vol. 54, pp. 1–28. Springer, Heidelberg (2014)
20. Park, K.S., Kim, S.H.: Fuzzy cognitive maps considering time relationships. Int. J. Hum. Comput. Stud. **42**(2), 157–168 (1995)
21. Pedrycz, W., Jastrzebska, A., Homenda, W.: Design of fuzzy cognitive maps for modeling time series. IEEE Trans. Fuzzy Syst. **24**(1), 120–130 (2016)
22. Peffers, K., Tuunanen, T., Rothenberger, M.A., Chatterjee, S.: A design science research methodology for information systems research. J. Manag. Inf. Syst. **24**(3), 45–77 (2007)
23. Poczta, K., Yastrebov, A.: Monitoring and prediction of time series based on fuzzy cognitive maps with multi-step gradient methods. In: Progress in Automation. Robotics and Measuring Techniques, pp. 197–206. Springer, Cham (2015)
24. Salmeron, J.L., Froelich, W.: Dynamic optimization of fuzzy cognitive maps for time series forecasting. Knowl.-Based Syst. **105**, 29–37 (2016)
25. Song, H., Miao, C., Roel, W., Shen, Z., Catthoor, F.: Implementation of fuzzy cognitive maps based on fuzzy neural network and application in prediction of time series. IEEE Trans. Fuzzy Syst. **18**(2), 233–250 (2010)
26. Song, H.J., Miao, C.Y., Shen, Z.Q., Roel, W., Maja, D.H., Francky, C.: Design of fuzzy cognitive maps using neural networks for predicting chaotic time series. Neural Networks **23**(10), 1264–1275 (2010)
27. Stach, W., Kurgan, L.A., Pedrycz, W.: Numerical and linguistic prediction of time series with the use of fuzzy cognitive maps. IEEE Trans. Fuzzy Syst. **16**(1), 61–72 (2008)
28. Stach, W., Kurgan, L., Pedrycz, W., Reformat, M.: Genetic learning of fuzzy cognitive maps. Fuzzy Sets Syst. **153**(3), 371–401 (2005)
29. Wang, D.J.: Pearson correlation coefficient. In: Dubitzky, W., Wolkenhauer, O., Cho, K.H., Yokota, H. (eds.) Encyclopedia of Systems Biology, pp. 1671–1671. Springer, New York (2013). doi:10.1007/978-1-4419-9863-7_372
30. Yastrebov, A., Piotrowska, K.: Synthesis and analysis of multi-step learning algorithms for fuzzy cognitive maps. In: Papageorgiou, E.I. (ed.) Fuzzy Cognitive Maps for Applied Sciences and Engineering. Intelligent Systems Reference Library, vol. 54, pp. 133–144. Springer, Heidelberg (2014)
31. Yesil, E., Urbas, L., Demirsoy, A.: FCM-GUI: A graphical user interface for big bang-big crunch learning of FCM. In: Papageorgiou, E.I. (ed.) Fuzzy Cognitive Maps for Applied Sciences and Engineering. Intelligent Systems Reference Library, vol. 54, pp. 177–198. Springer, Heidelberg (2014)

Applying Roughication to Support Establishing Intensive Insulin Therapy at Onset of T1D

Rafal Deja(✉)

Department of Computer Science, University of Dabrowa Gornicza, Cieplaka 1c, Dabrowa Gornicza, Poland
rdeja@wsb.edu.pl

Abstract. Determination of insulin therapy at onset of type 1 diabetes is a challenge. The doctor besides basic data and results is largely based on their own experience when prescribing initial dosage of insulin. The daily insulin dose is adjusted then while observing blood glucose level during hospitalization. In the paper we are constructing decision support tool that allows to establish proper insulin therapy faster based on real medical data. Using roughication for numerical data helps in obtaining higher classification accuracy and give more useful indications for the doctors.

1 Introduction

The diabetes mellitus is one of the civilization diseases. The number of cases is rapidly growing in the recent years [17]. In some countries it started to be a serious social and economic problem [1]. Thus there are many researches concerning different aspects of this disease. One of the most recent direction is about "closing the loop" [11,13]. The idea is to create such a solution that will allow to control the blood glucose level by automatically delivering insulin. There are number of problems concerning this approach such as the delay between the insulin application and its acting, the different absorption and the number of various factors that are influencing the blood glucose level [1,2]. One of the approach is to predict the insulin dosage based on most important factors [12,16]. The other approaches concentrate on building medical guidelines that can show the possible treatment paths based on current blood glucose levels and previous insulin doses [4–6].

It is particularly difficult to establish the treatment for a new onset of diabetes. Currently it is mainly based on physician's experience. The initial daily insulin dosage is proposed and then it is adjusted by observing the blood glucose level measured many times a day. Usually the proper dosage is obtain after several days of observation. In the paper we would like to propose the rule based tool that supports the physician in establishing the insulin therapy faster. The rule based classifier is mined from the real historical data. The attributes (factors) considered in the classifier are chosen be the doctor. The starting point for classification is applying rough set theory, which usefulness has been proven in many

I. Czarnowski et al. (eds.), *Intelligent Decision Technologies 2017*,
Smart Innovation, Systems and Technologies 72, DOI 10.1007/978-3-319-59421-7_25

medical problems [3,15]. The information system [8] is defined by $I = (U, A)$, where $u \in U$ is a set of patients, which are described by the set of attributes $a \in A$ (patient sex, age, weight, some medical examinations' results like C-peptyde, CRP etc.). It can be noticed that many of the attributes have values from the set of real numbers $a(u) \in R$. Because using the attributes with reals values is not suitable (there are many discernibility classes) the most often approach is using discretization technique [9]. The correspondence of the applied discretization is always under discussion, and is especially inconvenient in the case of a small number of subjects. In the paper we propose a solution to this problem by a method called "roughication" [14]. We are developing this approach by applying "roughication" also to decision attribute which is of real values as well. Since the decision class is also "roughied" we obtain non-deterministic rules. Applying "roughication" and using non-deterministic rules increases the quality of the tool.

2 Medical Problem

In the paper we are analyzing the patient with the onset of diabetes type 1. The onset is usually rapid and the patient has to be hospitalized. It takes several days in the hospital when child is first stabilized and then the insulin therapy is established. Generally the therapy consists of maintaining the blood glucose near to the normal level according to medical standard. Based on the factors like the patient weight, sex the state at admission, finally the size and number of the meals the physician is estimating the insulin daily dose. Daily insulin dose is usually split in a ratio of 30 to 70% into 2 parts. The first part is a long-acting insulin (up to 24 h), maintaining a reasonable level of glycemia at night and between meals. The second part of the insulin, is short acting one and it is delivered just before meals in the correct individual proportions calculated per each 100 kcal of food (or carbohydrate exchangers). Each day the insulin dosage is verified and adjusted by physician. The verification is done by the blood glucose level measured every 2–3 h.

Table 1. The attributes used in the study

Attribute	Medical meaning
Age	The patient age at onset
Sex	Male (1) or female (0)
Weight	The weight at onset
C-peptyde	Insulin secretion
CRP	Certificate of infection, 1 or 0
PH	ACID based balance
Glycemia	Measured every 3 h during a day
Insulin	Basal and pre-meal insulin doses

3 Medical Data

For the purpose of this study we gathered data of 102 cases of new onset of diabetes mellitus type 1 treated in Medical University of Silesia, diabetology department. The attributes considered in the study are presented in Table 1.

4 Roughication

Rough set theory developed by Pawlak [10] has been proven as an important tool for classificatory analysis of data tables also when dealing with uncertain knowledge. It has been used in many researches concerning medical problems [3,15]. Since in our medical problem the attributes are of real values we are proposing to apply roughication as the supplement of the rough set classification, and to avoid discretization of some of the attributes. The roughication has been proposed in [14], but we expand this method by allowing the decision attribute to be of real value as well. The roughication can be defined in the following way.

For the given information system with decision attribute $\mathbf{A} = (U, A \cup \{d\})$, we create the new information system $\mathbf{A}^* = (U^*, A^* \cup \{d^*\})$, where $U^* \equiv U \times U$ and $A^* \cup \{d^*\} \equiv A \cup \{d\}$ in the following way. For each attribute of real value $a \in A$, $V_a \in \mathbf{R}$ we create the new attribute $a^* \in A^*$, which values are from the set $\{$"$\geq a(x)$", "$< a(x)$"$\}$, where $x, y \in U$. More precisely the attribute $a^*(x, y)$ for each $x, y \in U$ obtains the value $a^*(x, y) =$ "$\geq a(x)$" when $a(y) \geq a(x)$ and $a^*(x, y) =$ "$< a(x)$" otherwise. For symbolic attributes $b^*(x, y) = b(y)$ i.e. $V_b^* = V_b$, and $b \in A$. It is worth to notice that the values from V_a^* in the following calculations are treated as symbolic values. Later on, after creating the classifier, the values from V_a^* are started to be treated as not symbolic again e.g. in the process of new case classification. We call this as deroughication (the naming convention refers to fuzzy set approach [18]). As can be noticed after deroughication we are obtaining non-deterministic rules. To be able to evaluate the rules the support and confidence coefficients are defined. Let $lh(r)$ denotes the conditional part of the rule and $rh(r)$ is the decision part. With $\|lh(r)\|_{\mathbf{A}}$ we denotes the set of all objects from decision table $\mathbf{A}$ that fits to conditions defined by $lh(r)$. For the rule r and decision table $\mathbf{A}$ the support is defined by:

$$supp(r) = \frac{card(\|lh(r)\| \cap \|rh(r)\|)}{card(U)} \tag{1}$$

and the confidence by:

$$conf(r) = \frac{card(\|lh(r)\| \cap \|rh(r)\|)}{card(\|lh(r)\|)} \tag{2}$$

Let us observe that support and confidence are defined against decision table $\mathbf{A}$ and not $\mathbf{A}^*$. After deroughication we consider the objects from original decision table $\mathbf{A}$ again. The support coefficient allows us to restrict the number of rules to most important, while the confidence helps to select the decision when rules are non-deterministic.

5 Therapy Supporting Tool

The idea of the tool we are building is to support the physician in establishing the insulin therapy for a new admitted patient. The supporting tool is made of the rule base classifier where the decision class is the insulin dose. The classifier is mined from the historical data describing the patient treatment. The assumption is made that similar patients should be treated in similar way. Thus based on the classifier the physician is proposed the initial insulin dose for a patient. Of course due to the huge number of potential patients' states the precise dose cannot be predicted. Typically such a real value like insulin dose is approximated to the range of values defined by discretization algorithm. Here instead we are proposing to use roughication approach. Using of the tool can be presented in the following steps:

- When admitting to the hospital the initial patient data are gathered.
- The rules based tool (classifier) is applied for that data. Since the rules are non-deterministic more than one decision is fired.
- The decisions with support and confidence grater than the given thresholds are presented. These decisions are analyzed to retrieved the smallest possible range of decision values.
- The physician is decided the proper insulin dosage considering suggested values.

5.1 Data Preparation

In the study we considered the attributes described in Table 1 for all 102 patients. The daily insulin dose for each patient has been calculated as the sum of basal insulin value and the insulin ratio. The insulin ratio is the number of insulin units taken for every 100 kcal of meal. The therapy changes during patients hospitalization were analyzed by an expert. Finally only these daily insulin doses were considered where corresponding glycemia was around normal level in most of the measurements. This has been reached usually at the end of hospitalization. Since the glycemia has been considered to be around normal state according to medical standards in all the cases we skipped this attribute in the following calculations.

6 Experiments and Results

The therapy support tool is built using real medical data as described in Sect. 3. Mainly it consists on creating the rule based classifier. The daily insulin dose is considered as the decision attribute. The fragment of this decision table is presented in Table 2. This decision table after roughication is partially presented in Table 3.

The rules has been deduced using LEM2 algorithm [7]. They are partially presented in Table 4 together with support and confidence values. If the rules are

Table 2. Fragment of decision table

Patient	Sex	Weight	C-peptyde	CRP	PH	Age	Dose
1	0	15	0	0	0	2	7
2	1	23	0	0	0	6	14
3	1	41	0	0	0	13	41
4	0	47	0	0	0	16	25
5	0	33	0	0	0	11	11
6	1	44	0	0	0	12	32

Table 3. Fragment of decision table after roughication

Patient	Sex	Weight	C-peptyde	CRP	PH	Age	Dose
(1, 1)	0	≥15	0	0	0	≥2	≥7
(1, 2)	1	≥15	0	0	0	≥2	≥7
(1, 3)	1	≥15	0	0	0	≥2	≥7
(1, 4)	0	≥15	0	0	0	≥2	≥7
(1, 5)	0	≥15	0	0	0	≥2	≥7
(1, 6)	1	≥15	0	0	0	≥2	≥7
(2, 1)	0	<23	0	0	0	<6	<14
(2, 2)	1	≥23	0	0	0	≥6	≥14
(2, 3)	1	≥23	0	0	0	≥6	≥14
(2, 4)	0	≥23	0	0	0	≥6	≥14
(2, 5)	0	≥23	0	0	0	≥6	<14
(2, 6)	1	≥23	0	0	0	≥6	≥14
(3, 1)	0	<41	0	0	0	<13	<41

inconsistent only the rules with higher confidence are considered. As the example of applying the tool let examine the potentially new case described with the following values: crp = 0 and c-peptyde = 0 and ph = 0 and sex = 1 and age = 9 and weight = 30. One can notice that first 7 rules will be triggered. Considering decisions with highest confidence value i.e. 0.91 the *dose* should be in the range $[10, 46)$, however considering a little smaller confidence level 0.73 the *dose* range in reduced to $[14, 22)$. This way the physician is obtaining quite useful treatment recommendation. Furthermore, the proposed values are in the range mined based on real examples in contrast to discretization which is based on predefined, often artificial intervals.

The usefulness of the proposed approach has been verified using cross - validation technique. We built the test set by randomly taken 20% of the cases. The rest of the data constitute the training set in the form of decision table. The training set was the subject of roughication, and afterwords the rule based

Table 4. The example of deduced rules (cpe stands for c-peptyde)

Rule	Supp.	Conf.
$(crp=0)\wedge(cpe=0)\wedge(ph=0)\wedge(sex=1)\wedge(age<17)\wedge(weight<51)\Rightarrow dose\geq 12$	0.27	0.87
$(crp=0)\wedge(cpe=0)\wedge(ph=0)\wedge(sex=1)\wedge(age<17)\wedge(weight<51)\Rightarrow dose\geq 13$	0.27	0.87
$(crp=0)\wedge(cpe=0)\wedge(ph=0)\wedge(sex=1)\wedge(age<13)\wedge(weight<53)\Rightarrow dose<46$	0.26	1
$(crp=0)\wedge(cpe=0)\wedge(ph=0)\wedge(sex=1)\wedge(age<13)\wedge(weight<53)\Rightarrow dose<31$	0.26	0.76
$(crp=0)\wedge(cpe=0)\wedge(ph=0)\wedge(sex=1)\wedge(age<10)\wedge(weight<42)\Rightarrow dose\geq 14$	0.19	0.73
$(crp=0)\wedge(cpe=0)\wedge(ph=0)\wedge(sex=1)\wedge(age<10)\wedge(weight<42)\Rightarrow dose<22$	0.19	0.82
$(crp=0)\wedge(cpe=0)\wedge(ph=0)\wedge(sex=1)\wedge(age<10)\wedge(weight<42)\Rightarrow dose\geq 10$	0.19	0.91
$(crp=0)\wedge(cpe=0)\wedge(ph=0)\wedge(sex=0)\wedge(age\geq 6)\wedge(weight\geq 31)\Rightarrow dose\geq 11$	0.19	1
$(crp=0)\wedge(cpe=0)\wedge(ph=0)\wedge(sex=0)\wedge(age\geq 6)\wedge(weight\geq 31)\Rightarrow dose\geq 16$	0.19	0.86
$(crp=0)\wedge(cpe=0)\wedge(ph=0)\wedge(sex=0)\wedge(age<17)\wedge(weight<51)\Rightarrow dose\geq 12$	0.19	0.95
$(crp=0)\wedge(cpe=0)\wedge(ph=0)\wedge(sex=0)\wedge(age<17)\wedge(weight<51)\Rightarrow dose\geq 13$	0.19	0.91

classifier has been deduced - it consists of more than 800 rules. After deroughication we calculated the support and confidence coefficients against the original decision table. We restricted the number of rules by considering the rules with minimum support equals 0.20 and minimum confidence equals 0.70. It is not surprised that coverage of the training set with the mentioned rules is 100%. Furthermore the quality of the classifier understood as a percentage of correctly classified objects is also 100%, however this understanding of quality could not fit directly to the doctor expectation. Infect there are usually many (up to 20) rules that are matching the single object from decision table. Again the highest the rule support is the rule is more general, and the support usually decreasing with more specific rule. Let us present this with an example case taken from the test set. It is described by the following values: $crp=0$ and $c\text{-}peptyde=1$ and $ph=0$ and $sex=0$ and $age=6$ and $weight=31$, with decision value $dose=21$.

Table 5. The rules decision for example object

Rule decision	Supp.	Conf.
$dose\geq 4$	0.93	1
$dose<54$	0.80	1
$dose<40$	0.35	1
$dose\geq 6$	0.27	0.87
$dose<32$	0.26	0.76
$dose\geq 12$	0.20	0.94
$dose<48$	0.20	0.76
$dose\geq 14$	0.20	0.73
$dose\geq 16$	0.20	0.80

The classifier for this case delivered the decisions values presented in Table 5. It can be observed that the proposed dose should be in the range [16, 32) with confidence 0.76. The real value in this case is just in the middle of this range.

7 Conclusions

The paper presents the construction of the decision support tool for patients with onset of type 1 diabetes. The tool is based on real medical data gathered during patients' hospitalization. Like the other medical data, the data is difficult to be analyzed due to the presence of many real-valued attributes, which can be often impacted by measurement accuracy. Therefore we propose to apply roughication method and rough set theory. We demonstrated they effectiveness in a situation where we have to deal with real-valued attributes and the number of objects is not large. The roughication method has been extended by applying it also to decision attribute. As the result the rule based classifier is based on non-deterministic rules. The quality of the classifier understood as a percentage of correctly classified new objects is very high. Finally the physician is obtaining the interval of possible values as the support for his/her decision. The interval is mined based on real examples in contrast to discretization which is based on predefined, often artificial partitions. The usefulness of the treatment recommendation given by presented tool has been positively verified by the physician.

References

1. ADA: American diabetes association. Standards of medical care in diabetes-2012. Diab. Care **35**, 11–63 (2012). doi:10.2337/dc12-s011
2. Deja, G., Jarosz-Chobot, P., Polanska, J.: The rate of improvement in metabolic control in children with diabetes mellitus type 1 on insulin glargine depends on age. Exp. Clin. Endocrinol. Diab. **115**(10), 662–668 (2007)
3. Deja, R.: Applying rough set theory to the system of type 1 diabetes prediction. In: Internet–Technical Development and Applications, pp. 119–127. Springer (2009)
4. Deja, R., Froelich, W., Deja, G.: Differential sequential patterns supporting insulin therapy of new-onset type 1 diabetes. Biomed. Eng. Online **14**(1), 13 (2015)
5. Deja, R., Froelich, W., Deja, G., Wakulicz-Deja, A.: Hybrid approach to the generation of medical guidelines for insulin therapy for children. Inf. Sci. **384**, 157–173 (2017)
6. Froelich, W., Deja, R., Deja, G.: Mining therapeutic patterns from clinical data for juvenile diabetes. Fundam. Informaticae **127**(1), 513–528 (2013)
7. Grzymala-Busse, J.W.: A new version of the rule induction system lers. Fundam. Informaticae **31**(1), 27–39 (1997)
8. Komorowski, J., Pawlak, Z., Polkowski, L., Skowron, A.: Rough sets: a tutorial. Rough fuzzy hybridization: a new trend in decision-making, pp. 3–98 (1999)
9. Liu, H., Hussain, F., Tan, C.L., Dash, M.: Discretization: an enabling technique. Data Min. Knowl. Discov. **6**(4), 393–423 (2002)
10. Pawlak, Z.: Rough Sets: Theoretical Aspects of Reasoning about Data, vol. 9. Springer Science & Business Media (2012)

11. Pérez-Gandía, C., Facchinetti, A., Sparacino, G., Cobelli, C., Gómez, E., Rigla, M., De Leiva, A., Hernando, M.: Artificial neural network algorithm for online glucose prediction from continuous glucose monitoring. Diab. Technol. Ther. **12**(1), 81–88 (2010)
12. Schlotthauer, G., Gamero, L.G., Torres, M.E., Nicolini, G.A.: Modeling, identification and nonlinear model predictive control of type i diabetic patient. Med. Eng. Phys. **28**(3), 240–250 (2006)
13. Shalitin, S., Phillip, M.: Closing the loop: combining insulin pumps and glucose sensors in children with type 1 diabetes mellitus. Pediatr. Diab. **7**(s4), 45–49 (2006)
14. Slezak, D., Wroblewski, J.: Roughfication of numeric decision tables: the case study of gene expression data. In: International Conference on Rough Sets and Knowledge Technology, pp. 316–323. Springer (2007)
15. Słowiński, K., Stefanowski, J., Siwiński, D.: Application of rule induction and rough sets to verification of magnetic resonance diagnosis. Fundam. Informaticae **53**(3–4), 345–363 (2002)
16. Stahl, F., Johansson, R.: Diabetes mellitus modeling and short-term prediction based on blood glucose measurements. Math. Biosci. **217**(2), 101–117 (2009)
17. WHO: Fact sheet no. 312 (2011). http://www.who.int/diabetes/en
18. Zadeh, L.A.: Fuzzy sets. Inf. Control **8**(3), 338–353 (1965)

Tips Service and Water Diary An Innovative Decision Support System for the Efficient Water Usage at Households

Ewa Magiera(✉), Tomasz Jach, and Lukasz Kurcius

Institute of Computer Science, University of Silesia, Katowice, Poland
ewa.magiera@us.edu.pl

Abstract. In the paper, we present the tips service and water diary, which are the modules of the mobile decision support system (DSS) for the efficient water usage at households. This system is a part of the Integrated Support System for Efficient Water Usage and resources management (ISS-EWATUS). The DSS at the household level is aimed at provoking the usage of the water more effective. Therefore, the tip service was designed and implemented. Tips are generated for each household as a result of the analysis of the previously gathered data concerning water consumption. In order to enhance and to personalise the influence of the DSS, the water diary was created. The water diary gives the possibility to members of the household to follow to identify their own water consumption measurement separately. Additionally, the water consumption data are presented to them as charts in the various views. The DSS was validated and accepted at 40 households in Poland and Greece.

Keywords: Mobile DSS · Water consumption · User behaviour

1 Introduction

Integrated Support System for Efficient Water Usage and resources management (ISS-EWATUS) is the outcome of the international research project entitled "Integrated Support System for Efficient Water Usage and resources management". The project was implemented by an international consortium and founded by the 7th Framework Programme [6, 7]. The ISS-EWATUS consist of four subsystems:

1. decision support system for the efficient water usage at households [2],
2. decision support system for efficient water management at municipal water company,
3. social-media platform: enabling and promoting water-saving behaviour, development and simulation of adaptive water price systems [10].

The ISS-EWATUS project involves the development of the DSS at the household level and its subsequent trial in a real-world setting.

I. Czarnowski et al. (eds.), *Intelligent Decision Technologies 2017*,
Smart Innovation, Systems and Technologies 72, DOI 10.1007/978-3-319-59421-7_26

At pilot stage of the project, sensors were attached to water appliances (such as kitchen taps and washing machines) in homes in two places, in Sosnowiec (Poland and Skiathos (Greece)). The installed home WiFi system sends data collected by these sensors to a remote server which records the water flow rate and water temperature associated with each appliance [1]. Each family was equipped with a tablet having the DSS application pre-installed This application provided access to the feedback on the household's water use; furthermore, other mobile devices or computers already possessed by users could be used. The application allowed users to observe their household's water consumption in the shape of different reports and charts. Along with the engineering part, the social studies measuring the impact and behavioural changes of users were conducted [8,9].

In this paper two DSS components related to provoking a change of water consumers behaviour are presented: tips service and water diary. Tips service and water diary are the embedded modules of DSS and are intended to be used on mobile devices.

2 Water User Behaviour

The study in [11] presents insights into water consumption behaviour at the household level on the base of the household water consumption surveys of the domestic water consumers in Greece and Poland. The paper [9] describes the results of investigating psychosocial and behavioural factors influencing consumers' intention to engage in everyday water-saving actions around the home. The source of the analysis and research were web surveys filled in by 174 individuals in Greece. These research were used in creating and designing the strategies implemented in the DSS. The strategies contain use cases (including tips service and water diary) related to DSS functionalities concerning the influence of water consumers at household level.

Tips service and water diary allow households members to follow their water consumption in near real-time. Additionally, the substantial feature of the DSS is the possibility of encouragement of water consumers (family's members) to change their behaviour in more effective water usage.

3 Tips Service

The Tips Service is a component of the DSS at household level enabling to generate water-related tips. The tips are predefined in the database and involve multiple behaviours of the users. The tips service contains:

- a Java-based application serving the purpose of a scheduler and tips generator,
- R-project [5] scripts implementing sophisticated water-usage models.

The Tips Service works as a service which, at specified periods of time, analyses the household's water usage users and provides tips. Tips are generated using a variety of forecasting algorithms, namely:

- SQL queries which analyse the water usage directly using the database;
- Bayesian [3,4];
- Linear Regression.

The latter two of these algorithms produce a forecast for the next period of time. Using the forecast, the user is given tips regarding their expected behaviour. For example, if the Linear Regression model predicts that expected water usage will dramatically increase in the relevant period, the user is presented with a warning which forms the tip.

The language of the generated tips is dynamically chosen for each user. The end user is only presented with tips which are relevant to their situation (e.g. if someone uses too much water during a shower, they will be presented with shower-related tips).

3.1 Tips Service Model

The Tip class serves as a part of Object-Relation-Mapping (ORM). It directly reflects the Tips table in the database. The Tip class is also provided with methods to conveniently manipulate the contents of the database from the Java source files. The JobAbstract class represents the different types of Tips Jobs (that is, different ways to generate tips). The LinearRegressionJob, LogicalRegressionJob, RscriptJob and SqlScriptJob classes provide specific methods and ways to generate tips from which methods their name derives. The PurgeJob is the special case of a job which purges the assignment of tips to users daily (in order to provide new tips every day) (Fig. 1).

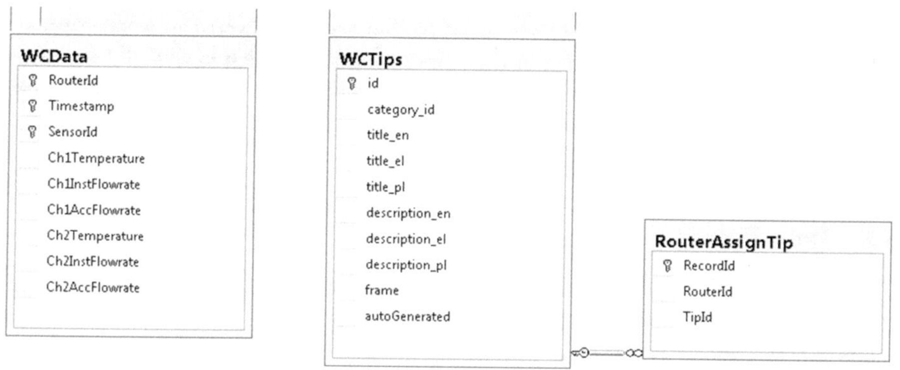

Fig. 1. Tips service database tables

3.2 Tips Service Database Structure

The Tips Service additional tables include:

1. WCTips table, which stores the tips itself. The tip is translated into three languages. Each tip has a title and a description. Additionally, each tip belongs to one of the predefined categories. A future plan is to use data mining techniques (e.g. clustering) in order to identify users who are then shown tips from a particular category.
2. WCData which is a common table for all the household sensor data.
3. RouterAssignTip is a table which stores the connection between Tips and Households. Every record stored here corresponds to one tip displayed on the tablet application in a particular household.

3.3 Tips Service Processing

The Tips Service uses the existing database to get current and historical water usage data for each and every household. These data are then analysed using methods described previously (R-scripts, Java-based regression models, SQL-based queries) in order to assign different tips to different household users. Because of the complexity of some of the computations, the R statistical language was used. The Fig. 2 shows the sequence diagram for this part of the DSS.

4 Water Diary

WaterDiary is a mobile application created for household members to identify their water readings. It enables the user to declare who used the water and for what purpose. This application allows users to identify which household member, and which activities, consume the largest volume of water. An additional feature of this application is a notification system which supports users in keeping their water diary.

4.1 Data Storage

The ISS-EWATUS spatio-temporal database contains data being gathered and transmitted by the water consumption monitoring system installed at 40 households. Each value of the water consumption is associated with the place (a bath, a toilet, a kitchen) and with the identifier of the household. The water diary requires the set of values related to each water consumption and the place, where this consumption has taken place. Hence, data, reflecting each water consumption, are retrieved from the remote ISS-EWATUS database and are published to the end user - each member of the household. The water diary database stores information about: water usage readings (dbo.HouseholdConsumption); households and household members (dbo.tblFamilies, dbo.tblFamiliesMembers); and what the water was used for (dbo.tblBehaviours) (Fig. 3).

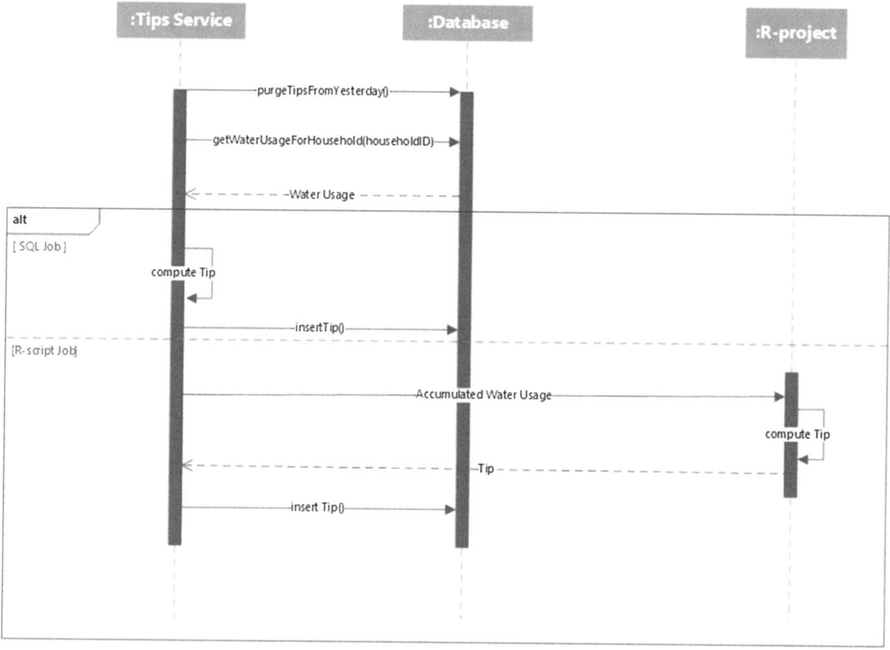

Fig. 2. Sequence diagram for tips service

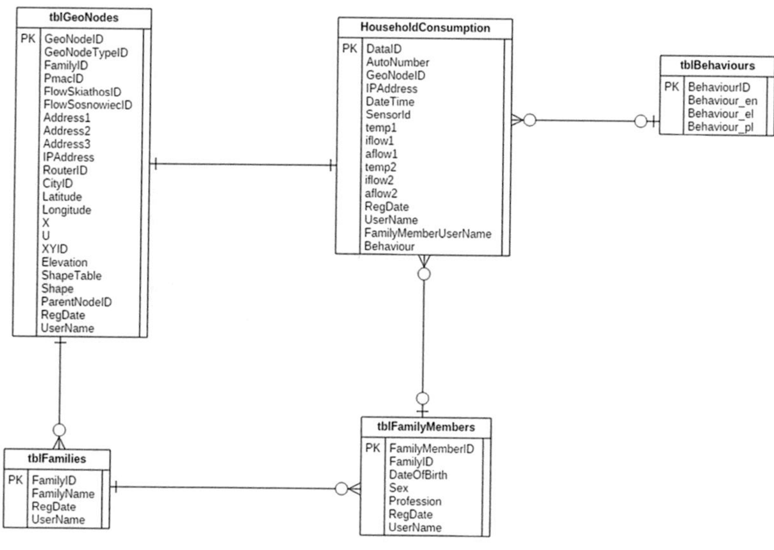

Fig. 3. Data schema of water diary

4.2 WaterDiary Application

The WaterDiary prompts household members to identify their personal water usage in a fixed range of time (e.g. daily or weekly). The need to identify usage is signalled by an alarm. In the first WaterDiary screen the user provides data about each water usage. In the column "Behaviour" the diary user specifies the purpose that the water was used for (behaviours are placed in a drop-down list). In the column "Family member" (family members are placed in drop-down list) the

← WaterDiary

ISS-EWATUS

From	To
2015-12-08	2015-12-09

Date	Behaviour	Family member	Volume
2015-12-08 00:01	Golenie	Anna	2.592
2015-12-08 00:02	Pranie	Adam	17.541
2015-12-08 00:03	Pranie	Katarzyna	31.164
2015-12-08 00:04	Pranie	Adam	2.693

Fig. 4. All readings in the chosen period

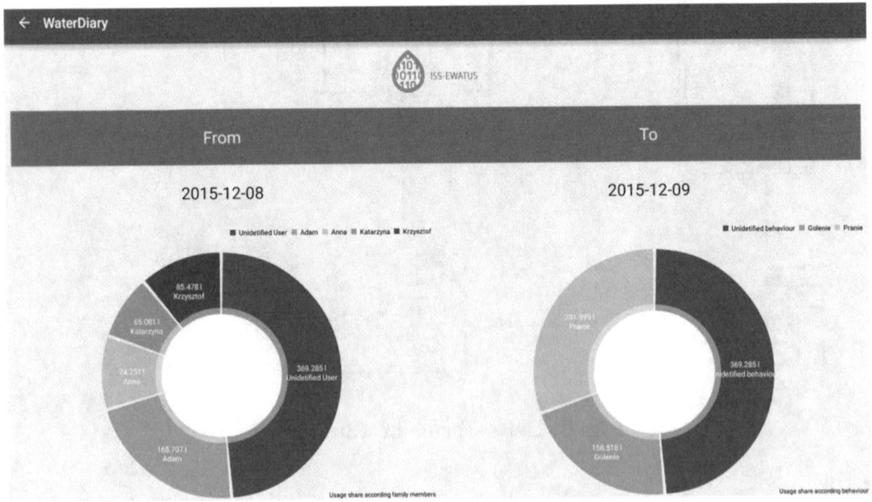

Fig. 5. Water usage charts

diary user specifies who used the water. The user is also able to select a period of time (dates below "From" and "To") to retrieve readings from a specified range of time (Fig. 4).

On the water usage screen, the user sees diagrams which present water usage in behaviour and family member contexts. The user is also able to select the period which should be presented.

The end user has the ability to submit these data after each and every water use from their home (in other words, the application works in near real time). If the user forgets to do so, once per day a notification is shown asking her or him to fill in the necessary data (Fig. 5).

5 Conclusions

In this we have presented, the tips service and the water diary, that are innovative solutions, parts of the mobile decision support system for the efficient water usage at households. Their implementation encouraged water household users to make more conscious decisions in a daily usage of water and to consume water in an efficient way. As the DSS was validated in 40 households in Poland and Greece, these positive results were observed and additionally, proven by conducted measurements of water consumption.

Acknowledgments. The research was undertaken as part of Integrated Support System for Efficient Water Usage and Resources Management (ISS-EWATUS) project, funded by European Union's Seventh Framework Programme for research, technological development and demonstration under grant agreement No. [619228].

References

1. Al-Hoqani, N., Yang, S.H.: Adaptive sampling for wireless household water consumption monitoring. Procedia Eng. **119**, 1356–1365 (2015)
2. Chen, X., Yang, S.H., Yang, L., Chen, X.: A benchmarking model for household water consumption based on adaptive logic networks. Procedia Eng. **119**, 1391–1398 (2015)
3. Froelich, W.: Forecasting daily urban water demand using dynamic Gaussian Bayesian network. In: International Conference: Beyond Databases, Architectures and Structures, pp. 333–342. Springer (2015)
4. Froelich, W., Magiera, E.: Forecasting domestic water consumption using bayesian model. In: Proceedings of the 8th KES International Conference on Intelligent Decision Technologies (KES-IDT 2016), pp. 337–346. Springer (2016)
5. Gentleman, R., Ihaka, R., Bates, D., et al.: The R project for statistical computing (1997). http://www.r-project.org
6. Magiera, E., Froelich, W.: Integrated support system for efficient water usage and resources management (ISS-EWATUS). Procedia Eng. **89**, 1066–1072 (2014)
7. Magiera, E., Froelich, W., Jach, T., Kurcius, Ł., Berbeka, K., Bhulai, S., Kokkinos, K., Papageorgiou, E., Laspidou, C., Yang, L., Perren, K., Yang, S.H., Capiluppi, A., El-Jamal, S., Wang, Z.: ISS-EWATUS an example of integrated system for efficient water management. In: Proceedings of the conference Computing and Control for Water Industry (CCWI), Amsterdam, pp. 1–10 (2016)

8. Perren, K., Yang, L., He, J., Yang, S.H., Shan, Y.: Incorporating persuasion into a decision support system: the case of the water user classification function. In: 2016 22nd International Conference on Automation and Computing (ICAC), pp. 429–434 (2016)
9. Perren, K., Yang, L.: Psychosocial and behavioural factors associated with intention to save water around the home: a Greek case study. Procedia Eng. **119**, 1447–1454 (2015)
10. Safa El-Jamal, A.C., Wang, Z.: A holistic dissemination strategy to deliver water conservation messages through gamification and social networks. In: Conference Water Efficency Network (WATEF), pp. 1–10 (2016)
11. Shan, Y., Yang, L., Perren, K., Zhang, Y.: Household water consumption: insight from a survey in Greece and Poland. Procedia Eng. **119**, 1409–1418 (2015)

Decision Making Theory for Economics

Strict and Strong Consistency in Pairwise Comparisons Matrix with Fuzzy Elements

Jaroslav Ramík(✉)

Faculty of Business Administration in Karviná,
Silesian University, Opava, Czech Republic
ramik@opf.slu.cz

Abstract. This paper forms both theoretical and practical innovation basis for decision making process in micro and macro economics. The decision making problem considered here is to rank n alternatives from the best to the worst, using the information given by the decision maker(s) in the form of an $n \times n$ pairwise comparisons matrix. Here, we deal with pairwise comparisons matrices with fuzzy elements. Fuzzy elements of the pairwise comparisons matrix are applied whenever the decision maker is uncertain about the value of his/her evaluation of the relative importance of elements in question. We investigate pairwise comparisons matrices with elements from abelian linearly ordered group (alo-group) over a real interval which is a generalization of traditional multiplicative or additive approaches. The concept of reciprocity and consistency of pairwise comparisons matrices with fuzzy elements is well known. Here, we extend these concepts, namely to the strict as well as strong consistency of pairwise comparisons matrices with fuzzy elements (PCF matrices). We derive the necessary and sufficient conditions for strict/strong consistency and investigate their properties as well as some consequences to the problem of ranking the alternatives. Illustrating examples are presented and discussed.

Keywords: Ranking alternatives · Pairwise comparisons matrix · Reciprocity · Consistency · Fuzzy elements

1 Introduction

Pairwise comparisons is a process of comparing entities in pairs to judge which of each entity is preferred, or has a greater amount of some quantitative property, or, whether or not the two entities are identical. The method of pairwise comparisons introduced by Thurstone in 1927, see [1], was a milestone in scientific studies of preferences, attitudes, voting systems, social choice, public choice, and multi-agent AI systems, etc. Mathematical analysis of pairwise comparisons is of considerable importance since this method has been used in projects of national importance, see e.g. [2], or, 150 applications of AHP [3].

A *decision making problem (DM problem)* which forms an application background in this paper can be formulated as follows:

I. Czarnowski et al. (eds.), *Intelligent Decision Technologies 2017*,
Smart Innovation, Systems and Technologies 72, DOI 10.1007/978-3-319-59421-7_27

Let $\mathfrak{A} = \{\mathfrak{a}_1, \mathfrak{a}_2, ..., \mathfrak{a}_n\}$ be a finite set of alternatives ($n > 1$). The decision maker's aim is to rank the alternatives from the best to the worst (or, vice versa), using the information given by the decision maker in the form of an $n \times n$ pairwise comparisons matrix. Alternatively, the problem can be formulated in a more complex way as the Multi-Criteria Decision Making problem (MCDM problem) with $\mathfrak{C} = \{\mathfrak{c}_1, \mathfrak{c}_2, ..., \mathfrak{c}_m\}$, a finite set of evaluation criteria ($m > 1$), see e.g. [4]. We obtain $m \times m$ pairwise comparisons matrix of the criteria evaluations.

Interval values and/or fuzzy values - the elements of the pairwise comparisons matrix can be applied whenever the decision maker is not sure about the preference degree of his/her evaluation of the pairs in question, see [5,6,8]. Fuzzy elements may be taken also as the aggregations of crisp pairwise comparisons of a group of DM in the group DM problem, see [7].

Usually, an ordinal *ranking* of alternatives is required to obtain the "best" alternative(s), however, it often occurs that the decision maker is not satisfied with the ordinal ranking among alternatives and a cardinal ranking with the help of so called weights, i.e. *rating* is then required.

The recent paper is in some sense a continuation of [9]. In comparison to [9], we extend our approach from fuzzy number entries to fuzzy intervals as the matrix entries of pairwise comparisons matrices (PCF matrices). We newly introduce the concepts of consistency, strong consistency and strict consistency of PCF matrices. Then we derive necessary and sufficient conditions for strict and/or strong consistent PCF matrices, which can be easily checked. Then we investigate some consequences of the new concepts to the problem of ranking the alternatives. Strict and strong consistency may be useful when constructing membership functions of fuzzy elements of PCF matrices, see e.g. [10]. Moreover, we also solve the problem of measuring the inconsistency of PCF matrices by defining corresponding indexes. We present several numerical examples in order to illustrate the proposed concepts and derived properties. Due to the space limitation the proofs of all propositions are omitted.

2 Preliminaries

For our approach, it will be useful to consider fuzzy sets as special nested families of subsets of a set, see [11].

A *fuzzy subset* of a nonempty set X (or a *fuzzy set* on X) is a family $\{F_\alpha\}_{\alpha\in[0,1]}$ of subsets of X such that $F_0 = X, F_\beta \subset F_\alpha$ whenever $0 \leq \alpha \leq \beta \leq 1$, and $F_\beta = \cap_{0\leq\alpha<\beta} F_\alpha$ whenever $0 < \beta \leq 1$. The *membership function of* F is the function μ_F from X into the unit interval $[0, 1]$ defined by $\mu_F(x) = \sup\{\alpha \mid x \in F_\alpha\}$. Given $\alpha \in]0, 1]$, the set $[F]_\alpha = \{x \in X \mid \mu_F(x) \geq \alpha\}$ is called the *α-cut of fuzzy set F*.

Let F be a subset of X and let $\{F_\alpha\}_{\alpha\in[0,1]}$ be the family of subsets of X defined by $F_0 = X$ and $F_\alpha = F$ for each $\alpha \in [0, 1]$. It can easily be seen that this family is a fuzzy set on X and that its membership function is equal to the characteristic function of F; we call it the *crisp fuzzy set* on X. If X is a nonempty subset of the n-dimensional Euclidean space $\mathbf{R}^n$, then a fuzzy set F

of X is called *closed, bounded, compact* or *convex* if the α-cut $[F]_\alpha$ is a closed, bounded, compact or convex subset of X for every $\alpha \in]0,1]$, respectively. The case $\alpha = 0$ will be dealt with below.

We say that a fuzzy subset F of $\mathbf{R}^* = \mathbf{R} \cup \{-\infty\} \cup \{+\infty\}$ is a *fuzzy interval* whenever F is normal and its membership function μ_F satisfies the following condition: there exist $a, b, c, d \in \mathbf{R}^*$, $-\infty \le a \le b \le c \le d \le +\infty$, such that

$$\begin{aligned} \mu_F(t) &= 0 \text{ if } t < a \text{ or } t > d, \\ \mu_F &\text{ is strictly increasing and continuous on } [a,b], \\ \mu_F(t) &= 1 \text{ if } b \le t \le c, \\ \mu_F &\text{ is strictly decreasing and continuous on } [c,d]. \end{aligned} \tag{1}$$

A fuzzy interval F is *bounded* if $[a,d]$ is a compact interval. Moreover, for $\alpha = 0$ we define the *zero-cut* of F as $[F]_0 = [a,d]$. A bounded fuzzy interval F is the *fuzzy number*, if $b = c$.

An *abelian group* is a set, G, together with an operation $\odot$ (read: operation *odot*) that combines any two elements $a, b \in G$ to form another element in G denoted by $a \odot b$, see [12,13]. The symbol $\odot$ is a general placeholder for a concretely given operation. $(G, \odot)$ satisfies the following requirements known as the *abelian group axioms*, particularly: *commutativity (abelian)*, *associativity*, there exists an *identity element* $e \in G$ and for each element $a \in G$ there exists an element $a^{(-1)} \in G$ called the *inverse element to a*.

The *inverse operation* $\div$ to $\odot$ is defined for all $a, b \in G$ as follows

$$a \div b = a \odot b^{(-1)}. \tag{2}$$

An ordered triple $(G, \odot, \le)$ is said to be *abelian linearly ordered group, alo-group* for short, if $(G, \odot)$ is a group, $\le$ is a linear order on G, and for all $a, b, c \in G$

$$a \le b \text{ implies } a \odot c \le b \odot c. \tag{3}$$

If $\mathcal{G} = (G, \odot, \le)$ is an alo-group, then G is naturally equipped with the order topology induced by $\le$ and $G \times G$ is equipped with the related product topology. We say that $\mathcal{G}$ is a *continuous alo-group* if $\odot$ is continuous on $G \times G$.

By definition, an alo-group $\mathcal{G}$ is a lattice ordered group. Hence, there exists $\max\{a, b\}$, for each pair $(a, b) \in G \times G$. Nevertheless, a nontrivial alo-group $\mathcal{G} = (G, \odot, \le)$ has neither the greatest element nor the least element.

Because of the associative property, the operation $\odot$ can be extended by induction to n-ary operation.

$\mathcal{G} = (G, \odot, \le)$ is *divisible* if for each positive integer n and each $a \in G$ there exists the (n)-th root of a denoted by $a^{(1/n)}$, i.e. $\left(a^{(1/n)}\right)^{(n)} = a$.

Let $\mathcal{G} = (G, \odot, \le)$ be an alo-group. Then the function $\|.\| : G \to G$ defined for each $a \in G$ by

$$\|a\| = \max\{a, a^{(-1)}\} \tag{4}$$

is called a *$\mathcal{G}$-norm*.

The operation $d : G \times G \to G$ defined by $d(a, b) = \|a \div b\|$ for all $a, b \in G$ is called a *$\mathcal{G}$-distance*. The well known examples of alo-groups are [13] or [9].

Example 1: Additive alo-group $\mathcal{R} = (]-\infty, +\infty[, +, \leq)$ is a continuous alo-group with: $e = 0$, $a^{(-1)} = -a$, $a^{(n)} = a + a + ... + a = n.a$.

Example 2: Multiplicative alo-group $\mathcal{R}^+ = (]0, +\infty[, \bullet, \leq)$ is a continuous alo-group with: $e = 1$, $a^{(-1)} = a^{-1} = 1/a$, $a^{(n)} = a^n$. Here, by $\bullet$ we denote the usual operation of multiplication.

Example 3: Fuzzy additive alo-group $\mathcal{R}_a = (]-\infty, +\infty[, +_f, \leq)$, see [9], is a continuous alo-group with: $a +_f b = a + b - 0.5$, $e = 0.5, a^{(-1)} = 1 - a, a^{(n)} = n.a - (n-1)/2$.

Example 4: Fuzzy multiplicative alo-group $]\mathbf{0}, \mathbf{1}[_\mathbf{m} = (]0, 1[, \bullet_f, \leq)$, see [13], is a continuous alo-group with:

$$a \bullet_f b = \frac{ab}{ab + (1-a)(1-b)}, e = 0.5, a^{(-1)} = 1 - a.$$

3 PCF Matrices

In this paper we shall investigate an $n \times n$ pairwise comparisons matrix $C = \{\tilde{c}_{ij}\}$ with elements $\tilde{c}_{ij}$ being bounded fuzzy intervals of the alo-group $\mathcal{G}$ over an interval G of the real line $\mathbf{R}$. We call it shortly the *PCF matrix*. Moreover, we assume that all diagonal elements of this PCF matrix are crisp, particularly they are isomorfic to the identity element of $\mathcal{G}$.

By this general approach based on alo-group various approaches known from the literature can be unified. This fact has been demonstrated on the previous 4 examples, where the well known alo-groups are shown.

It was proven, see [9], that the above alo-goups are isomorfic to each other.

Notice that elements of PCF matrices may be crisp and/or fuzzy numbers, also crisp and/or fuzzy intervals, fuzzy intervals with bell-shaped membership functions, triangular fuzzy numbers, trapezoidal fuzzy numbers etc. Such fuzzy elements may be either evaluated by individual decision makers, or, they may be made up of crisp pairwise evaluations of decision makers in a group DM problem, see e.g. [10].

3.1 Consistency of PCF Matrices

Rationality and compatibility of a decision making process can be achieved by the consistency property of pairwise comparisons matrices, here, PCF matrices.

Let $\mathcal{G} = (G, \odot, \leq)$ be a continuous alo-group, $C = \{\tilde{c}_{ij}\}$ be a crisp PCF matrix, where $c_{ij} \in G \subset \mathbf{R}$ for all $i, j \in \{1, 2, ..., n\}$. The following definition is well known, see e.g. [13].

A crisp PCF matrix $C = \{c_{ij}\}$ is $\odot$-*consistent* if for all $i, j, k \in \{1, 2, ..., n\}$

$$c_{ik} = c_{ij} \odot c_{jk}. \tag{5}$$

The following equivalent condition for consistency of PCF matrices is popular e.g. in AHP, see [13].

Proposition 1. *A crisp PC matrix* $C = \{c_{ij}\}$ *is* $\odot$*-consistent if and only if there exists a vector* $w = (w_1, w_2, ..., w_n)$, $w_i \in G$ *such that*

$$w_i \div w_j = c_{ij} \text{ for all } i, j \in \{1, 2, ..., n\}. \tag{6}$$

Now, we extend the definition to PCF matrices as follows, see also [9]. Let $\alpha \in [0, 1]$. A PCF matrix $C = \{\tilde{c}_{ij}\}$ is said to be α-$\odot$-*consistent*, if the following condition holds:
For every $i, j, k \in \{1, 2, ..., n\}$, there exist $c_{ik} \in [\tilde{c}_{ik}]_\alpha$, $c_{ij} \in [\tilde{c}_{ij}]_\alpha$ and $c_{jk} \in [\tilde{c}_{jk}]_\alpha$ such that (5) is satisfied.
The matrix C is said to be $\odot$-*consistent*, if C is α-$\odot$-consistent for all $\alpha \in [0, 1]$. If for some $\alpha \in [0, 1]$ the matrix C is not α-$\odot$-consistent, then C is called α-$\odot$-*inconsistent.*

Remark 1. If C is crisp, then the previous definitions of consistency, α-$\odot$-consistency and $\odot$-consistency of a PCF matrix are equivalent.

Remark 2. Let $\alpha, \beta \in [0, 1], \alpha \geq \beta$. Evidently, if $C = \{\tilde{c}_{ij}\}$ is α-$\odot$-consistent, then it is β-$\odot$-consistent.

In order to extend Proposition 1 to the non-crisp case we define the notion of consistent vector with respect to a PCF matrix. Let $\alpha \in [0, 1]$, $C = \{\tilde{c}_{ij}\}$ be a PCF matrix. A vector $w = (w_1, w_2, ..., w_n)$, $w_i \in G$ for all $i \in \{1, 2, ..., n\}$, is an α-$\odot$-*consistent vector with respect to* C if for every $i, j \in \{1, 2, ..., n\}$ there exists $c_{ij} \in [\tilde{c}_{ij}]_\alpha$ such that (6) is satisfied.
The next proposition gives 3 necessary and sufficient conditions for a PCF matrix to be α-$\odot$-consistent.

Proposition 2. *Let* $\alpha \in [0, 1]$, $C = \{\tilde{c}_{ij}\}$ *be a PCF matrix, denote* $[c^L_{ij}(\alpha), c^R_{ij}(\alpha)] = [\tilde{c}_{ij}]_\alpha$. *The following four conditions are equivalent.*

(i) $C = \{\tilde{c}_{ij}\}$ *is* α-$\odot$*-consistent.*
(ii) *There exists a vector* $w = (w_1, w_2, ..., w_n)$ *with* $w_i \in G, i \in \{1, 2, ..., n\}$, *such that for all* $i, k \in \{1, 2, ..., n\}$ *it holds*

$$c^L_{ik}(\alpha) \leq w_i \div w_k \leq c^R_{ik}(\alpha). \tag{7}$$

(iii) *For all* $i, j, k \in \{1, 2, ..., n\}$, *it holds*

$$[c^L_{ik}(\alpha), c^R_{ik}(\alpha)] \cap [c^L_{ij}(\alpha) \odot c^L_{jk}(\alpha), c^R_{ij}(\alpha) \odot c^R_{jk}(\alpha)] \neq \emptyset. \tag{8}$$

(iv) *For all* $i, j, k \in \{1, 2, ..., n\}$, *it holds*

$$c^L_{ik}(\alpha) \leq c^R_{ij}(\alpha) \odot c^R_{jk}(\alpha), \tag{9}$$

$$c^R_{ik}(\alpha) \geq c^L_{ij}(\alpha) \odot c^L_{jk}(\alpha). \tag{10}$$

Remark 3. Property *(iv)* in Proposition 2 is very useful for checking α-$\odot$-consistency of PCF matrices. For a given PCF matrix $C = \{\tilde{c}_{ij}\}$ it can be easily calculated whether inequalities (9) and (10) are satisfied or not.

3.2 Strong Consistency of PCF Matrices

In this section the consistency property of PCF matrices investigated in the previous section will be strengthen. We define strong α-$\odot$-consistent PCF matrices and derive their properties.

Let $\alpha \in [0,1]$. A PCF matrix $C = \{\tilde{c}_{ij}\}$ is said to be *strong* α-$\odot$-*consistent*, if the following condition holds:
For every $i,j,k \in \{1,2,...,n\}$, and for every $c_{ij} \in [\tilde{c}_{ij}]_\alpha$, there exist $c_{ik} \in [\tilde{c}_{ik}]_\alpha$ and $c_{jk} \in [\tilde{c}_{jk}]_\alpha$ such that (5) is satisfied.

The matrix C is said to be *strong* $\odot$-*consistent*, if C is strong α-$\odot$-consistent for all $\alpha \in [0,1]$.

Remark 4. Again, if C is crisp, then the definitions of consistency, strong α-$\odot$-consistency and strong $\odot$-consistency of a PCF matrix are equivalent.

Remark 5. Evidently, each strong α-$\odot$-consistent PCF matrix is α-$\odot$-consistent.

Remark 6. Notice that Remark 2 is not true for strong consistency. Particularly, if $C = \{\tilde{c}_{ij}\}$ is a strong α-$\odot$-consistent PCF matrix, $\alpha, \beta \in [0,1], \beta < \alpha$, then $C = \{\tilde{c}_{ij}\}$ need not be strong β-$\odot$-consistent. However, it must be β-$\odot$-consistent.

Remark 7. If for a PCF matrix $C = \{\tilde{c}_{ij}\}$, property (5) is valid only for all $i,j,k \in \{1,2,...,n\}$ such that $i = k$ (i.e. not necessarily for all $i,j,k \in \{1,2,...,n\}$), then we say that the matrix C is α-$\odot$-*reciprocal.* Hence, each α-$\odot$-consistent PCF matrix is α-$\odot$-reciprocal. The opposite assertion is, however, not true.

In the following proposition we formulate two necessary and sufficient conditions for strong α-$\odot$-consistency of a PCF matrix. This property may be useful for checking strong consistency of PCF matrices.

Proposition 3. *Let $\alpha \in [0,1]$, $C = \{\tilde{c}_{ij}\}$ be a PCF matrix, $[c_{ij}^L(\alpha), c_{ij}^R(\alpha)] = [\tilde{c}_{ij}]_\alpha$. The following three conditions are equivalent.*

(i) *$C = \{\tilde{c}_{ij}\}$ is strong α-$\odot$-consistent.*
(ii) *For all $i,j,k \in \{1,2,...,n\}$, it holds*

$$[c_{ik}^L(\alpha), c_{ik}^R(\alpha)] \cap [c_{ij}^R(\alpha) \odot c_{jk}^L(\alpha), c_{ij}^L(\alpha) \odot c_{jk}^R(\alpha)] \neq \emptyset. \tag{11}$$

(iii) *For all $i,j,k \in \{1,2,...,n\}$, it holds*

$$c_{ik}^L(\alpha) \leq c_{ij}^L(\alpha) \odot c_{jk}^R(\alpha), \tag{12}$$

$$c_{ik}^R(\alpha) \geq c_{ij}^R(\alpha) \odot c_{jk}^L(\alpha). \tag{13}$$

Remark 8. Property (*iii*) in Proposition 3 is useful for checking strong α-$\odot$-consistency of PCF matrices. For a given PCF matrix $C = \{\tilde{c}_{ij}\}$ it can be easily calculated whether inequalities (12) and (13) are satisfied or not.

Example 5: Consider the additive alo-group $\mathcal{R} = (\mathbf{R}, \odot, \leq)$ with $\odot = +$, see Example 1. Let PCF matrices $A = \{\tilde{a}_{ij}\}$ be given as follows:

$$A = \begin{bmatrix} (0;0;0) & (1;2;3) & (3.5;6;8) \\ (-3;-2;-1) & (0;0;0) & (2.5;4;5) \\ (-8;-6;-3.5) & (-5;-4;-2.5) & (0;0;0) \end{bmatrix}.$$

Here, A is a 3×3 PCF matrix, particularly a PCF matrix with triangular fuzzy number elements and the usual "linear" membership functions. Checking inequalities (12) and (13), we obtain that A is strong α-$\odot$-consistent for $\alpha = 0$ and $\alpha = 1$. As the membership functions of all elements of A are triangular piecewise linear functions, we obtain that A is strong α-$\odot$-consistent PCF matrix for all α, $0 \leq \alpha \leq 1$, hence, A is strong $\odot$-consistent.

3.3 Strict Consistency of PCF Matrices

In this section the strong consistency property of PCF matrices will be more strengthen. Similarly to α-$\odot$-strong consistency defined earlier, we define strict α-$\odot$-consistent PCF matrices and derive their properties.

Let $\alpha \in [0,1]$. A PCF matrix $C = \{\tilde{c}_{ij}\}$ is said to be *strict α-$\odot$-consistent*, if the following condition holds:
For every $i, j, k \in \{1, 2, ..., n\}$, and for every $c_{ik} \in [\tilde{c}_{ik}]_\alpha$, and every $c_{ij} \in [\tilde{c}_{ij}]_\alpha$ there exists $c_{jk} \in [\tilde{c}_{jk}]_\alpha$ such that $c_{ik} = c_{ij} \odot c_{jk}$.

The matrix C is said to be *strict $\odot$-consistent*, if C is strict α-$\odot$-consistent for all $\alpha \in [0,1]$.

Remark 9. Again, if C is crisp, then all the above stated definitions of consistency are equivalent. Evidently, each strict α-$\odot$-consistent PCF matrix is strong α-$\odot$-consistent.

Now we formulate two necessary and sufficient conditions for strict α-$\odot$-consistency of a PCF matrix. This property is useful for checking strict consistency of PCF matrices.

Proposition 4. *Let $\alpha \in [0,1]$, $C = \{\tilde{c}_{ij}\}$ be a PCF matrix. The following three conditions are equivalent.*

(i) *$C = \{\tilde{c}_{ij}\}$ is strict α-$\odot$-consistent.*
(ii) *For all $i, j, k \in \{1, 2, ..., n\}$, it holds*

$$[c_{ik}^L(\alpha), c_{ik}^R(\alpha)] \subseteq [c_{ij}^L(\alpha) \odot c_{jk}^L(\alpha), c_{ij}^R(\alpha) \odot c_{jk}^R(\alpha)]. \tag{14}$$

(iii) *For all $i, j, k \in \{1, 2, ..., n\}$, it holds*

$$c_{ik}^L(\alpha) \geq c_{ij}^L(\alpha) \odot c_{jk}^L(\alpha), \tag{15}$$

$$c_{ik}^R(\alpha) \leq c_{ij}^R(\alpha) \odot c_{jk}^R(\alpha). \tag{16}$$

3.4 Priority Vectors, Inconsistency of PCF Matrices

Now, the definition of the priority vector for ranking the alternatives will be based on the concept of consistency (the "weakest" one), i.e. Proposition 2, (*ii*), particularly on the optimal solution of the following optimization problem:

(P1) $\alpha \longrightarrow \max;$

subject to

$$c_{ij}^L(\alpha) \le w_i \div w_j \le c_{ij}^R(\alpha) \text{ for all } i,j \in \{1,2,...,n\}, \tag{17}$$

$$\bigodot_{i=1}^{n} w_i = e,\ 0 \le \alpha \le 1,\ w_k \in G \text{ for all } k \in \{1,2,...,n\}. \tag{18}$$

If optimization problem (P1) has a feasible solution, i.e. system of constraints (17), (18) has a solution, then (P1) has also an optimal solution. Let α^* and $w^* = (w_1^*, ..., w_n^*)$ be an optimal solution of problem (P1). Then α^* is called the $\odot$*-consistency grade of* C, denoted by $g_\odot(C)$, i.e.

$$g_\odot(C) = \alpha^*. \tag{19}$$

Here, $w^* = (w_1^*, ..., w_n^*)$ is called the $\odot$*-priority vector of* C. This vector is associated with the ranking of alternatives $\mathfrak{a}_i, \mathfrak{a}_j$ from the set of alternatives $\mathfrak{A}$ as follows:

$$\text{If } w_i^* \ge w_j^* \text{ then } \mathfrak{a}_i \succeq \mathfrak{a}_j,$$

where $\succeq$ stands for "is not worse then".

If optimization problem (P1) has no feasible solution, then we define

$$g_\odot(C) = 0. \tag{20}$$

In that case the priority vector will be defined later in Sect. 3.4.

Proposition 5. *Let $C = \{\tilde{c}_{ij}\}$ be a PCF matrix, where all entries $\tilde{c}_{ij}$ are fuzzy numbers. If $w^* = (w_1^*, ..., w_n^*)$ is an optimal solution of (P1), i.e. $\odot$-priority vector of C, then w^* is unique.*

Remark 10. The optimal solution α^* and $w^* = (w_1^*, ..., w_n^*)$ of problem (P1) should be unique as decision makers usually ask for unique decision, or, a unique ranking of the alternatives in X. A sufficient condition for uniqueness of the priority vector $w^* = (w_1^*, ..., w_n^*)$ is formulated in Proposition 5. However, this is not the case of PCF matrices where the entries are fuzzy intervals (i.e. trapezoidal fuzzy numbers). Then the uniqueness is not secured and multiple solutions of (P1) can occur. In practical decision making problems such cases usually require reconsidering evaluations of some elements of the PCF matrix.

Let $C = \{\tilde{c}_{ij}\}$ be a PCF matrix, $\alpha \in [0,1]$. If there exist elements $i,j,k \in \{1,2,...,n\}$ such that for any $c_{ij} \in [\tilde{c}_{ij}]_\alpha$, any $c_{ik} \in [\tilde{c}_{ik}]_\alpha$, and any $c_{kj} \in [\tilde{c}_{kj}]_\alpha$:

$$c_{ik} \ne c_{ij} \odot c_{jk}, \tag{21}$$

then the PCF matrix C is α-$\odot$*-inconsistent*. If for all $\alpha \in [0,1]$ the PCF matrix C is α-$\odot$-inconsistent, then we say that C is $\odot$*-inconsistent*. By this definition, for a given PCF matrix C and given $\alpha \in [0,1]$, C is either α-$\odot$-consistent, or, C is α-$\odot$-inconsistent.

Notice, that for a PCF matrix C problem (P1) has no feasible solution, iff C is $\odot$-inconsistent, i.e. C is α-$\odot$-inconsistent for all $\alpha \in [0,1]$.

The $\odot$-inconsistency of C will be measured by the minimum of the distance of the "ratio" matrix $W = \{w_i \div w_j\}$ to the "left" matrix $C^L = \{c_{ij}^L(0)\}$ and "right" matrix $C^R = \{c_{ij}^R(0)\}$, as follows. For $w = (w_1, ..., w_n), w_i \in G$, $i, j \in \{1, ..., n\}$, denote

$$\begin{aligned} d_{ij}(C,w) &= e \text{ if } c_{ij}^L(0) \leq w_i \div w_j \leq c_{ij}^R(0), \\ &= \min\{\|c_{ij}^L(0) \div (w_i \div w_j)\|, \|c_{ij}^R(0) \div (w_i \div w_j)\|\}, \text{ otherwise.} \end{aligned} \tag{22}$$

Here, by $\|.\|$ we denote the norm defined in Sect. 2. We define the maximum deviation to the matrix $W = \{w_i \div w_j\}$ as follows

$$I_\odot(C,w) = \max\{d_{ij}(C,w) | i,j \in \{1,...,n\}\}. \tag{23}$$

Now, consider the following optimization problem.

(P2) $\quad I_\odot(C,w) \longrightarrow \min;$

subject to

$$\bigodot_{i=1}^{n} w_i = e, \; w_k \in G \text{ for all } k \in \{1,2,...,n\}. \tag{24}$$

The *$\odot$-inconsistency index of PCF matrix C*, $I_\odot(C)$, is defined as

$$I_\odot(C) = \inf\{I_\odot(C,w) | w_k \text{ satisfies (24)}\}. \tag{25}$$

If $w^* = (w_1^*, ..., w_n^*)$ is an optimal solution of (P2), then

$$I_\odot(C) = I_\odot(C, w^*).$$

If there exists a feasible solution of (P1), then $\odot$-inconsistency index of PCF matrix C, $I_\odot(C)$, is equal to e, i.e. $I_\odot(C) = e$.

Now, we define a priority vector also in case of $g_\odot(C) = 0$, i.e. if no feasible solution of (P1) exists. Here, in contrast to the case of $g_\odot(C) > 0$, the priority vector cannot become an α-$\odot$-consistency vector of C for some $\alpha > 0$.

Let C be an $\odot$-inconsistent PCF matrix. The optimal solution $w^* = (w_1^*, ..., w_n^*)$ of (P2) will be called the *$\odot$-priority vector of C*.

Let us summarize the obtained results.

Let $C = \{\tilde{c}_{ij}\}$ be a PCF matrix. Then exactly one of the following two cases occurs:

- Problem (P1) has a feasible solution. Then consistency grade $g_\odot(C) = \alpha$, for some α, $0 \leq \alpha \leq 1$, $I_\odot(C) = e$. The $\odot$-priority vector of C is the optimal solution of problem (P1).
- Problem (P1) has no feasible solution. Then consistency grade $g_\odot(C) = 0$, C is $\odot$-inconsistent, and $I_\odot(C) > e$. The $\odot$-priority vector of C is the optimal solution of problem (P2).

4 Conclusion

This paper forms both theoretical and practical innovation basis for decision making process in micro and macro economics. We deal with pairwise comparisons matrices with fuzzy elements. Fuzzy elements of the pairwise comparison matrix are usually applied whenever the decision maker is not sure about the value of his/her evaluation of the relative importance of elements in question. In comparison with pairwise comparisons matrices investigated in the literature, here we investigated pairwise comparisons matrices with elements of abelian linearly ordered group (alo-group) over a real interval (PCF matrices). In the future research we shall investigate interdependences among the elements of PCF matrices, e.g. by applying Choquet integral.

Acknowledgment. This paper was supported by the Ministry of Education, Youth and Sports Czech Republic within the Institutional Support for Long-term Development of a Research Organization in 2017.

References

1. Thurstone, L.L.: A law of comparative judgement. Psychol. Rev. **34**, 278–286 (1927)
2. Saaty, T.L.: Scaling method for priorities in hierarchical structures. J. Math. Psychol. **15**(3), 234–281 (1977)
3. Vaidya, O.S., Sushil, K.: Analytic hierarchy process: an overview of applications. Eur. J. Oper. Res. **169**, 1–29 (2006)
4. Kou, G., Ergu, D., Lin, C.S., Chen, Y.: Pairwise comparison matrix in multiple criteria decision making. Technol. Econ. Dev. Econ. **22**(5), 738–765 (2016)
5. Entani, T., Sugihara, K.: Uncertainty index based interval assignment by Interval AHP. Eur. J. Oper. Res. **219**(2), 379–385 (2012)
6. Ohnishi, S., Yamanoi, T., Imai, H.: A weights representation for fuzzy constraint-based AHP. In: IPMU 2008 (2008). http://www.gimac.uma.es/ipmu08/proceedings/papers/036-OhnishiYamanoiImai.pdf
7. Zhang, H.: Group decision making based on multiplicative consistent reciprocal preference relations. Fuzzy Sets Syst. **282**, 31–46 (2016)
8. Ramik, J., Korviny, P.: Inconsistency of pairwise comparison matrix with fuzzy elements based on geometric mean. Fuzzy Sets Syst. **161**, 1604–1613 (2010)
9. Ramik, J.: Pairwise comparison matrix with fuzzy elements on alo-group. Inf. Sci. **297**, 236–253 (2015)
10. Bilgic, T., Turksen, I.B.: Measurement of membership functions: theoretical and empirical work. In: Dubois, D., Prade, H. (eds.) Fundamentals of Fuzzy Sets, pp. 195–227. Kluwer Academic Publ., New York (2000)
11. Ramik, J., Vlach, M.: Generalized concavity in optimization and decision making. Kluwer Academic Publishers, Boston-Dordrecht-London (2001)
12. Bourbaki, N.: Algebra II. Springer, Heidelberg (1998)
13. Cavallo, B., D'Apuzzo, L.: Reciprocal transitive matrices over abelian linearly ordered groups: characterizations and application to multi-criteria decision problems. Fuzzy Sets Syst. **266**, 33–46 (2015)

Adjustment from Inconsistent Comparisons in AHP to Perfect Consistency

Kazutomo Nishizawa(✉)

Nihon University, 1-2-1 Izumicho, Narashino, Chiba 275-8575, Japan
nishizawa.kazutomo@nihon-u.ac.jp

Abstract. In this study, for inconsistent pairwise comparisons in Analytic Hierarchy Process (AHP), a method for improving consistency is proposed. The pairwise comparisons constructed by the decision maker, usually include inconsistent comparisons. In the traditional AHP, the consistency of comparisons is judged by the value of the consistency index (CI). Another consistency judgment is based on the directed graph of comparisons. The purpose of this study is to adjust an inconsistent comparison matrix to a perfect consistency matrix. The proposed method is based on adjusting all the elements of the comparison matrix. The idea of the proposed method is developed based on a previous study of estimating unknown or missing comparisons. By applying the proposed method to an inconsistent example, the process of improvement is illustrated.

Keywords: AHP · Consistency · Comparisons · Graph · Cycles

1 Introduction

In this study, for inconsistent pairwise comparisons in Analytic Hierarchy Process (AHP)[4], a method for improving consistency is proposed. In a previous study, the proposed method did not adjust the comparison matrix [3]. This proposed method adjusts all the elements of the comparison matrix. The idea of the proposed method is developed based on a previous study of estimating unknown or missing comparisons [2].

In AHP, the weights of alternatives are obtained from the eigenvector of the pairwise comparison matrix. The principal eigenvalue and its corresponding eigenvector of the matrix are usually calculated by the power method. In this study, we assume that the pairwise comparison matrix $\boldsymbol{A}$, constructed by the decision maker, consists of $n \times n$ elements and $\boldsymbol{A}$ is a complete matrix of comparisons.

In the traditional AHP, the consistency of $\boldsymbol{A}$ was judged based on the value of the Consistency Index (CI). CI is calculated by Eq. (1),

$$CI = (\lambda_{max} - n)/(n - 1), \tag{1}$$

where λ_{max} is the principal eigenvalue of $\boldsymbol{A}$. In general, if $CI > 0.1$ then we judge $\boldsymbol{A}$ as inconsistent. In the case, we obtain $CI = 0$, that is $\lambda_{max} = n$,

I. Czarnowski et al. (eds.), *Intelligent Decision Technologies 2017*,
Smart Innovation, Systems and Technologies 72, DOI 10.1007/978-3-319-59421-7_28

it is called perfectly consistent in this paper. Unfortunately, we cannot obtain $CI = 0$, because comparisons by the decision maker usually include inconsistent comparisons.

Another consistency judgment is based on the directed graph of $\boldsymbol{A}$ [1]. If the directed graph of $\boldsymbol{A}$ has no cycles, that is the Transitive Law holds, then we judge $\boldsymbol{A}$ as consistent. If there are cycles in the directed graph, that is the Circulation Law holds, then we judge it an inconsistent. In the case of complete comparisons, cycles of various lengths always include cycles of length 3, so we only check for the existence of cycles of length 3.

In this study, adjusting each element of $\boldsymbol{A}$, we would like to obtain a perfect consistency matrix $\boldsymbol{A}'$. $\boldsymbol{A}'$ has $\lambda_{max} = n$ and no cycles in its directed graph. In addition, in the case of perfect consistency, its principal eigenvector is easily obtained. Each column vector of a perfect consistency matrix coincides with its principal eigenvector. So we do not need to use the power method.

This paper consists of the following sections. In Sect. 2, the proposed method is described. In Sect. 3, for an inconsistent example, the results obtained by the proposed method are illustrated and discussed. And finally, in Sect. 4, we conclude this study.

2 Proposed Method

In this section, the proposed method is explained. The idea of this method was developed in a previous study on the estimation of unknown or missing pairwise comparisons [2]. Assume that the number of alternatives is n, and the pairwise comparison matrix $\boldsymbol{A}$ consists of complete comparisons.

As it is well known, in the perfectly consistent $\boldsymbol{A}$, the element of $\boldsymbol{A}$, that is a_{ij} ($i = 1$ to n and $j = 1$ to n), Eq. (2) holds. And $\boldsymbol{A}$ has $\lambda_{max} = n$.

$$a_{ij} = w_i/w_j, \tag{2}$$

where w_i is the eigenvector of alternative ai. And based on Eq. (2), for any k ($k = 1$ to n), Eq. (3) is obtained.

$$a_{ij} = a_{kj}/a_{ki} \tag{3}$$

In the imperfectly consistent case, Eq. (2) does not hold. Therefore based on Eq. (3) and using the geometric mean, the first adjusted element a'_{ij} is obtained by Eq. (4).

$$a'_{ij} = \sqrt[n]{\prod_{k=1}^{n} (a_{kj}/a_{ki})} \tag{4}$$

The second adjusted element a''_{ij} is calculated by (a'_{kj}/a'_{ki}), instead of (a_{kj}/a_{ki}) in Eq. (4). Repeating Eq. (4) for $i = 1$ to n and $j = 1$ to n, the consistency of $\boldsymbol{A}$

can be improved. After some repetitions, adjusted elements converge. Then the adjusted comparison matrix $\boldsymbol{A}'$ is obtained. As a result, $\boldsymbol{A}'$ has $\lambda_{max} = n$ and becomes perfectly consistent.

The procedure of the proposed method, consisting of P1 to P4, is summarized as follows.

P1: From comparison matrix $\boldsymbol{A}$, calculate CI, and draw a directed graph. Then find cycles of length 3.
P2: Calculate the adjusted comparison elements by Eq. (4).
P3: Repeat Eq. (4) until adjusted comparison elements have converged. This gives us the adjusted comparison matrix $\boldsymbol{A}'$.
P4: The eigenvector of $\boldsymbol{A}'$ is obtained from any column of $\boldsymbol{A}'$.

Using the proposed method, the principal eigenvalue and eigenvector of the comparison matrix are easily obtained. Equation (2) holds for the obtained $\boldsymbol{A}'$. For any k ($k = 1$ to n), if $a'_{kk} = 1.0$ and is the maximum element of a'_{ik} ($i = 1$ to n), then the elements of i-th column of $\boldsymbol{A}'$ coincide with the eigenvector of $\boldsymbol{A}'$ calculated by the power method. There is no need to use the power method.

3 Example

In this section, using the proposed method, an inconsistent example is adjusted. This example consists of three criteria and five alternatives. To simplify this study, assume that the priority of criteria $\boldsymbol{v}$ are equally important. Therefore $\boldsymbol{v}^T = [1, 1, 1]$. First, in this study, each consistency of the comparisons of this example is judged by the values of CI and the directed graphs. Next, each value of the comparisons is adjusted by the proposed method. And finally, the overall evaluation of the five alternatives is calculated.

For three criteria, in this example, corresponding three comparison matrices, $\boldsymbol{A_1}$ to $\boldsymbol{A_3}$ (Eqs. (5) to (7)), are shown below. Each matrix consists of five alternatives, a1 to a5.

$$\boldsymbol{A_1} = \begin{bmatrix} 1 & 6 & 5 & 1/2 & 7 \\ 1/6 & 1 & 8 & 1/4 & 1/2 \\ 1/5 & 1/8 & 1 & 1/4 & 1/3 \\ 2 & 4 & 4 & 1 & 2 \\ 1/7 & 2 & 3 & 1/2 & 1 \end{bmatrix} \tag{5}$$

$$\boldsymbol{A_2} = \begin{bmatrix} 1 & 1/4 & 1/6 & 1/2 & 1/7 \\ 4 & 1 & 1/2 & 1/3 & 1/3 \\ 6 & 2 & 1 & 1/2 & 2 \\ 2 & 3 & 2 & 1 & 1/2 \\ 7 & 3 & 1/2 & 2 & 1 \end{bmatrix} \tag{6}$$

$$\boldsymbol{A_3} = \begin{bmatrix} 1 & 2 & 1/4 & 5 & 1/3 \\ 1/2 & 1 & 5 & 2 & 3 \\ 4 & 1/5 & 1 & 4 & 5 \\ 1/5 & 1/2 & 1/4 & 1 & 2 \\ 3 & 1/3 & 1/5 & 1/2 & 1 \end{bmatrix} \tag{7}$$

By procedure P1, using the power method, the principal eigenvalue and corresponding eigenvector of three matrices are obtained. The results are summarized in Table 1.

Table 1. Eigenvalue, Eigenvector and CI by the power method

Comparison matrix		$\boldsymbol{A_1}$	$\boldsymbol{A_2}$	$\boldsymbol{A_3}$
Eigenvalue		5.9048	5.5905	7.2565
Eigenvector	a1	1.000000	0.178776	0.577184
	a2	0.293815	0.391607	1.000000
	a3	0.113145	0.932448	0.865291
	a4	0.864540	0.848993	0.253770
	a5	0.306271	1.000000	0.377978
CI		0.2262	0.1476	0.5641

As a result, in Table 1, each value of CI is greater than 0.1. Therefore we can judge $\boldsymbol{A_1}$, $\boldsymbol{A_2}$ and $\boldsymbol{A_3}$ as inconsistent.

On the other hand, we can draw the directed graphs of $\boldsymbol{A_1}$, $\boldsymbol{A_2}$ and $\boldsymbol{A_3}$.

First, the directed graph of $\boldsymbol{A_1}$ is shown in Fig. 1. There are no cycles in Fig. 1, so we can judge $\boldsymbol{A_1}$ as consistent.

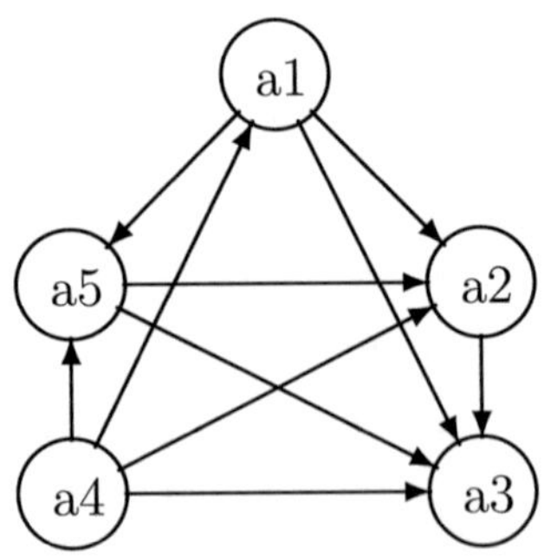

Fig. 1. Directed graph of $\boldsymbol{A_1}$

Next, the directed graph of $\boldsymbol{A_2}$ is shown in Fig. 2. There is one cycle in Fig. 2, (a3 → a5 → a4), so we can judge $\boldsymbol{A_2}$ as inconsistent.

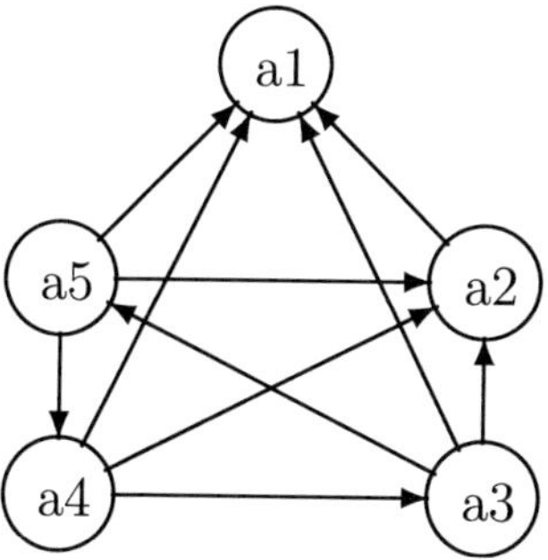

Fig. 2. Directed graph of $\boldsymbol{A_2}$

And the directed graph of $\boldsymbol{A_3}$ is shown in Fig. 3. There are three cycles in Fig. 3. These cycles are (a1 → a2 → a3), (a1 → a2 → a5) and (a1 → a4 → a5). So we can judge that $\boldsymbol{A_3}$ is also inconsistent.

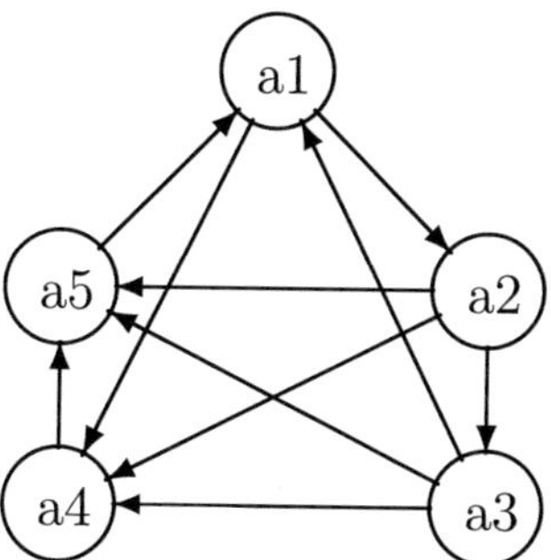

Fig. 3. Directed graph of $\boldsymbol{A_3}$

Based on each CI value and the directed graph, mentioned above, $\boldsymbol{A_1}$, $\boldsymbol{A_2}$ and $\boldsymbol{A_3}$ are judged inconsistent. In this inconsistent example, the overall evaluation vector of alternatives, $\boldsymbol{w}$, is obtained by $\boldsymbol{w} = \boldsymbol{W}\boldsymbol{v}$. Where $\boldsymbol{W}$ is the weight matrix of alternatives in Table 1.

$$\boldsymbol{w} = \begin{bmatrix} 1.000000 & 0.178776 & 0.577184 \\ 0.293815 & 0.391607 & 1.000000 \\ 0.113145 & 0.932448 & 0.865291 \\ 0.864540 & 0.848993 & 0.253770 \\ 0.306271 & 1.000000 & 0.377978 \end{bmatrix} \begin{bmatrix} 1 \\ 1 \\ 1 \end{bmatrix} = \begin{bmatrix} 0.892572 \\ 0.856717 \\ 0.971321 \\ 1.000000 \\ 0.856121 \end{bmatrix} \tag{8}$$

As a result, in this inconsistent example, the order of priority of alternatives is a4 > a3 > a1 > a2 > a5.

Next, using the proposed method, as mentioned the previous section, adjusted comparison matrices are obtained.

By procedure P2 and P3, based on Eq. (4), the results of adjusted matrices $\boldsymbol{A}'_1$, $\boldsymbol{A}'_2$ and $\boldsymbol{A}'_3$ are shown below.

$$\boldsymbol{A}'_1 = \begin{bmatrix} 1.000000 & 3.629678 & 8.719390 & 1.104083 & 3.004922 \\ 0.275507 & 1.000000 & 2.402249 & 0.304182 & 0.827876 \\ 0.114687 & 0.416277 & 1.000000 & 0.126624 & 0.344625 \\ 0.905729 & 3.287504 & 7.897402 & 1.000000 & 2.721644 \\ 0.332787 & 1.207911 & 2.901703 & 0.367425 & 1.000000 \end{bmatrix} \tag{9}$$

$$\boldsymbol{A}'_2 = \begin{bmatrix} 1.000000 & 0.422060 & 0.190062 & 0.218324 & 0.169937 \\ 2.369329 & 1.000000 & 0.450320 & 0.517282 & 0.402637 \\ 5.261434 & 2.220643 & 1.000000 & 1.148698 & 0.894113 \\ 4.580344 & 1.933182 & 0.870551 & 1.000000 & 0.778371 \\ 5.884529 & 2.483627 & 1.118427 & 1.284735 & 1.000000 \end{bmatrix} \tag{10}$$

$$\boldsymbol{A}'_3 = \begin{bmatrix} 1.000000 & 0.560978 & 0.553783 & 1.755374 & 1.528142 \\ 1.782602 & 1.000000 & 0.987175 & 3.129135 & 2.724070 \\ 1.805761 & 1.012991 & 1.000000 & 3.169786 & 2.759459 \\ 0.569679 & 0.319577 & 0.315479 & 1.000000 & 0.870551 \\ 0.654389 & 0.367098 & 0.362390 & 1.148698 & 1.000000 \end{bmatrix} \tag{11}$$

Using the power method, the results of Eqs. (9) to (11) are shown in Table 2. This table shows each principal eigenvalue, eigenvector and CI for $\boldsymbol{A}'_1$, $\boldsymbol{A}'_2$ and $\boldsymbol{A}'_3$. Each matrix has $\lambda_{max} = 5$ and $CI = 0$.

Table 2. Eigenvalue, Eigenvector and CI by the power method

Comparison matrix		$\boldsymbol{A}'_1$	$\boldsymbol{A}'_2$	$\boldsymbol{A}'_3$
Eigenvalue		5.0000	5.0000	5.0000
Eigenvector	a1	1.000000	0.169937	0.553783
	a2	0.275507	0.402637	0.987175
	a3	0.114687	0.894113	1.000000
	a4	0.905729	0.778371	0.315479
	a5	0.332787	1.000000	0.362390
CI		0	0	0

By procedure P4, each column of $\boldsymbol{A}'_1$ (as same as $\boldsymbol{A}'_2$ and $\boldsymbol{A}'_3$) is its eigenvector. The first column of $\boldsymbol{A}'_1$, the fifth column of $\boldsymbol{A}'_2$ and third column of $\boldsymbol{A}'_3$ coincide with the eigenvector obtained by the power method in Table 2.

As a result, the order of alternatives between $\boldsymbol{A}_1$ and $\boldsymbol{A}'_1$, and $\boldsymbol{A}_2$ and $\boldsymbol{A}'_2$ do not change. However, the order in $\boldsymbol{A}_3$ and $\boldsymbol{A}'_3$ is different.

On the other hand, based on Eqs. (9), (10) and (11), the directed graphs for $\boldsymbol{A}'_1$, $\boldsymbol{A}'_2$ and $\boldsymbol{A}'_3$ are illustrated in Figs. 4, 5, and 6.

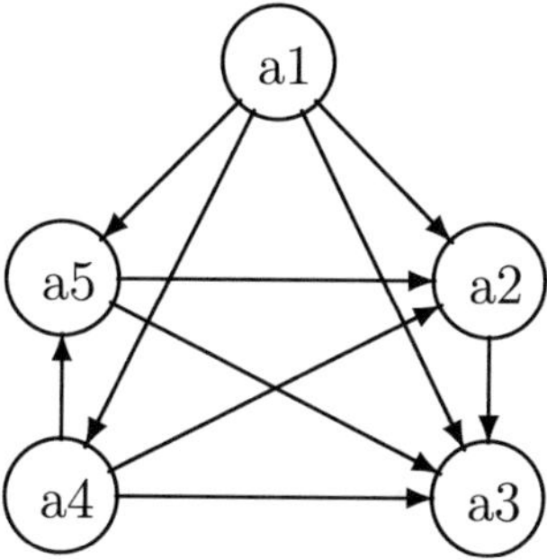

Fig. 4. Directed graph of $\boldsymbol{A}_1'$

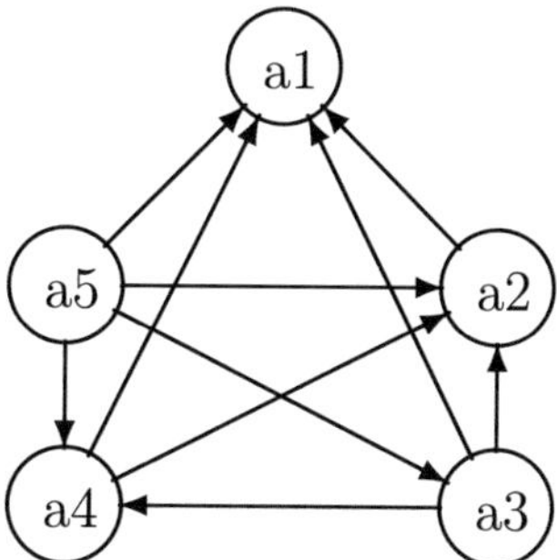

Fig. 5. Directed graph of $\boldsymbol{A}_2'$

In Fig. 4, there are no cycles, however, so comparison between a1 and a4 is different from Fig. 1. A pairwise comparison between a1 and a4 reveals the cause of the inconsistency.

In Fig. 5, there are also no cycles. However, two comparisons are different from Fig. 2. The differences are between a3 and a4 and between a3 and a5.

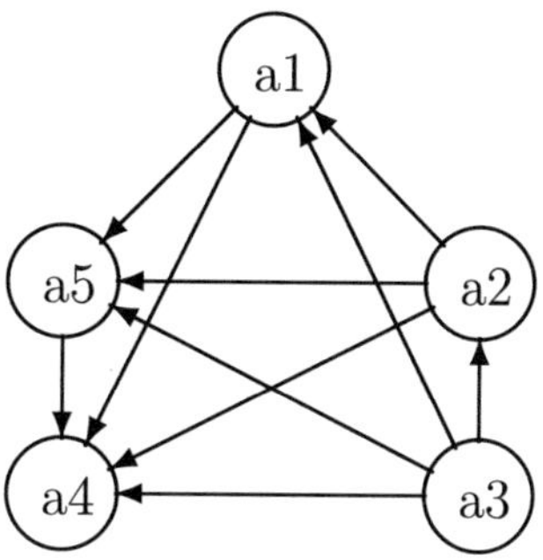

Fig. 6. Directed graph of $\boldsymbol{A}_3'$

In Fig. 6, there are also no cycles. However, four comparisons are different from Fig. 3. The differences are between a1 and a2, a1 and a5, a2 and a3, a4 and a5.

The adjusted overall evaluation vector of alternatives is obtained by $\boldsymbol{w}' = \boldsymbol{W}'\boldsymbol{v}$, where $\boldsymbol{W}'$ is the adjusted weight matrix of alternatives in Table 2.

$$\boldsymbol{w}' = \begin{bmatrix} 1.000000 & 0.169937 & 0.553783 \\ 0.275507 & 0.402637 & 0.987175 \\ 0.114687 & 0.894113 & 1.000000 \\ 0.905729 & 0.778371 & 0.315479 \\ 0.332787 & 1.000000 & 0.362390 \end{bmatrix} \begin{bmatrix} 1 \\ 1 \\ 1 \end{bmatrix} = \begin{bmatrix} 0.858085 \\ 0.829012 \\ 1.000000 \\ 0.995409 \\ 0.843876 \end{bmatrix} \quad (12)$$

As a result, the order of priority of alternatives, adjusted by the proposed method, is a3 > a4 > a1 > a5 > a2. Before adjustment, the order was a4 > a3 > a1 > a2 > a5.

4 Conclusion

In this paper, a method for improving consistency for inconsistent pairwise comparisons in AHP, was proposed. This method was to adjust all elements of the comparison matrix based on the geometric mean. Applying the proposed method to an inconsistent example, the following was obtained. Consistency improvement effects were confirmed by the value of CI and cycles of the directed graph of comparisons.

1. By the proposed method, $CI = 0$ and no cycles in the directed graph of comparisons were obtained.
2. Based on the directed graph of comparison matrix and its adjusted directed graph, the reasons for inconsistency were revealed.
3. To obtain the principal eigenvector, by the proposed method, the power method is not use. Each column of the adjusted comparison matrix coincides with the principal eigenvector.

References

1. Nishizawa, K.: A consistency improving method in binary AHP. J. Oper. Res. Soc. Jpn. **38**, 21–33 (1995)
2. Nishizawa, K.: Estimation of unknown comparisons in incomplete AHP and it's compensation. Rep. Res. Inst. I. T. Nihon Univ. **77**, 1–8 (2005)
3. Nishizawa, K.: The improvement of pairwise comparison method of the alternatives in the AHP. In: Intelligent Decision Technologies. SIST, vol. 39, pp. 483–491 (2015)
4. Saaty, T.L.: The Analytic Hierarchy Process. McGraw-Hill, New York (1980)

A Macroeconomic Model for Service Science Capitalism: Thetical and Antithetical Economics

Takafumi Mizuno(✉) and Eizo Kinoshita

School of Urban Science, Meijo University, Gifu, Japan
{tmizuno,kinoshit}@meijo-u.ac.jp

Abstract. In capitalist societies whose businesses are almost all service business, measurements of values of products are important. Researchers anticipate that Big Data technologies will enable them. A model for economic agents provided in the article is suitable for the Big Data technologies. The model represents behavioral principles of agents by using OR techniques. The principles have mathematical representation with high affinity of correlation deduced from Big Data. And we provide an explanation of macroeconomic states of Japan, EU, and United States since 1970 as examples of use of the model.

Keywords: Macroeconomics · Service science

1 Introduction

A major premise of macroeconomics is that our world is capitalism. If the world is not capitalism, then every theory of macroeconomics will lost its senses. Researchers of macroeconomics must consider whether the world is capitalism at first.

The most important concept of capitalism is fixed price sales. It means that one good has one price, and one service has one price. Fixed price sales enable managers to run their planned business and guarantee value of capitals.

To enforce fixed price sales without any contradictions on businesses, mangers must measure values of their products precisely. In a word, precise measurements of products provide bases of every index about economics and managements in the capitalist world.

For any goods products, we can measure its values relatively easily. Because the goods have physical entities and properties, we can reduce eventually their values to their length, weight, temperature, velocity, or entropy.

On the other hand, we cannot measure values of service products easily. Service products often stand on relations between goods and goods, or between services and services. Relationship is combinations of products, and increasing the number of the combinations makes measurements of values of the products complex. As service products consist of some lower level services, they are developed in high abstraction level far from physical goods products. To overcome

I. Czarnowski et al. (eds.), *Intelligent Decision Technologies 2017*,
Smart Innovation, Systems and Technologies 72, DOI 10.1007/978-3-319-59421-7_29

the complexity and the distance abstraction level, we need much knowledge of many fields.

In 1980s, researchers of macroeconomics recognized difference between goods products and service products, and they have tried to define what service products are. Now, service products are defined as products that have properties: intangibility, immediacy, variability, perishability, and customer's high satisfaction. In early 2000s, IBM researchers advocated a necessity of service science which is a new research filed to construct knowledge systems for service products.

We refer to a society in which almost all employees work for service industry as service science capitalism society. In the society, every price value has large amount of information in the background of the value, and the value is detected in high abstraction level far from its physical entity. To fill the gap between abstraction levels, we must learn techniques which reduce from large amounts of data.

Recent years, we face to floods of large data. They are impossible to treat by use of traditional technologies, and are provided by information communication technologies. We call them Big Data. They have 3 V properties: large in quantity (Volume), traded high speed (Velocity), and formed in many style (Variety).

Big Data provide us new measurements for service products, and enable us to classify service products into three services: stock service, flow service, and rate-of-flow-change service. Stock service is construction of social infrastructures or information infrastructures. Flow service is ordinary everyday service which provided by government administrators and private companies. Rate-of-flow-change service is unusual service. The classification depends on that we can trace changes of values of service products every times. It corresponds to time derivative in physics. Big Data technologies provide us feasibility of the classification.

Let us consider a mathematical model with any parameters. We can specialize the mathematical model by fill the parameters with any values. The specialized mathematical model is referred to as hypothesis. In statistical fields, targets of tests are the hypothesis.

$$\text{model} \quad + \quad \text{specific parameters} \quad = \quad \text{hypothesis} \tag{1}$$

We can test hypotheses by concrete procedures. A hypothesis which tolerate against various tests are called theory. We can use Big Data for constructing hypotheses, and for testing the hypotheses. In other words, Big Data can make hypotheses. And the hypotheses are representations of system that we want to understand.

We can understand a system by combining data and model, not only seeing data. Arising Big Data and Big Data systems reduces costs of data, and reduces values of data itself. We treat Big Data as assets. It means, to be exact, our assets are Big Data with model, and hypotheses made by Big Data.

When we treat Big Data sufficiently, correlation plays important roles in any analyses of economics. So we must build macroeconomic models which we can construct by detecting parameters from correlation deduced from Big Data.

2 A Model for Decision Making in Macroeconomic Issues

Kinoshita provides a macroeconomic model which is referred to as Thetical economics and Antithetical economics. That is a rearrangement of theories of macroeconomics into two set; a set of them is Thetical economics and another set is Antithetical economics. If Say's law is valid in an economic phase in an economic cycle, then the Thetical economics dominates the phase. We feel that we are in normal economy and economic growth in the phase. While if the Keynes's effective demand is effective in an economic phase, then the Antithetical economics dominates the phase. We feel that we are in depressed economy in the phase. Easy to say, Thetical economics represents what prosperity is, while Antithetical economics represents what recession is.

With the macroeconomic model, we can provide behavioral principles of economic agents such as corporations and governments as follows:

A principle of corporations under Thetical economics

Objective function (maximize profits)

$$\max \sum_{j=1}^{n} c_j x_j \tag{2}$$

Constraint condition

$$\sum_{j=1}^{n} a_{ij} x_j \leq b_i, \quad i = 1, \cdots, m \tag{3}$$

A principle of corporations under Antithetical economics

Objective function (minimize debts)

$$\min \sum_{i=1}^{m} u_i b_i \tag{4}$$

Constraint condition

$$\sum_{i=1}^{m} u_i a_{ij} \leq c_i, \quad j = 1, \cdots, n \tag{5}$$

Following list is correspondence of variables and its meanings.

- x_j The number of units of a product j made by the corporation.
- c_j The amount of profits of one unit of a product j; $P_j - (1 + r)h_j$, where P_j is price of the product j, r is interest rate, and h_j is cost of the product j.
- a_{ij} Costs in an account subject i to produce the product j for one unit.
- b_i The amount debts of an account subject i.
- u_i Unpaid balance rate for the accounting subject i; $u_i = 1-$amortization_rate.

A principle of governments under Thetical economics

Objective function (fiscal reconstruction)

$$\min \sum_{j=1}^{N} G_j K_j \tag{6}$$

Constraint condition

$$\sum_{j=1}^{N} A_{ij} K_j \leq B_i, \quad i = 1, \cdots, M \tag{7}$$

A principle of governments under Antithetical economics

Objective function (fiscal stimulus)

$$\max \sum_{i=1}^{M} Y_i B_i \tag{8}$$

Constraint condition

$$\sum_{i=1}^{M} Y_i A_{ij} \leq c_i, \quad j = 1, \cdots, N \tag{9}$$

Following list is correspondence of variables and its meanings.

- K_j A rate of the remainder of national loans for an administrative service j. Increasing the rate increases expenses of the service j.
- G_j Demand for funds as national loans for an administrative service j.
- A_{ij} Satisfaction of a resident i when the government gives the resident one unit of costs of a service j.
- B_i A desiring level of total services of the government for a resident i.
- Y_i The amount of public money to increase satisfaction by one unit for a resident i.

In usual studies of the macroeconomics, economic agents, such as customer, corporations, and governments, are modeled simply. All agents expand their profits, they are well-disciplined, they can acquire all information of markets, and their behavior is rational. The principles, which we provide, give a concrete mathematical model of the rationality.

The behavioral principle is linear equation system. Construction the principle is detecting parameters of the equations. So, the model has high affinity with correlation obtained from Big Data.

3 Examples: Japan, EU, and United States

For examples, we provide an explanation of macroeconomic states in Japan, EU, and United States since 1970 with the model.

Let us see Fig. 1. This is transitions of gross capital formation of non-financial corporations and GDP (Gross Domestic Products) of Japan. Roughly speaking, gross capital formation represents investments; sum of gross capital formation and inventory equals to investment. GDP is a macroeconomic index which represents business conditions of the nation.

Japan is dominated by Thetical economics before 1995, and is dominated by Antithetical economics after 1995. Before 1995, corporations increase investments. It is an evidence of behavior of maximization of their profits; the Japanese economy was dominated by Thetical economics. At February 1990, Heisei bubble collapsed. Five years later, in 1995, Japanese economy was into recession. Since the year, Japanese corporations decrease investments. They do not try to expand their profits. And GDP does not increase from then. Figure 2 shows that they reduce their debt since the year. Net worth, which is assert minus liability, of non-financial corporations is plotted in the figure. We can see a change of behavioral principle of corporations at the year. The economy is dominated by Antithetical economics.

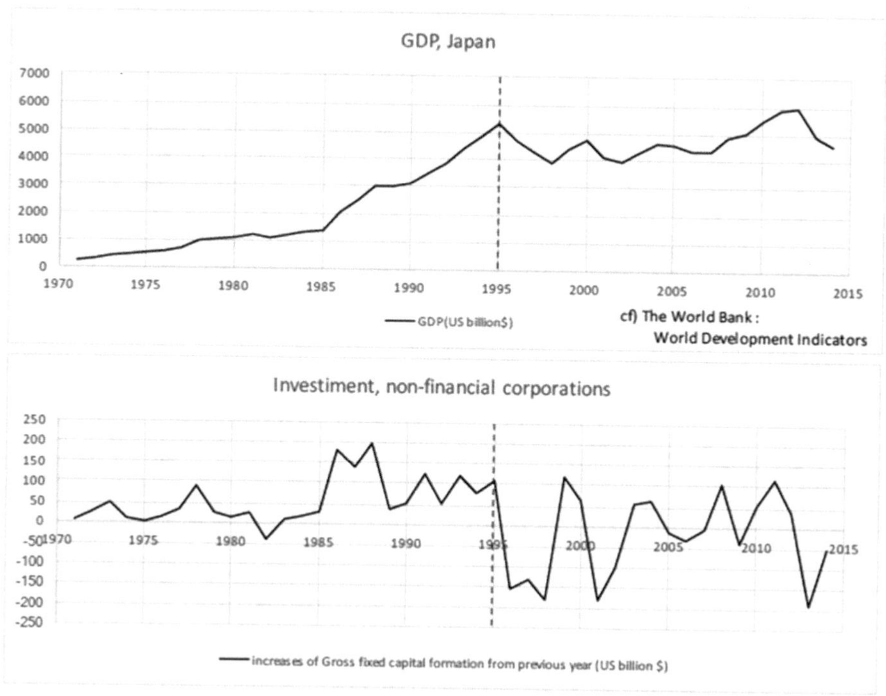

Fig. 1. GDP (upside) and gross capital formation of non-financial corporations (downside) in Japan since 1970.

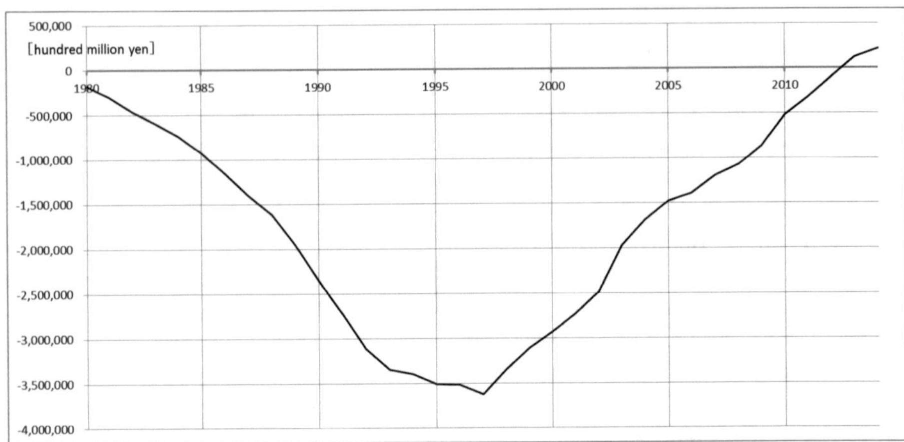

Fig. 2. Net worth of non-financial corporations in Japan since 1980. The data are from Japan Bank.

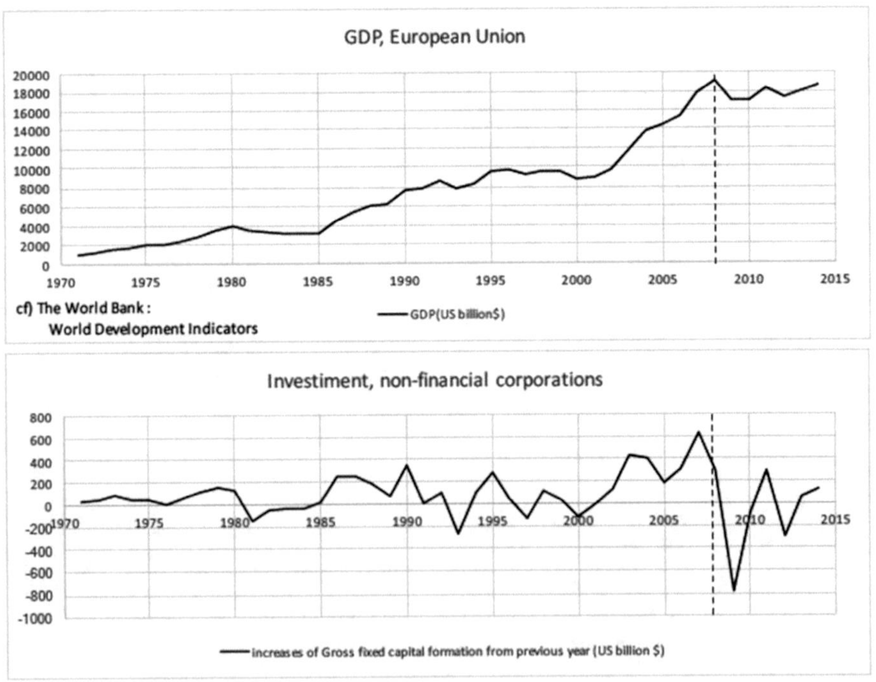

Fig. 3. GDP (upside) and gross capital formation of non-financial corporations (downside) in EU since 1970.

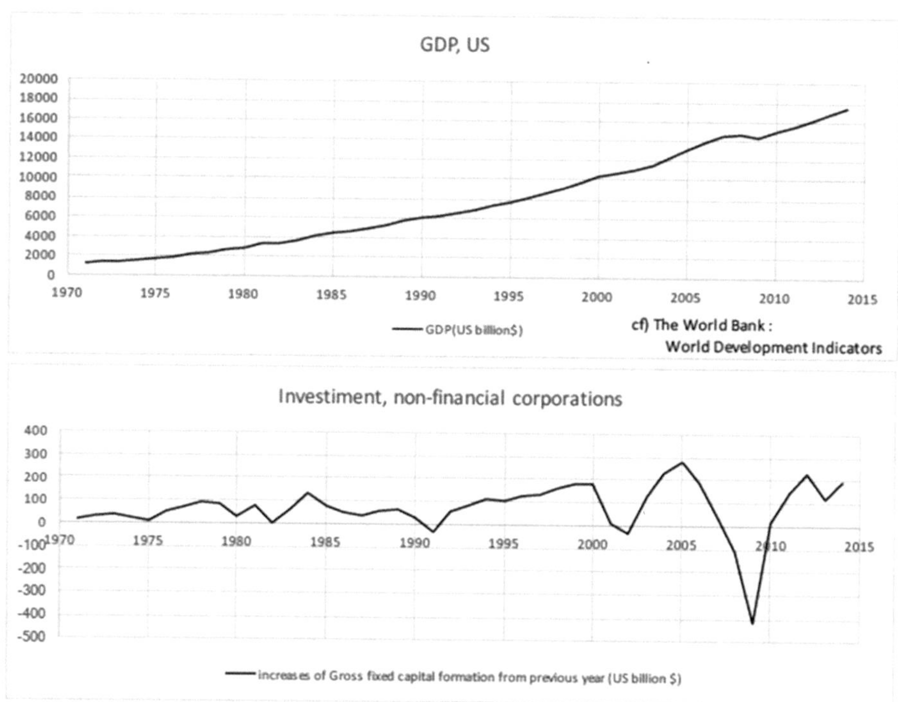

Fig. 4. GDP (upside) and gross capital formation of non-financial corporations (down-side) in United States since 1970.

For EU area and United States, their GDP and gross capital formations are in Figs. 3 and 4. In EU area, a turning point is 2008. EU has suffered effects of subprime-loan problem from United States. Before the year, EU is in Thetical economics. Since the year, corporations in EU have reduced investiments. We must check whether EU is in Antithetical economics. While EU reduces their investiments, the United States keeps increasing their investiments. The United states has been in Thetical economics.

4 Conclusions

We provide a macroeconomic model of behavioral principles of economic agents. The principles have mathematical representation with high affinity of correlation deduced from Big Data. As examples of use of the model, we describe macro-economic states in Japan, EU, and US since 1970.

References

1. Kinoshita, E.: A proposal of primal and dual problems in macro-economics. China-USA Bus. Rev. **10**(2), 115–124 (2011)
2. Kinoshita, E.: Why bubble economy occurs and crashes? - Repeated history of economic growth and collapse. Chin. Bus. Rev. **10**(2), 102–111 (2011)
3. Kinoshita, E.: A proposal of thetical economy and antithetical economy-mechanism of occurrence and collapse of bubble economy. J. Bus. Econ. **3**(2), 117–130 (2012). Academic Star Publishing Company, USA. ISSN 2155-7950
4. Kinoshita, E., Mizuno, T.: Analysing mechanism of an economic phase. China-USA Bus. Rev. **12**(11), 1025–1032 (2013). ISSN 1537-1514
5. Kinoshita, E.: Thetical and antithetical business management. J. Bus. Econ. **6**(6), 1086–1096 (2015)

Managerial Decisions Modelling for the Company Development Strategy

Irina Kalinina, Valery Maslennikov, and Marina Kholod(✉)

Chair of Management and Business Technologies, Plekhanov Russian University of Economics, 36 Stremyanny Lane, 117997 Moscow, Russia
{kalinina.ia, maslennikov.vv, kholod.mv}@rea.ru

Abstract. In this paper we investigate the subject of strategies development by Russian companies through the managerial decision modelling. In the traditional economic model it is assumed that the company exists in order to maximize profits in the long term. Profit determines the viability of a business, and protects the organization from the threat of bankruptcy. We develop strategy development matrix, based on the sales profitability metrics. Furthermore, we apply this methodology to the sample of leading companies of Russia in order to propose the necessary managerial decision in order to pursue the strategy of development of a company.

Keywords: Development strategy · Sales profitability · Development strategy matrix · Managerial decision making · Success strategy · Fragile balance strategy · Development dilemma · Crisis

1 Introduction

In the traditional economic model it is assumed that the company exists in order to maximize profits in the long term. Profit determines the viability of a business, and protects the organization from the threat of bankruptcy. Thus the rational use of profit leads to higher capacity of the organization. If the maximum sustainable growth of the potential is considered to be the global strategic goal of the company, then, the achievement of such growth is possible, if the profit is used rationally [1].

In order to get profit, you need to strive to increase revenue and/or reduce costs. The increase in revenue (sales) can be achieved by entering the market with products, for which demand is unsatisfied, to use methods of sales promotion, for example, the preferential party, i.e. by expanding market shares of competitive products.

In a market economy the volume of production resources depends on the number of sales of the enterprise products produced in previous periods. Therefore, ensuring sales is so important for the development of the organization.

Cost reduction is possible through rational use of available resources, allowing you to obtain additional funds to finance ongoing activities in the organization and for development.

At certain points, companies go for the partial decline in profit to increase its market share or retain existing market share. In addition, getting a profit does not guarantee its

I. Czarnowski et al. (eds.), *Intelligent Decision Technologies 2017*,
Smart Innovation, Systems and Technologies 72, DOI 10.1007/978-3-319-59421-7_30

further effective use, because profit may be diverted for non-productive needs, and not to financing the development of the organization.

2 Development Strategies Matrix

In real economic life, some companies work more successfully than others, getting a lower rate and mass of profit. This possibility is explained by the transactional theory of the firm, in which under the success of the firm should be understood the organization's expansion. The expansion of the organization refers to its ability to increase the share in the developed markets or develop new markets (to provide new services). In other words, the success of economic organizations is measured by the increase in the scale of its activities and therefore its ability to replace the market, reducing transaction costs [2].

However, the effectiveness of the organization development in the dynamics is characterized by the growth of its operations that is strategically not less important than getting profit. Moreover, from the point of view of development possibility of getting profit is a subordinate task compared to the purpose of growth, because the mass of profit largely depends on the volume of products sold (work performed, services rendered).

Thus, for the purpose of development of the company's strategy, it is necessary to fulfill two conditions:

- to increase market share, which would increase revenues (or sales) [3];
- to manage the self-financing of development, acquiring the necessary resources through the use of additional revenue.

2.1 Sales Profitability as the Main Indicator for Managerial Decision-Making

The increase in sales by gaining market share characterizes the ability of the organization to function effectively in the external environment, and getting profit for self-financing of expanded reproduction through the rational management of the resources provides internal capability of the development.

The generalized indicator characterizing level of use of capacities, as a result of the operation from the point of view of resource management and market position is the comparison of profit and sales. One of such indicators is the indicator of sales profitability (or return on sales), defined as the ratio of profit obtained in a given period to sales volume earned during the same period (formula 1):

$$P = \frac{\Pi}{V} \tag{1}$$

where P - profit made in a given period;

V - volume of the sales in this period.

Depending on the use in the numerator of the various indicators of profit - net profit or retained earnings, the sales profitability indicator represents its various types:

economic profitability of turnover or profit per 1 ruble of products sold (in foreign sources, return on sales).

If we consider the development purposes of the organization, not by absolute values of current profits and sales volume, but by increase in their values, it is possible to evaluate the result of development from the point of view of the growth of the organization.

2.2 The Change in the Sales Profitability as the Metrics of a Company's Development

The change in the profitability of sales can be mathematically written as follows:

$$\Delta P = \Delta(\Pi : V) = \Delta\Pi : \Delta V \tag{2}$$

$\Delta\Pi$ – the change in profit made in a given period;
ΔV – the change volume of the sales in this period.

Let us analyze the possible situations related to the measure of the marginal profitability of the organization (formula 2).

In practice, the change in profit can have a positive ("+") or negative sign ("–") value. If the profit in this period increased compared to the previous period, the change will have positive value (sign "+"), if the profit in this period decreased, the change in profits will have a negative value (sign "–"). The same applies to changes in sales.

As a result, let's define 4 cases as the aggregation of possible signs of changes in $\Delta\Pi$ and ΔV:

1. ΔV - "+"; $\Delta\Pi$ - "+";
2. ΔV - "–"; $\Delta\Pi$ - "–";
3. ΔV - "–"; $\Delta\Pi$ - "+";
4. ΔV - "+"; $\Delta\Pi$ - "–";

In each of the four possible cases of the ratio, ΔV and $\Delta\Pi$ can have different rates of change of $\Delta\Pi$ relative to ΔV. For example, $\Delta\Pi$ can grow faster or slower than ΔV.

The situation at the company is better in the case if the rate of growth of $\Delta\Pi$ is greater than the rate of growth of ΔV. This means that without increasing profit as an internal source of formation of resources of the organization it is impossible to provide the resources for the organization's activities in the case of increasing sales (ceteris paribus). Therefore, in each of the four cases two areas can be determined:

- $\Delta\Pi$ grows faster than ΔV – the area with higher level of the use of potential;
- $\Delta\Pi$ grows slower than ΔV – the area with lower level of the use of potential.

In order to control the potential of the entrepreneurial firms, let's build a model describing the change in the level of use of the potential depending on various factors [4]. The model of evaluation of use of the firm's potential is constructed depending on the changes in the market of goods (works, services) - rate of change in sales volume (ΔV), as well as the internal capabilities of the firm's financing of its development - the rate of change of profit ($\Delta\Pi$).

2.3 Graphical Interpretation of the Dynamics of Change in the Sales Profitability

Graphical interpretation of the dynamics of ΔV and $\Delta \Pi$ is represented as a matrix (Fig. 1). The essence of the matrix is that the position of the firm in economic space is determined, by marked quadrants and depending on this, one of the strategies is recommended. The minimum number of fields (quadrants) is 4, the maximum is theoretically unlimited, but in practice the big amount of them is not used [5].

Using the matrix, the task of developing a company development strategy gets solved: to win a bigger share of the market, or at least to keep the previously won positions.

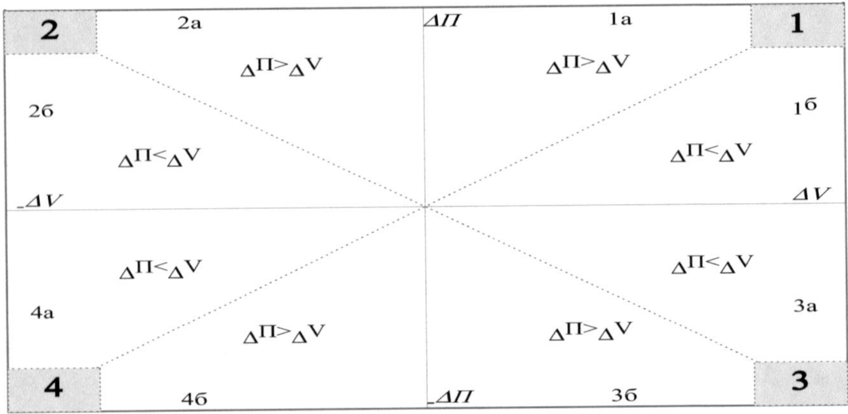

Fig. 1. Matrix assessment of the level of use of potential of the company.

The matrix describes not only existing at the time of analyzing the state of the company and the level of utilization of its capacity, but also allows you to specify the directions of developing the strategic actions to achieve the desired level of utilization of capacity, i.e., the matrix allows to control the potential.

In the matrix on the y-axis pending change in profit for a given period in percent, i.e. change the internal capabilities of the firm's financing of its development; on the X - axis is the change in sales volume over a specified period in percent, i.e., changes occurring in the market of goods (works, services).

As a result, the matrix consists of four quadrants. In one of the four quadrants the company is in accordance with actually received for the period changes in profits and sales. Each quadrant describes a specific state of the organization at present and the level of use potential.[1]

[1] Each quadrant depending on the level of use of capacities with a brief motto.

3 Interpretation of the Development Strategies Matrix

Entrepreneurial firms, in terms of the level of use of growth potential (Fig. 2) is characterized by the following possible provisions:

Unstable Equilibrium $\Delta \Pi/\Delta V \to min$	Success $\Delta \Pi/\Delta V \to max$
Crisis $\Delta \Pi/\Delta V \to min$	Dilemma $\Delta \Pi/\Delta V \to max$

Fig. 2. The definition of the type positions of the company's strategy

In the First Quadrant: ΔV "+"; $\Delta \Pi$ "+"; ΔP "+".
If a positive change $\Delta \Pi$ increases the opportunity of firms to finance their own development. Positive growth ΔV allows you to increase the turnover of produced goods. As a result, the company has a high level of integrated use of capabilities and can use significant internal and external conditions for sustainable development.

The total conclusion for quadrant 1: firm sustainable currently and effectively using the potential.

The integral index of the level of use of growth potential in the first quadrant to strive for the best. The motto of the position of potential: success.

In the Second Quadrant: ΔV "–"; $\Delta \Pi$ "+"; ΔP "–".
Positive ΔP shows that there is a growing opportunity for self-financing. Negative change ΔV shows a decline in sales. The ratio of the two components of the marginal profitability leads to the conclusion that in the future the company has internal sources that are generated due to the increase ΔP, which allows it to finance its own development.

Positive ΔP shows that there is a growing opportunity for self-financing. Negative change ΔV shows a decline in sales. The ratio of the two components of the marginal profitability leads to the conclusion that in the future the company has internal sources that are generated due to the increase ΔP, which allows it to Finance its own development.

The total conclusion for quadrant 2: firm stable at the moment, however not effectively using the potential. The result of correlation between changes in profits and changes in sales can distinguish two particular cases. The motto of position of potential: a delicate balance.

In the Third Quadrant: ΔV "+"; $\Delta \Pi$ "–"; ΔP "–".
Negative change $\Delta \Pi$ reduces the ability of the company to finance its development. The positive change ΔV shows the growth in sales. The ratio of the two components of the marginal profitability leads to the conclusion that in the future the company will spend more money for increasing and sustaining growth in sales. This leads to the fact that the firm has exhausted the internal resources for self-financing its development.

The total conclusion for quadrant 3: firm stable at the moment, but not effectively using the potential. The motto of the position of potential: the development dilemma.

In the Fourth Quadrant: ΔV "–"; $\Delta \Pi$ "–"; ΔP "–".
Negative $\Delta \Pi$ decreases the ability of the company to finance its own development. Negative ΔV means reducing the turnover of produced goods. In the very near future this will lead to the fact that the company will not be able to recoup the resources (i.e., to provide self-sufficiency), but also to carry out the expansion of production, social goals, etc., i.e. to provide self-financing. As a result, the company does not use its capacity and cannot use internal and external conditions for sustainable development.

The total conclusion for quadrant 4: the firm has a stable position currently, and inefficient use of capacity.

The integral index of the level of use of growth potential in the fourth quadrant tends to minimize. The motto of the position of potential: crisis of development.

The management of potential of the organization is based on our proposed matrix (Fig. 1), which allows us to formulate the potential change of the organization, focusing on the organization's location in a particular quadrant. Indeed, by acting on the factors affecting the change in the rate of sales or profits, we influence the level of these indicators and respectively, by use of the economic potential of the organization.

Let us analyze the possible situations in each of the quadrants of the matrix to define the different possibilities for organization, as well as the favorable and unfavorable impact of certain economic decisions and activities.

Quadrant 1 "Success"

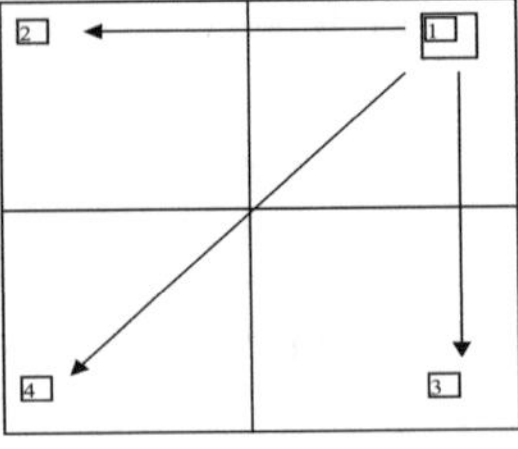

The level of use of growth potential of the organization high. If there is a decrease in sales, but does not change a profit, it automatically moves to quadrant 2.

If a slight change in the volume of sales is the decline in profits, the company moved to quadrant 3, where the utilization of capacity is low. If the decline in profits will occur in the future, the firm risks being in quadrant 4.

Quadrant 2 "Fragile Balance"

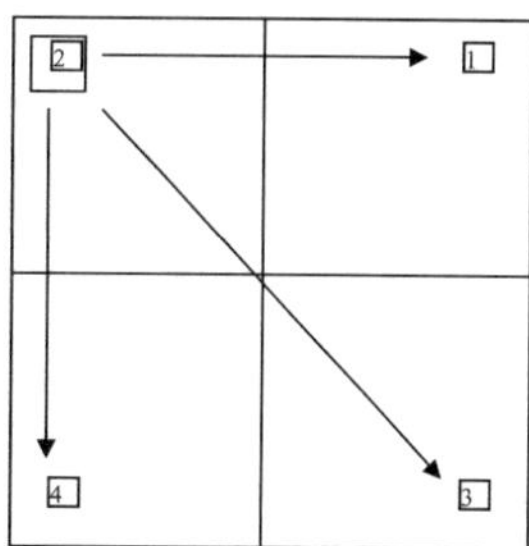

In this quadrant, the firm level of use of growth potential is quite high. To move into quadrant 1, where a value of the level of utilization of the growth potential of the maximum, you need to increase sales. When you fall in profits the firm cannot avoid quadrant 4 or 3.

Quadrant 3 "Development Dilemma"

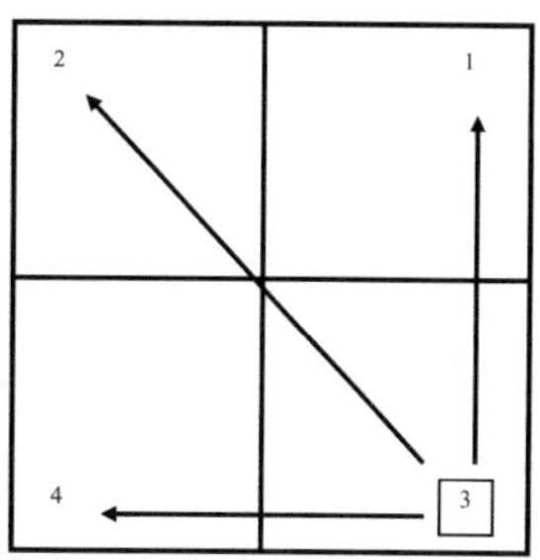

In quadrant 3 the level of use of growth potential is low. If you manage to increase profits, the company moves into quadrant 1, where the level of use of growth potential at its maximum. However, if the sales volume will fall (with little change in profit, of course), the firm is in quadrant 4.

Quadrant 4 "Crisis"

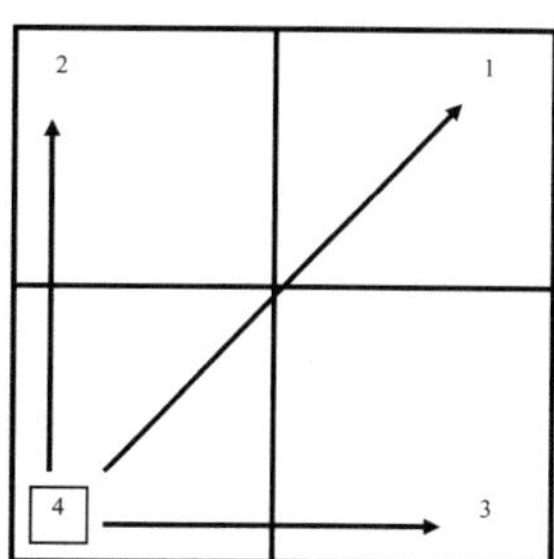

As the level of use of the potential of the situation is the worst. To improve the situation and move into quadrant 3 need to increase sales. If you manage to increase profits, the firm will move to quadrant 2.

Thus, to determine the estimated condition level of use of capacities, we can consider various management decisions, based on the fact that the level of use of capacity should be located in quadrant 1, 2, 3 or 4. In case of violation of sequence of actions or the detection that the solution to the problem of the utilization of capacity has deteriorated, it is possible quickly enough to change the aim, strategy, or suggest other tasks.

4 Data Analysis and Results

We use in assessing the potential capabilities of the organization we selected the integral index of the level of use of growth potential is confirmed by the data taken from the magazine "Expert", including the list of 200 largest Russian companies.

For our purposes, the data in both rankings were transformed according to the following criteria:

1. From the raw data both ratings are excluded for various kinds of financial-credit institutions (banks, investment companies, various insurance companies), a company specializing in real estate.
2. Due to the nature of the trade-wholesalers and the impossibility of their Association with production companies, both ratings are excluded from all trade and wholesale companies.
3. To achieve greater comparability of the ratings of both the excluded companies in which there was no data on the volume of sales and profit during the period.
4. In mind the particular situation in the economy is excluded company "Gazprom".

The rate of change of profit was calculated as the difference of profit margins in 2015 and 2016. Accordingly, the rate of change of sales was calculated.

In the end, after the above-mentioned transformations, a list of 36 of the largest Russian companies was generated for testing the methodology of the study (Appendix A).

According to the list of 36 major Russian companies, we built a graphical model of the economic potential utilization (Fig. 3).

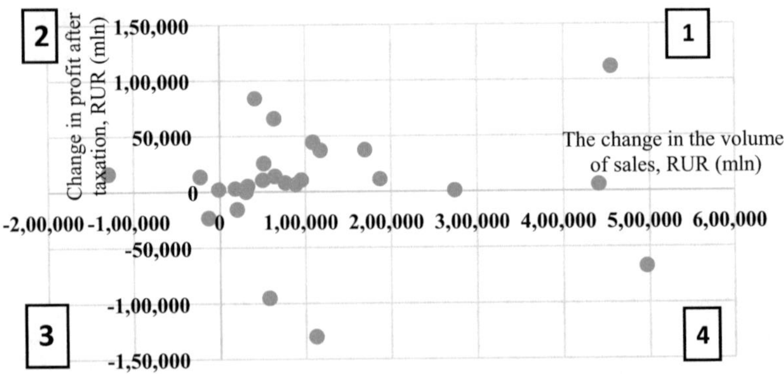

Fig. 3. The distribution of surveyed companies by the ratio of "change in sales profitability"

On the y-axis pending change indicator profits in 2016, compared to 2016 in %; X-axis - rate of change of sales volume in 2016, compared to 2016 in %. Based on the analysis of the obtained graphic model of the utilization of economic potential include the following:

1. Even visually assessing the location of the largest Russian companies can say that they are grouped more densely in the first quadrant, which indicates a good financial position of most enterprises and the efficient use of capacity.
2. If the sales capacity of the organization is located in quadrant - a crisis in the market value of this organization has declined compared to last year.

In 90% of cases, if a company from the rating of the largest Russian companies as potential was in the crisis quadrant, place the company on rating market value decreased.

5 Conclusion

The proposed matrix, as an integral instrument for showcasing the economic potential, allows to consider in more details the parameters of the potential group of companies. This requires us to change the scale on axes X and Y, by setting the desired interval scales.

Thus, for the simulation of management decisions for the development strategy of the company management assessment of the potential of the company according to the proposed method allows to obtain predictions and to make decisions for the company's development strategy.

Acknowledgments. Marina V. KHOLOD acknowledges the previous financial support from MEXT.KAKENHI 24730373 and "Support Project for Strategic Collaboration between Private Universities": Matching Fund Subsidy from MEXT (Ministry of Education, Culture, Sports, Science and Technology), 2015–2019.

A Appendix. A List of 36 of the Largest Russian Companies for Testing the Methodology of the Study

Place in the ranking of the «Expert»	Company	Change in the volume of sales, in RUR (mln)	Change in profit before taxation, in RUR (mln)	Change in profit after taxation, in RUR (mln)	The ratio of increase of profit and growth in sales	Matrix quadrant	Strategy
1	«Gazprom»	376 995	618 418	648 007	Outlier in the Sample		
2	NK «Lukoil»	455 241	128 912	111 625	Revenue is growing, profit is growing	quadrant 1 "Success"	Keep investment strategy
3	NK «Rosneft»	441 000	−18 000	6 000	Revenue is growing, profit is growing	quadrant 1 "Success"	Keep investment strategy
4	Sberbank of Russia	496 300	−43 000	−67 400	Revenue grows, profit falls	quadrant 4 "Crisis of development"	Review investment strategy

(*continued*)

(*continued*)

Place in the ranking of the «Expert»	Company	Change in the volume of sales, in RUR (mln)	Change in profit before taxation, in RUR (mln)	Change in profit after taxation, in RUR (mln)	The ratio of increase of profit and growth in sales	Matrix quadrant	Strategy
5	Russian Railways	109 028	64 199	44 396	Revenue is growing, profit is growing	quadrant 1 "Success"	Keep investment strategy
6	VTB Group	273 900	−20 500	900	Revenue is growing, profit is growing	quadrant 1 "Success"	Keep investment strategy
7	«Surgutneftegaz»	112 031	−148 614	−130 106	Revenue grows, profit falls	quadrant 4 "Crisis of development"	Review investment strategy
8	«Magnit»	187 086	−692 331	11 375	Revenue is growing, profit is growing	quadrant 1 "Success"	Keep investment strategy
9	AK «Transneft»	41 272	69 110	83 927	Revenue is growing, profit is growing	quadrant 1 "Success"	Keep investment strategy
11	Group «InterRAO»	64 243	9 867	14 162	Revenue is growing, profit is growing	quadrant 1"Success"	Keep investment strategy
15	«Tatneft»	76 352	13 655	8 095	Revenue is growing, profit is growing	quadrant 1 "Success"	Keep investment strategy
16	«Evraz»	32 592	−1 448	5 274	Revenue is growing, profit is growing	quadrant 1 "Success"	Keep investment strategy
17	United Company «Rusal»	169 603	40 863	37 511	Revenue is growing, profit is growing	quadrant 1 "Success"	Keep investment strategy
18	«Bashneft»	−129 641	19 639	16 543	Revenue grows, profit falls	Quadrant 2 "Delicate Balance"	Strategy effectiveness
19	GMK «Norilsk Nickel»	50 127	15 886	10 656	Revenue is growing, profit is growing	quadrant 1"Success"	Keep investment strategy
20	Group NLMK	88 749	12 506	6 557	Revenue is growing, profit is growing	quadrant 1 "Success"	Keep investment strategy
21	«NovaTAK»	117 682	40 098	37 204	Revenue is growing, profit is growing	quadrant 1 "Success"	Keep investment strategy

(*continued*)

(*continued*)

Place in the ranking of the «Expert»	Company	Change in the volume of sales, in RUR (mln)	Change in profit before taxation, in RUR (mln)	Change in profit after taxation, in RUR (mln)	The ratio of increase of profit and growth in sales	Matrix quadrant	Strategy
22	«Aeroflot—Russian Airlines»	95 402	14 792	10 652	Revenue is growing, profit is growing	quadrant 1 "Success"	Keep investment strategy
24	«Severstal'»	63 462	−7 140	65 724	Revenue is growing, profit is growing	quadrant 1 "Success"	Keep investment strategy
27	«Vimpelcom»	20 188	−21 203	−15 819	Revenue grows, profit falls	quadrant 4 "Crisis of development"	Review investment strategy
28	Magnitogorsk Metal Factory	51 377	38 029	25 769	Revenue is growing, profit is growing	quadrant 1 "Success"	Keep investment strategy
29	United Aviation Corporation	57 304	−97 102	−95 113	Revenue is growing, profit is growing	quadrant 1 "Success"	Keep investment strategy
30	«Rusgidro»	17 952	5 487	3 028	Revenue is growing, profit is growing	quadrant 1 "Success"	Keep investment strategy
31	«Ashan»	30 569	−740	−49	Revenue grows, profit falls	quadrant 4 "Crisis of development"	Review investment strategy
33	ГК «Megafon»	−1 592	1 182	2 216	Revenue grows, profit falls	Quadrant 2 "Delicate Balance"	Strategy effectiveness
35	«Rostelecom»	−13 562	−30 927	−23 416	Worst	Quadrant 3 «Development Dilemma»	To choose the direction of development – the growth of sales or profits
36	«Stroygazmontazh»	−23 221	16 872	13 780	Revenue drops, profit grows	Quadrant 2 "Delicate Balance"	Strategy effectiveness

References

1. Maslennikov, V.V., Lyandau, Y.V.: Balanced Scorecard Systems-Based Strategy Formaliztion. RuScience, Moscow (2016)
2. Kalinina, I.A.: Economic University Potential Realization. Vestn. PRUE, vol. 6, no. 90, pp. 20–30 (2016)

3. Kholod, M., Yada, K.: An examination of the impact of neurophysiologic and environmental variables on shopping behavior of customers in a grocery store in Japan. In: Advances in Knowledge-Based and Intelligent Information and Engineering Systems, vol. 243, pp. 2099–2103. IOS Press (2012)
4. De Melo, F., et al.: Quantitative analysis in economics based on wavelet transform: a new approach. Asian Soc. Sci. **11**(20), 66–73 (2015)
5. Kholod, M., Nakahara, T., Azuma, H., Yada, K.: The influence of shopping path length on purchase behavior in grocery store. In: 14th International Conference KES 2010, Knowledge-Based and Inteligent, Information and Enineering Systems, LNAI 6278, vol. III, pp. 273–280. Springer, Heidelberg (2010)

Super Pairwise Comparison Matrix in the Dominant AHP with Hierarchical Criteria

Takao Ohya[1(✉)] and Eizo Kinoshita[2]

[1] School of Science and Engineering, Kokushikan University, Tokyo, Japan
takaohya@kokushikan.ac.jp
[2] Faculty of Urban Science, Meijo University, Gifu, Japan
kinoshit@urban.meijo-u.ac.jp

Abstract. We have proposed a super pairwise comparison matrix (SPCM) to express all pairwise comparisons in the evaluation process of the dominant analytic hierarchy process (D-AHP) or the multiple dominant AHP (MDAHP) as a single pairwise comparison matrix. This paper shows calculations of super pairwise comparison matrix in D-AHP with hierarchical criteria.

Keywords: Super pairwise comparison matrix · The dominant AHP · The multiple dominant AHP · Logarithmic least square method

1 Introduction

AHP (the Analytic Hierarchy Process) proposed by Saaty [1] enables objective decision making by top-down evaluation based on an overall aim.

In actual decision making, a decision maker often has a specific alternative (regulating alternative) in mind and makes an evaluation on the basis of the alternative. This was modeled in D-AHP (the dominant AHP), proposed by Kinoshita and Nakanishi [2].

If there are more than one regulating alternatives and the importance of each criterion is inconsistent, the overall evaluation value may differ for each regulating alternative. As a method of integrating the importance in such cases, CCM (the concurrent convergence method) was proposed. Kinoshita and Sekitani [3] showed the convergence of CCM.

Ohya and Kinoshita [4] proposed an SPCM (Super Pairwise Comparison Matrix) to express all pairwise comparisons in the evaluation process of the dominant analytic hierarchy process (AHP) or the multiple dominant AHP (MDAHP) as a single pairwise comparison matrix.

Ohya and Kinoshita [5] showed, by means of a numerical counterexample, that in MDAHP an evaluation value resulting from the application of the logarithmic least squares method (LLSM) to an SPCM does not necessarily coincide with that of the evaluation value resulting from the application of the geometric mean multiple D-AHP (GMMDAHP) to the evaluation value obtained from each pairwise comparison matrix by using the geometric mean method.

I. Czarnowski et al. (eds.), *Intelligent Decision Technologies 2017*,
Smart Innovation, Systems and Technologies 72, DOI 10.1007/978-3-319-59421-7_31

Ohya and Kinoshita [6] showed, using the error models, that in D-AHP an evaluation value resulting from the application of the logarithmic least squares method (LLSM) to an SPCM necessarily coincide with that of the evaluation value resulting obtained by using the geometric mean method to each pairwise comparison matrix.

Ohya and Kinoshita [7] showed the treatment of hierarchical criteria in D-AHP with super pairwise comparison matrix.

Ohya and Kinoshita [8] showed the example of using SPCM with the application of LLSM for calculation of MDAHP.

Ohya and Kinoshita [9] showed that the evaluation value resulting from the application of LLSM to an SPCM agrees with the evaluation value determined by the application of D-AHP to the evaluation value obtained from each pairwise comparison matrix by using the geometric mean.

This paper shows calculations of SPCM in D-AHP with hierarchical criteria.

2 D-AHP and SPCM

This section explains D-AHP, GMMDAHP and a SPCM to express the pairwise comparisons appearing in the evaluation processes of D-AHP and MDAHP as a single pairwise comparison matrix. Section 2.1 outlines D-AHP procedure and explicitly states pairwise comparisons, and Sect. 2.2 explains the SPCM that expresses these pairwise comparisons as a single pairwise comparison matrix.

2.1 Evaluation in D-AHP

The true absolute importance of alternative $a(a = 1,\ldots,A)$ at criterion $c(c = 1,\ldots,C)$ is v_{ca}. The final purpose of the AHP is to obtain the relative value between alternatives of the overall evaluation value $v_a = \sum_{c=1}^{C} v_{ca}$ of alternative a. The procedure of D-AHP for obtaining an overall evaluation value is as follows:

D-AHP

Step 1: The relative importance $u_{ca} = \alpha_c v_{ca}$ (where α_c is a constant) of alternative a at criterion c is obtained by some kind of methods. In this paper, u_{ca} is obtained by applying the pairwise comparison method to alternatives at criterion c.

Step 2: Alternative d is the regulating alternative. The importance u_{ca} of alternative a at criterion c is normalized by the importance u_{cd} of the regulating alternative d, and $u^d_{ca}(= u_{ca}/u_{cd})$ is calculated.

Step 3: With the regulating alternative d as a representative alternative, the importance w^d_c of criterion c is obtained by applying the pairwise comparison method to criteria, where, w^d_c is normalized by $\sum_{c=1}^{C} w^d_c = 1$.

Step 4: From u^d_{ca}, w^d_c obtained at Steps 2 and 3, the overall evaluation value $t_a = \sum_{c=1}^{C} w^d_c u^d_{ca}$ of alternative a is obtained. By normalization at Steps 2 and 3, $u_d = 1$. Therefore, the overall evaluation value of regulating alternative d is normalized to 1

2.2 SPCM

The relative comparison values $r^{ca}_{c'a'}$ of importance v_{ca} of alternative a at criteria c as compared with the importance $v_{c'a'}$ of alternative a' in criterion c', are arranged in a (CA × CA) or (AC × AC) matrix. This is proposed as the SPCM $R = (r^{ca}_{c'a'})\, or\, (r^{ac}_{a'c'})$.

In a (CA × CA) matrix, index of alternative changes first. In a (CA × CA) matrix, SPCM's $(A(c-1)+a, A(c'-1)+a')$ th element is $r^{ca}_{c'a'}$.

In a (AC × AC) matrix, index of criteria changes first. In a (AC × AC) matrix, SPCM's $(C(a-1)+c, C(a'-1)+c')$ th element is $r^{ac}_{a'c'}$.

In a SPCM, symmetric components have a reciprocal relationship as in pairwise comparison matrices. Diagonal elements are 1 and the following relationships are true:

If $r^{ca}_{c'a'}$ exists, then $r^{c'a'}_{ca}$ exists and

$$r^{c'a'}_{ca} = 1/r^{ca}_{c'a'} \tag{1}$$

$$r^{ca}_{ca} = 1 \tag{2}$$

Pairwise comparison at Step 1 of D-AHP consists of the relative comparison value $r^{ca}_{ca'}$ of importance v_{ca} of alternative a, compared with the importance $v_{ca'}$ of alternative a' at criterion c.

Pairwise comparison at Step 3 of D-AHP consists of the relative comparison value $r^{cd}_{c'd}$ of importance v_{cd} of alternative d at criterion c, compared with the importance $v_{c'd}$ of alternative d at criterion c', where the regulating alternative is d.

SPCM of D-AHP or MDAHP is an incomplete pairwise comparison matrix. There-fore, the LLSM based on an error model or an eigenvalue method such as the Harker method [10] or two-stage method is applicable to the calculation of evaluation values from an SPCM.

3 Numerical Example of Using SPCM for Calculation of D-AHP

Let us take as an example the hierarchy shown in Fig. 1. Three alternatives from 1 to 3 and seven criteria from I to VI, and S are assumed, where Alternative 1 is the regulating alternative. Criteria IV to VI are grouped as Criterion S, where Criterion IV is the regulating criterion.

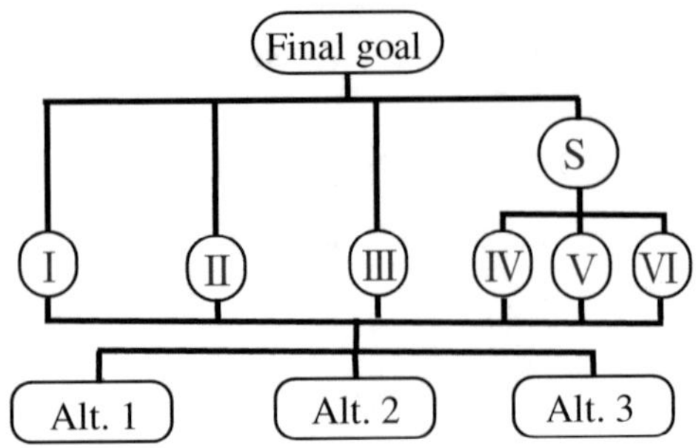

Fig. 1. The hieratical structure

As the result of pairwise comparison between alternatives at criteria c (c = I,…,VI), the following pairwise comparison matrices $R_c^A, c = \mathrm{I}, \ldots, \mathrm{VI}$ are obtained:

$$R_{\mathrm{I}}^A = \begin{pmatrix} 1 & 1/3 & 5 \\ 3 & 1 & 3 \\ 1/5 & 1/3 & 1 \end{pmatrix}, \quad R_{\mathrm{II}}^A = \begin{pmatrix} 1 & 7 & 3 \\ 1/7 & 1 & 1/3 \\ 1/3 & 3 & 1 \end{pmatrix},$$

$$R_{\mathrm{III}}^A = \begin{pmatrix} 1 & 1/3 & 1/3 \\ 3 & 1 & 1/3 \\ 3 & 3 & 1 \end{pmatrix}, \quad R_{\mathrm{IV}}^A = \begin{pmatrix} 1 & 3 & 5 \\ 1/3 & 1 & 1 \\ 1/5 & 1 & 1 \end{pmatrix},$$

$$R_{\mathrm{V}}^A = \begin{pmatrix} 1 & 1/3 & 3 \\ 3 & 1 & 5 \\ 1/3 & 1/5 & 1 \end{pmatrix}, \quad R_{\mathrm{VI}}^A = \begin{pmatrix} 1 & 1/5 & 3 \\ 5 & 1 & 7 \\ 1/3 & 1/7 & 1 \end{pmatrix}.$$

With regulating alternative 1 as the representative alternative, and regulating criterion IV as the representative alternative, importance between criteria was evaluated by pairwise comparison. As a result, the following pairwise comparison matrices R_{I}^C, R_1^S are obtained:

$$R_1^C = \begin{array}{c} \\ \mathrm{I} \\ \mathrm{II} \\ \mathrm{III} \\ \mathrm{IV} \end{array} \begin{array}{c} \begin{array}{cccc} \mathrm{I} & \mathrm{II} & \mathrm{III} & \mathrm{IV} \end{array} \\ \begin{pmatrix} 1 & 1/3 & 3 & 1/3 \\ 3 & 1 & 3 & 1 \\ 1/3 & 1/3 & 1 & 1/3 \\ 3 & 1 & 3 & 1 \end{pmatrix} \end{array}, \quad R_1^S = \begin{array}{c} \\ \mathrm{IV} \\ \mathrm{V} \\ \mathrm{VI} \end{array} \begin{array}{c} \begin{array}{ccc} \mathrm{IV} & \mathrm{V} & \mathrm{VI} \end{array} \\ \begin{pmatrix} 1 & 1/3 & 3 \\ 3 & 1 & 5 \\ 1/3 & 1/5 & 1 \end{pmatrix} \end{array}$$

The (CA × CA) order SPCM for this example is

$$R_{(CA\times CA)} =$$
$$\begin{pmatrix}
1 & 1/3 & 5 & 1/3 & & & 3 & & & 1/3 & & & & & & & & \\
3 & 1 & 3 & & & & & & & & & & & & & & & \\
1/5 & 1/3 & 1 & & & & & & & & & & & & & & & \\
3 & & & 1 & 7 & 3 & 3 & & & 1 & & & & & & & & \\
 & & & 1/7 & 1 & 1/3 & & & & & & & & & & & & \\
 & & & 1/3 & 3 & 1 & & & & & & & & & & & & \\
1/3 & & & 1/3 & & & 1 & 1/3 & 1/3 & 1/3 & & & & & & & & \\
 & & & & & & 3 & 1 & 1/3 & & & & & & & & & \\
 & & & & & & 3 & 3 & 1 & & & & & & & & & \\
3 & & & 1 & & & 3 & & & 1 & 3 & 5 & 1/3 & & & 3 & & \\
 & & & & & & & & & 1/3 & 1 & 1 & & & & & & \\
 & & & & & & & & & 1/5 & 1 & 1 & & & & & & \\
 & & & & & & & & & 3 & & & 1 & 1/3 & 3 & 5 & & \\
 & & & & & & & & & & & & 3 & 1 & 5 & & & \\
 & & & & & & & & & & & & 1/3 & 1/5 & 1 & & & \\
 & & & & & & & & & 1/3 & & & 1/5 & & & 1 & 1/5 & 3 \\
 & & & & & & & & & & & & & & & 5 & 1 & 7 \\
 & & & & & & & & & & & & & & & 1/3 & 1/7 & 1
\end{pmatrix}$$

For pairwise comparison values in an SPCM, an error model is assumed as follows:

$$r^{ca}_{c'a'} = \varepsilon^{ca}_{c'a'} \frac{v_{ca}}{v_{c'a'}} \tag{3}$$

Taking the logarithms of both sides gives

$$\log r^{ca}_{c'a'} = \log v_{ca} - \log v_{c'a'} + \log \varepsilon^{ca}_{c'a'} \tag{4}$$

To simplify the equation, logarithms will be represented by overdots as $\dot{r}^{ca}_{c'a'} = \log r^{ca}_{c'a'}, \dot{v}_{ca} = \log v_{ca}, \dot{\varepsilon}^{ca}_{c'a'} = \log \varepsilon^{ca}_{c'a'}$. Using this notation, Eq. (4) becomes

$$\dot{r}^{ca}_{c'a'} = \dot{v}_{ca} - \dot{v}_{c'a'} + \dot{\varepsilon}^{ca}_{c'a'}, c, c' = 1, \ldots, C, a, a' = 1, \ldots, A \tag{5}$$

From Eqs. (1) and (2), we have

$$\dot{r}^{c'a'}_{ca} = -\dot{r}^{ca}_{c'a'} \tag{6}$$

$$\dot{r}^{ca}_{ca} = 0 \tag{7}$$

If $\dot{\varepsilon}^{ca}_{c'a'}$ is assumed to follow an independent probability distribution of mean 0 and variance σ^2, irrespective of c, a, c', a', the least squares estimate gives the best estimate for the error model of Eq. (5) according to the Gauss Markov theorem.

There are two types of pairwise comparison in D-AHP: $r^{ca}_{ca'}$ at Step 1 and $r^{cd}_{c'd}$ at Step 3. Then Eq. (5) comes to following Eq. (8) by vector notation.

$$\dot{\mathbf{Y}} = \mathbf{S}\dot{\mathbf{x}} + \dot{\epsilon} \tag{8}$$

Where

$$\dot{\mathbf{x}} = (\dot{\mathbf{v}}_{\mathrm{I2}}\ \dot{\mathbf{v}}_{\mathrm{I3}}\ \dot{\mathbf{v}}_{\mathrm{II1}}\ \dot{\mathbf{v}}_{\mathrm{II2}}\ \dot{\mathbf{v}}_{\mathrm{II3}}\ \dot{\mathbf{v}}_{\mathrm{III1}}\ \dot{\mathbf{v}}_{\mathrm{III2}}\ \dot{\mathbf{v}}_{\mathrm{III3}}\ \dot{\mathbf{v}}_{\mathrm{IV1}} \bullet \bullet \bullet\ \dot{\mathbf{v}}_{\mathrm{VI2}}\ \dot{\mathbf{v}}_{\mathrm{VI3}})^{\mathrm{T}},$$

$$\dot{\mathbf{Y}} = \begin{pmatrix} \dot{\mathbf{r}}^{\mathrm{I1}}_{\mathrm{I2}} \\ \dot{\mathbf{r}}^{\mathrm{I1}}_{\mathrm{I3}} \\ \dot{\mathbf{r}}^{\mathrm{I1}}_{\mathrm{II1}} \\ \dot{\mathbf{r}}^{1}_{\mathrm{III1}} \\ \dot{\mathbf{r}}^{1}_{\mathrm{IV1}} \\ \dot{\mathbf{r}}^{\mathrm{I2}}_{\mathrm{I3}} \\ \dot{\mathbf{r}}^{\mathrm{II1}}_{\mathrm{II2}} \\ \dot{\mathbf{r}}^{\mathrm{II1}}_{\mathrm{II3}} \\ \dot{\mathbf{r}}^{\mathrm{II1}}_{\mathrm{III1}} \\ \dot{\mathbf{r}}^{\mathrm{II1}}_{\mathrm{IIV1}} \\ \dot{\mathbf{r}}^{\mathrm{II2}}_{\mathrm{II3}} \\ \dot{\mathbf{r}}^{\mathrm{III1}}_{\mathrm{III2}} \\ \dot{\mathbf{r}}^{\mathrm{III1}}_{\mathrm{III3}} \\ \dot{\mathbf{r}}^{\mathrm{III1}}_{\mathrm{IV1}} \\ \dot{\mathbf{r}}^{\mathrm{III2}}_{\mathrm{III3}} \\ \bullet \\ \bullet \\ \bullet \\ \dot{\mathbf{r}}^{\mathrm{VI2}}_{\mathrm{VI3}} \end{pmatrix} = \begin{pmatrix} \log(1/3) \\ \log 5 \\ \log(1/3) \\ \log 3 \\ \log(1/3) \\ \log 3 \\ \log 7 \\ \log 3 \\ \log 3 \\ \log 1 \\ \log(1/3) \\ \log 1 \\ \log(1/3) \\ \log(1/3) \\ \log(1/3) \\ \bullet \\ \bullet \\ \bullet \\ \log 7 \end{pmatrix}, \mathbf{S} = \begin{pmatrix} -1 & & & & & & & & & & & \\ & -1 & & & & & & & & & & \\ & & -1 & & & & & & & & & \\ & & & & & -1 & & & & & & \\ & & & & & & & & -1 & & & \\ 1 & -1 & & & & & & & & & & \\ & & 1 & -1 & & & & & & \bullet\ \bullet\ \bullet & & \\ & & 1 & & -1 & & & & & & & \\ & & 1 & & & -1 & & & & & & \\ & & 1 & & & & & & -1 & & & \\ & & & 1 & -1 & & & & & & & \\ & & & & & 1 & -1 & & & & & \\ & & & & & 1 & & -1 & & & & \\ & & & & & 1 & & & -1 & & & \\ & & & & & & 1 & -1 & & & & \\ & & & & & \bullet & & & & \bullet & & \\ & & & & & \bullet & & & & & \bullet & \\ & & & & & \bullet & & & & & & \bullet \\ & & & & & & & & & & 1 & -1 \end{pmatrix}$$

To simplify calculations, $v_{11} = 1$ that is $\dot{v}_{11} = 0$. The least squares estimates for formula (8) are calculated by $\hat{\dot{\mathbf{x}}} = (\mathbf{S}^{\mathrm{T}}\mathbf{S})^{-1}\mathbf{S}^{\mathrm{T}}\dot{\mathbf{Y}}$.

Table 1 shows the evaluation values obtained from the SPCM for this example.

Table 1. Evaluation values obtained by SPCM + LLSM

Criterion	I	II	III	IV	V	VI	Overall evaluation value
Alternative 1	1	2.280	0.577	2.280	2.280	5.622	14.038
Alternative 2	1.754	0.299	1.201	0.641	5.622	21.803	31.320
Alternative 3	0.342	0.826	2.498	0.541	0.924	2.416	7.547

4 Conclusion

SPCM of D-AHP is an incomplete pairwise comparison matrix. Therefore, the LLSM based on an error model or an eigenvalue method such as the Harker method or two-stage method is applicable to the calculation of evaluation values from an SPCM. In this paper, we showed the way of using SPCM with the application of LLSM for calculation of D-AHP with hierarchical criteria.

References

1. Saaty, T.L.: The Analytic Hierarchy Process. McGraw-Hill, New York (1980)
2. Kinoshita, E., Nakanishi, M.: Proposal of new AHP model in light of dominative relationship among alternatives. J. Oper. Res. Soc. Jpn. **42**, 180–198 (1999)
3. Kinoshita, E., Sekitani, K., Shi, J.: Mathematical properties of dominant AHP and concurrent convergence method. J. Oper. Res. Soc. Jpn. **45**, 198–213 (2002)
4. Ohya, T., Kinoshita, E.: Proposal of super pairwise comparison matrix. In: Watada, J., et al. (eds.) Intelligent Decision Technologies, pp. 247–254. Springer, Heidelberg (2011)
5. Ohya, T., Kinoshita, E.: Super pairwise comparison matrix in the multiple dominant AHP. In: Watada, J., et al. (eds.) Intelligent Decision Technologies. Smart Innovation, Systems and Technologies 15, vol. 1, pp. 319–327. Springer, Heidelberg (2012)
6. Ohya, T., Kinoshita, E.: Super pairwise comparison matrix with the logarithmic least squares method. In: Neves-Silva, R., et al. (eds.) Intelligent Decision Technologies. Frontiers in Artificial Intelligence and Applications, vol. 255, pp. 390–398. IOS press, (2013)
7. Ohya, T., Kinoshita, E.: The treatment of hierarchical criteria in dominant AHP with super pairwise comparison matrix. In: Neves-Silva, R., et al. (eds.) Smart Digital Futures 2014, pp. 142–148. IOS press (2014)
8. Ohya, T., Kinoshita, E.: Using super pairwise comparison matrix for calculation of the multiple dominant AHP. In: Neves-Silva, R., et al. (eds.) Intelligent Decision Technologies, Smart Innovation, Systems and Technologies 39, pp. 493–499. Springer, Heidelberg (2015)
9. Ohya, T., Kinoshita, E.: Super pairwise comparison matrix in the dominant AHP. In: Czarnowski, I., et al. (eds.) Intelligent Decision Technologies 2016. Smart Innovation, Systems and Technologies, vol. 57, pp. 407–414. Springer, Heidelberg (2016)
10. Harker, P.T.: Incomplete pairwise comparisons in the analytic hierarchy process. Math. Model. **9**, 837–848 (1987)

Author Index

I. Czarnowski et al. (eds.), *Intelligent Decision Technologies 2017*,
Smart Innovation, Systems and Technologies 72, DOI 10.1007/978-3-319-59421-7

Printed in the United States
By Bookmasters